Molecular & Cell Biology FOR DUMMIES®

Molecular & Cell Biology For Dummies®

Published by
Wiley Publishing, Inc.
111 River St.
Hoboken, NJ 07030-5774
www.wiley.com

Published by Wiley Publishing, Inc., Indianapolis, Indiana

Published simultaneously in Canada

For general information on our other products and services, please contact our Customer Care Department within the U.S. at 877-762-2974, outside the U.S. at 317-572-3993, or fax 317-572-4002.

For technical support, please visit www.wiley.com/techsupport.

Wiley also publishes its books in a variety of electronic formats. Some content that appears in print may not be available in electronic books.

Library of Congress Control Number: 2009926359

ISBN: 978-0-470-43066-8

Manufactured in the United States of America

10 9 8 7 6 5 4 3 2 1

About the Author

René Fester Kratz, PhD, grew up near the ocean in Rhode Island. From a young age, she wanted to be a teacher (because she loved her teachers at school) and a biologist (because her dad was one). She graduated from Warwick Veterans Memorial High School and then went on to major in biology at Boston University. As a freshman and sophomore at BU, René got excited by subjects other than biology and even considered changing her major. Then, she met and studied under Lynn Margulis, who reignited René's love of biology and introduced her to the world of microbes. René graduated with a bachelor's degree in Biology from BU and then went on to get a master's and a doctorate degree in botany from the University of Washington. At UW, René studied reproductive onset in *Acetabularia acetabulum,* a marine green alga that grows as single cells big enough to pick up with your fingers.

René currently teaches biology and general science classes at Everett Community College in Everett, Washington. René spends most of her time introducing students to the wonders of cells and microbes as she teaches cellular biology and microbiology. René also has a strong interest in science education and science literacy for everyone. As a member of the North Cascades and Olympic Science Partnership, she helped create inquiry-based science courses for future teachers that are based on research on human learning. René loves teaching these courses because they make science accessible for all kinds of people. In the summer, René enjoys working with K–12 teachers on the improvement of science education in the public schools. René also enjoys writing about science and is the author of *Microbiology The Easy Way* for Barron's Educational Press.

René loves living in the Pacific Northwest because she is near the ocean and her daffodils start blooming in February (when her family back East is still shoveling snow). She doesn't mind the rain and thinks the San Juan Islands are one of the most beautiful places on Earth. Her husband, two sons, and two very bad dogs help her remember what is truly important and her "sisters" help keep her sane. René loves to scrapbook, quilt, stitch, and read.

Dedication

To my husband, Dan, and my sons, Hueston and Dashiel. You are the center of my world.

To my mom, Annette. It may be corny, but you are the "wind beneath my wings."

To my dad, James. I wanted to be a biologist because of you.

To Lynn: Thank you for inspiring a lifelong love of learning and microbes.

Author's Acknowledgments

Thanks to Matt Wagner of Fresh Books, Inc., for helping me find the opportunity to write this book. And thanks to all the great people at Wiley who made it happen: my editor, Kelly Ewing, who was always helpful and upbeat; the acquisitions editor, Stacy Kennedy, who helped get me started on the project; Alicia South, who coordinated the art; and Barry Ludvik, my technical reviewer. Thanks also to Patrick Redmond, the project coordinator, and those who worked on the art: Kitty Auble, Kathryn Born, Ana Carillo, Rhonda David-Burroughs, Brooke Graczyk, Gary Hunt, Ashley Layfield, Shelley Lea, Beth Morgan, Andew Recher, Heidi Richter, Simon Shak, Melissa K. Smith, Alicia South, Ron Terry, Janet Wahlfeldt, and Tobin Wilkerson.

On the home front, thanks to my husband, Dan, for all his love and support. To Hueston and Dashiel for once again being patient when Mommy was glued to her computer. To my sister, Alyson, and my friend, Julie, for reading chapters and sharing their nonscientist perspectives. To Staci, for helping me walk off some stress. And a big thanks always to my mother, Annette, for just being so supportive of everything I do.

Thanks to all my students at Everett Community College for your enthusiasm and hard work. You have all inspired me to keep doing what I do. Thanks also to my dean, Al Friedman, for letting me teach a reduced load for one quarter so that I'd have more time to write.

Publisher's Acknowledgments

We're proud of this book; please send us your comments through our Dummies online registration form located at http://dummies.custhelp.com. For other comments, please contact our Customer Care Department within the U.S. at 877-762-2974, outside the U.S. at 317-572-3993, or fax 317-572-4002.

Some of the people who helped bring this book to market include the following:

Acquisitions, Editorial, and Media Development

Project Editor: Kelly Ewing

Acquisitions Editor: Stacy Kennedy

Assistant Editor: Erin Calligan Mooney

Editorial Program Coordinator: Joe Niesen

General Reviewer: Barry Ludvik

Senior Editorial Manager: Jennifer Ehrlich

Editorial Supervisor and Reprint Editor: Carmen Krikorian

Editorial Assistant: Jennette ElNaggar

Art Coordinator: Alicia B. South

Cover Photos: © Colin Anderson/Brand X/ Corbis

Cartoons: Rich Tennant (www.the5thwave.com)

Composition Services

Project Coordinator: Patrick Redmond

Layout and Graphics: Samantha Allen, Reuben W. Davis, Andrea Hornberger, Melissa Jester, Christin Swinford

Special Art: Kitty Auble, Kathryn Born, Ana Carillo, Rhonda David-Burroughs, Brooke Graczyk, Gary Hunt, Ashley Layfield, Shelley Lea, Beth Morgan, Andrew Recher, Heidi Richter, Simon Shak, Melissa K. Smith, Ron Terry, Janet Wahlfeldt, Tobin Wilkerson

Proofreader: Christine Sabooni

Indexer: Potomac Indexing, LLC

Publishing and Editorial for Consumer Dummies

Diane Graves Steele, Vice President and Publisher, Consumer Dummies

Kristin Ferguson-Wagstaffe, Product Development Director, Consumer Dummies

Ensley Eikenburg, Associate Publisher, Travel

Kelly Regan, Editorial Director, Travel

Publishing for Technology Dummies

Andy Cummings, Vice President and Publisher, Dummies Technology/General User

Composition Services

Debbie Stailey, Director of Composition Services

Contents at a Glance

Table of Contents

Introduction

Molecular and cellular biology isn't just something that happens in a lab; it reaches out and touches your life in many ways, seen and unseen. Genetically modified organisms, designer cancer drugs, forensic science, and even home pregnancy tests are all applications of the science and techniques of molecular and cellular biology. Gaining an understanding of molecular and cellular biology can help you make informed decisions about your lifestyle and health.

Understanding cells and how they function is fundamental to all other fields of biology, including medicine. All living things are made of cells, and scientists can trace every response, every function of larger organisms back to the structure and function of cells. Genetic diseases are a dramatic example of how important one cell type, one protein, and one gene can be to an organism. Exploring this connection between genes, proteins, and cellular function is at the heart of molecular and cellular biology.

As you take your own journey into the inner space of the cell, I hope this book will act as your guidebook, pointing out landmarks and signposts, and translating the sometimes complicated language of the local inhabitants!

About This Book

Molecular & Cellular Biology For Dummies is an overview of the fundamentals of molecular and cellular biology. My goal is to explain each topic in a clear and straightforward fashion, keeping scientific jargon to a minimum. I want this book to be understandable by anyone who picks it up, even if they don't have a science background. To help you understand what is sometimes a complex science, I share every analogy, funny story, and memory trick that I've gathered in my ten years of teaching this subject. These types of gimmicks help my students wrap their brains around the fundamental principles of molecular and cellular biology, and I hope these strategies will help you, too.

In *Molecular & Cellular Biology For Dummies*, I emphasize the main concepts and fundamental processes that are at the heart of molecular and cellular biology. When I had to make a choice between getting the main idea across or including every molecular detail of a process, I chose the main idea. I think that once you understand the main concept or the big events of a process, you can later add in details fairly easily. However, if you try to tackle a complicated process and every little detail at the same time, you can hit information overload and not really understand anything at all. My emphasis on main ideas and events will make the subject of molecular and cellular biology easier, but that doesn't mean the topic will be easy. The world of the cell is complex and busy with detailed processes, so understanding molecular and cellular biology is a challenge for most people. I hope that this book will help you succeed in that challenge.

Conventions Used in This Book

In order to explain things as clearly as possible, I keep scientific jargon to a minimum and present information in straightforward, linear style. I break dense information into main concepts and divide complicated processes into steps.

To help you find your way through the subjects in this book, I use the following style conventions:

- *Italic* is used for emphasis and to highlight new words or terms that are defined in the text.
- **Boldface** is used to indicate key words in bulleted lists or the action parts of numbered steps.
- Web addresses are written in monofont so that you can easily recognize them.

What You're Not to Read

Sidebars are shaded gray boxes that include stories or information related to the main topic, but not necessary to your understanding. You can skip the sidebars if you want, but they contain some pretty fun and interesting information so I'm guessing you'll read them anyway.

You can also skip any information marked with a Technical Stuff icon (see "Icons Used in This Book" later in this Introduction) without hurting your understanding of the main concepts. For someone who wants or needs to pick up all the details of a process, the Technical Stuff provides a more in-depth explanation.

Foolish Assumptions

As I wrote this book, I tried to imagine who you might be and what you may need in order to understand molecular and cellular biology. Here's who I pictured:

- You're a student in a molecular and cellular biology class who is having trouble understanding everything and keeping up with the pace of the class. For you, I present the topics in a straightforward way with an emphasis on the most important concepts and processes. By reading this book before you go to lecture, you may have an easier time understanding what your professor is talking about.
- You're a student in a molecular and cellular biology class who is determined to get an *A,* and you want to gather all possible resources to help you in your goal. For you, I make studying more efficient by presenting the core concepts of molecular and cellular biology in straightforward bulleted lists. These lists can supplement your own notes, making sure that you've nailed the big ideas.
- You're someone who wants to know more about the science behind the stories you hear in the news and see on TV. Maybe you're interested in forensic science and want a better understanding of what they're talking about on CSI and the Discovery Channel. Or maybe you're worried about the potential impacts of genetically modified organisms or genetic screening on our society, and you want to know more about the science behind these topics. For you, I try to keep terminology to a minimum and include lots of analogies to help you relate the science to your everyday life.

How This Book Is Organized

This book is divided into seven parts, with each part containing related subjects. Like all *For Dummies* books, each chapter is self-contained, so you can pick up whenever you need it and jump into the topic you're working on. Once I explain a subject, I use that information in later topics. So, if you don't read the book in order, you may occasionally want to refer to another section for background information. In those cases, I refer you to the appropriate chapter.

Part I: The World of the Cell

All living things are made of cells. In this part, I introduce the fundamental structure and function of cells. I also introduce viruses, microscopic particles that attack and destroy cells.

Part II: Molecules: The Stuff of Life

Living things are made of cells and cells are made of molecules. In this part, I explain the fundamental cellular chemistry that is necessary to understand the molecular nature of cells.

Part III: The Working Cell

The cell is the fundamental unit of life and possesses all the characteristics of living things: Cells require food, reproduce themselves, respond to signals, and exchange materials. In this part, I describe how cells function, including how they communicate, obtain matter and energy, and reproduce.

Part IV: Genetics: From One Generation to the Next

In living things that reproduce sexually, including humans, parents pass the instructions for life to their offspring. In this part, I describe how parental cells organize their DNA during sexual reproduction and demonstrate how scientists can predict and analyze inheritance patterns using the principles of Mendelian genetics.

Part V: Molecular Genetics: Reading the Book of Life

The DNA code is the underlying programming for how cells function and develop. In this part, I explain the essential core of molecular biology, including how DNA is copied and read by cells, how it determines the traits of organisms, and how DNA is regulated by the cell.

Part VI: Tools of Molecular Biology: Harnessing the Power of DNA

Powerful tools have enabled scientists to explore and manipulate the DNA code, opening a new frontier in biological science. In this part, I describe how scientists can use the tools of molecular biology to explore genomes and apply biological knowledge to solve current world problems.

Part VII: The Part of Tens

Like all *For Dummies* books, this book contains a Part of Tens where I include lists of fun and interesting topics related to molecular and cellular biology. In this part, I include ten fundamental rules that govern the behavior of cells and ten tips for improving your grade!

Icons Used in This Book

All *For Dummies* books use icons to help identify particular types of information. Here's the list of icons I use in this book and what they all mean:

I use this icon to emphasize main ideas that you should definitely keep in mind.

I use this icon to present study tips or other information that can help you navigate through difficult material.

I use this icon to flag detailed information that isn't essential to the main concept or process being presented. If you're not a student in a molecular and cellular biology class, you can definitely skip this material.

I use this icon to flag potentially confusing ideas or common wrong ideas that people typically have about how something works. I know about these danger spots from my years of teaching, and I've flagged them to help you avoid these pitfalls.

Where to Go from Here

With *Molecular & Cellular Biology For Dummies,* you can start anywhere in the book that you want. If you're reading this book for general interest, you'll probably find it best to begin at the beginning with the chapter on cells and then move to whatever interests you next from there. If you're currently having trouble in a molecular and cellular biology class, jump right into the subject that's confusing you. If you're using the book as a companion to a molecular and cellular biology class that is just beginning, the book follows the organization of most college classes with one exception — most college classes work from the smallest to the largest, beginning with molecules then moving on to cells. I prefer to start with cells to give you a sense of context, an idea of where everything is happening, and then move on to the molecules.

Whatever your circumstance, the Table of Contents and Index can help you find the information you need. Best wishes from me to you as you begin your journey into the marvelous world of the cell.

Part I
The World of the Cell

"Okay—now that the paramedic is here with the defibrillator and smelling salts, prepare to learn about covalent bonds."

In this part . . .

Molecular and cellular biology looks at life on the smallest level, from the microscopic cells that make up living things to the mysterious molecules within those cells that contain the programming for how life functions. Cells are the smallest living things and they have all the properties of life, including reproduction, response to environmental signals, a need for energy, and the release of waste products.

Viruses are very small parasites of cells that have the ability to attack cells and convert them into factories for viral reproduction. In this part, I explain the science of molecular and cellular biology and present the basic structure and function of cells and viruses.

Chapter 1

Exploring the World of the Cell

In This Chapter

- Discovering the microscopic world
- Getting matter and energy
- Reading the genetic code

Molecular and cellular biology is about studying cell structure and function down to the level of the individual molecules that make up the cell. The most famous molecule in cells is DNA, and much of molecular biology focuses on this molecule — reading DNA, working with DNA, and understanding how cells use DNA.

In this chapter, I present an overview of molecular and cellular biology and how it relates to your life. My goal is to illustrate the importance of molecular and cellular biology and to give you a preview of the topics I explore in more depth in the later chapters of this book.

Cells and Viruses: Discovering the Inhabitants of the Microscopic World

If you were alive just 400 years ago, you would've had no idea that germs can spread diseases, that your blood contains cells that carry oxygen around your body, or that new people are made when sperm cells join with egg cells. Four hundred years ago, no one had any idea that there was an entire world just beyond the power of the human eye. A Dutch cloth merchant named Antony van Leeuwenhoek changed all that when he used small, hand-held microscopes to peer beyond the known world into the world of the cell.

In 1676, van Leeuwenhoek used his microscopes to look into a drop of lake water — water that appeared clear to his eyes — and was astounded to see tiny creatures swimming around in it. van Leeuwenhoek was the first to see bacteria, blood cells, and sperm cells fertilizing an egg. Along with Robert Hooke, who observed the first plant cells, van Leeuwenhoek laid the foundation for the development of cell biology and microbiology and began new chapters in the sciences of anatomy, physiology, botany, and zoology.

You: On the cellular level

Imagine your eyes have super powers, and you're staring at your own skin, revealing a patchwork of thin, flaky cells. These skin cells are just one type of more than 200 types of cells found in your body — cells that make up your tissues, organs, and organ systems (see Figure 1-1). Increase the power of your eyes, and you can zoom in on your chromosomes, which are made of DNA (see Chapter 7) and contain the instructions for your traits (see Chapter 15).

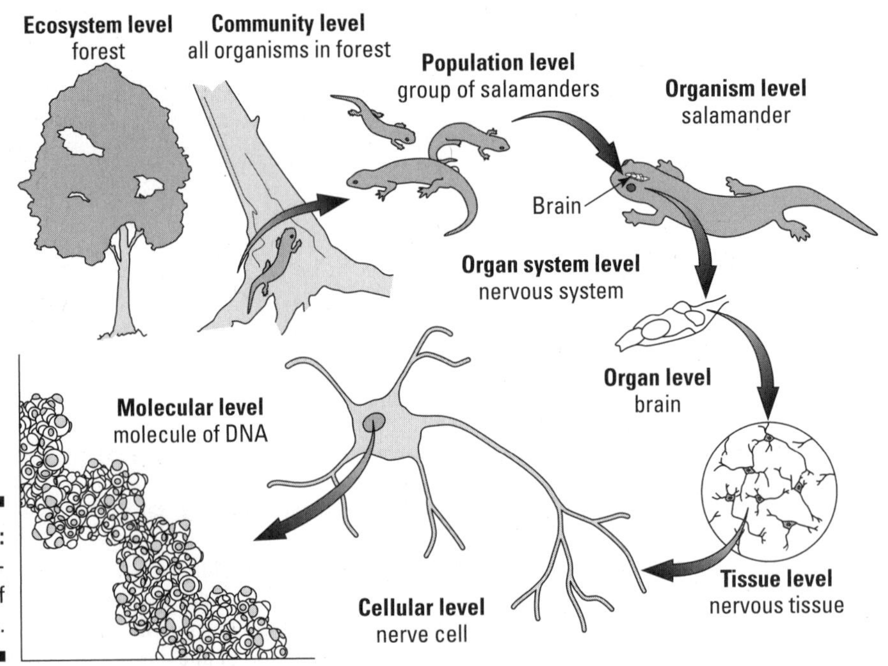

Figure 1-1: The organization of living things.

Them: Bacteria and Viruses

If you looked at your body with super-powered eyes, your cells aren't the only cells you'd see. All over your body and, in fact, everywhere on Earth you look, you can see another type of cell — the prokaryotic cell (see Chapter 2). Prokaryotic cells come in two types:

- **Bacteria** are probably most familiar to you because they can make you sick, but bacteria do many good things, too. The bacteria that live all over your body actually help keep you from getting sick, and many of the foods you eat, such as yogurt, owe their flavors to bacteria.
- **Archaea** are just as common as bacteria but are usually less familiar to people because they aren't known for causing human disease, and they're still being studied by scientists. On a microscope (or with super-powered eyes), archaea look just like bacteria, so scientists didn't realize archaea existed until around 40 years ago when improvements in molecular biology made their discovery possible.

Your super-powered eyes could also show you another type of alien creature, even smaller than the cells of bacteria and archaea — viruses (see Chapter 3). Viruses really are like little alien ships that land on your cells and take them over, enslaving the molecules within your cells and making them work to build more viruses. Your cells don't work for you anymore, and you feel the effects — your throat gets sore, your nose runs, or you ache all over. Fortunately, your immune system comes to the rescue, sending in white blood cells to fight off the invading viruses.

Because bacteria and viruses both make people sick, they often get confused — even in the news media! However, bacteria and viruses have very different structures — bacteria are cells, and viruses are not — which makes a big difference when it comes to medicine. Antibiotics target bacterial cells, and they *don't* work on viruses!

Speaking the language of cells

If you want to learn about cells, you need to speak their chemical chemistry. Cells are made of molecules, they communicate through molecules, and they respond to signals by changing existing molecules or making new ones. The DNA code (see Chapter 7), written in the chemical letters A, T, C, and G, is used by your body to create cellular workers like proteins (see Chapter 6) that control how your cells function. DNA and proteins, along with carbohydrates (see Chapter 5) and lipids (see Chapter 8), are the fundamental building blocks that make up your cells and thus your entire body.

The Life of a Cell: How Cells Get What They Need to Survive and Reproduce

Your cells are the smallest piece of you that is alive. All the things that you can think of that you need to do to keep your body alive — get energy from food, take in oxygen, and release wastes — are also true for your cells:

- When you eat food, you take in a source of energy and matter for your cells that you process with your cellular metabolism (see Chapter 10).
- Your cells do cellular respiration (see Chapter 11), using oxygen to transfer energy out of food into a form that they can use to do work.
- Cells can also use the energy and molecules from food to grow and make new cells (see Chapter 13).

Ultimately, you can trace all the food that you eat back to cells, like those of plants, that make food through photosynthesis (see Chapter 12). In fact, life on Earth couldn't even exist without the organisms that make food, because they capture the energy and matter that all cells need to survive.

Sexual Reproduction: Shuffling the Genetic Deck for the Next Generation

You began life as a single cell, when a sperm cell from your dad combined with an egg cell from your mom. Your parents made these special reproductive cells through a special type of cell division called meiosis (see Chapter 14). Each cell from your parents donated half of your genetic information — 23 chromosomes from Mom and 23 from Dad — for a total of 46 chromosomes in each of your cells. What you look like and much of how you behave is a result of the interaction between the genes you got from Mom and the genes you got from Dad.

Tracking the inheritance of genes and how they interact to determine traits is part of the science of genetics (see Chapters 15 and 16). Through genetics, you can understand things like why your eyes are a certain color or why some traits seem to run in families.

DNA to Protein: Following the Instructions in the Genetic Code

The instructions for your traits, from the level of the cell to the level of the whole you, are encoded in your DNA. Whenever your cells divide to make new cells, they must copy your DNA through DNA replication (see Chapter 17) so that each new cell gets a set of instructions. The working cells of your body are constantly reading the DNA code and using the instructions to build molecules, such as proteins, that they need to do their jobs for the body. Proteins are constructed by the combined efforts of two processes, called transcription and translation (see Chapter 18).

Signals, such as hormones, can tell your working cells that they need to change their behavior. To change their behavior, your cells may need to change their tools. Gene regulation (see Chapter 19) allows your cells to turn off some genes for proteins and turn others on. In fact, how your cells use your DNA is just as important as what your code actually says!

DNA Technology: Tackling the World's Problems

You've probably heard a lot about the impacts of biotechnology — genetically modified organisms (GMOs), DNA fingerprinting, the Human Genome Project, and gene therapy are just some of the topics that regularly appear in the news.

A revolution in biology has occurred over the past 50 years or so, a revolution based on scientists' ability to read and manipulate the genetic code of life. Scientists can extract, snip, copy, read, modify, and place DNA from cells into different cells using recombinant DNA technology (see Chapter 20). New technologies developed in the last 20 years allow scientists to read the entire genetic code, or *genome,* of organisms (see Chapter 21), essentially opening up the book of life for everyone to read.

New branches of biology are growing to study all this new information and present many opportunities for future careers:

- **Bioinformatics** is a science that blends computing, biology, and information technology to organize and analyze the large amounts of information that are being generated by biologists all around the world.
- **Genomics** is the study of entire genomes of organisms. By studying all of the DNA sequence of a cell, scientists are discovering new proteins and new understandings of how DNA is regulated in cells.
- **Proteomics** is the study of the entire body of proteins in a cell and how they interact with each other. The types of proteins found in different cells are compared in order to look for patterns common to certain cell types.

Molecular biology has spread throughout the older branches of biology as well. Botany, zoology, ecology, physiology — every "ology" you can think of, really — now has a molecular component. Living things are studied down to the level of the cell and the molecules, such as DNA and proteins, that make up the cell.

Even medicine is becoming increasingly molecular — Departments of Molecular Medicine are popping up all over — as doctors and scientists seek to understand and treat disease at the level of the cell and molecule. Designer drugs that specifically target the molecular defect of a particular disease are already in the works.

Molecular and cellular biology already impacts your life in many ways and will almost certainly become more important in your future.

Chapter 2

Take a Tour Inside the Cell

In This Chapter

- Comparing life on Earth
- Exploring the eukaryotic cells of plants, animals, and fungi
- Getting to know bacteria and other prokaryotic cells

All living things are made of cells. All cells are built out of the same materials and function in similar ways, showing the relationship of all life on Earth. Eukaryotic cells, such as those of plants and animals, are structurally complex. Prokaryotic cells, such as those of bacteria, have a simpler organization. In this chapter, I present cell structures and their functions for both eukaryotic and prokaryotic cells.

Admiring the Unity and Diversity of Cells

The unity among cells on Earth is truly amazing. All cells have DNA as the genetic material, use the same processes to make proteins, and follow the same basic metabolic principles as other cells. So, on the most fundamental level, cells on Earth show their unity and their relationship to each other.

Beyond the fundamentals, however, cells have fantastic variations. Cells differ in size, from the neurons of giant squid to tiny bacteria. They differ in function, from free-living amoebae to muscle cells in an animal to sperm cells inside the pollen grain of a plant. Cells also differ in their role in the environment, from food makers to predators to decomposers that eat the dead.

Based on their basic chemistry, structure, and hereditary material, all cells on Earth fall into one of three groups, as if the family tree of life on Earth split into three main branches, called *domains*:

- **Eucarya:** Plants, animals, fungi, and protists
- **Bacteria:** Familiar, single-celled microorganisms, some of which are useful to humans and some of which cause human diseases
- **Archaea:** Single-celled microorganisms found in all types of environments, but first discovered in extreme environments, such as hot springs

Cells of the Eucarya are structurally distinct from cells of the Bacteria and the Archaea. Cells of Eukarya are *eukaryotic*, while cells of the Bacteria and Archaea are *prokaryotic*.

- **Eukaryotic cells** have a *nucleus,* a chamber within the cell that is separated by a membrane and contains the DNA. They also have *organelles,* membrane-enclosed structures inside the cell that perform various functions for the cell. Finally, eukaryotic cells are typically much larger than prokaryotic cells, on average about ten times larger. (For more on these cells, see the section "Your Body, Your Cells: Eukaryotic Cells," later in this chapter.)
- **Prokaryotic cells** don't have a nucleus; their DNA is contained within the cytoplasm of the cell. They also don't have any membrane-enclosed organelles, and they're typically much smaller than eukaryotic cells. (See the upcoming section "Tiny but Mighty: Prokaryotic Cells" for more on prokaryotic cells.)

The root *eu* means true and *karyon* means seed, so eukaryotic cells are true-seeded cells because the nucleus looks a little bit like a seed inside the cell. On the other hand, *pro* means before, so prokaryotes are before seed cells because they don't have a nucleus.

Finding Common Ground: Structures in All Cells

Every living thing on Earth, including animals, plants, bacteria, yeast, and mold, is made of cells. Some living things, such as animals and plants, are *multicellular;* their bodies are made of many cells. Other living things, such as bacteria and yeast, are *unicellular* — made of just one cell. But whether a cell is one of many or the only one making up a living thing, all cells have certain things in common:

- Just like you have skin that covers your body, all cells have a boundary that separates them from their environment. The boundary of a cell is called the *plasma membrane* (or *cytoplasmic membrane*).
- The area inside all cells is called the *cytoplasm*.
- All cells contain *deoxyribonucleic acid* (DNA), which contains the plans for how the cell is built and how it functions.
- All cells make proteins to help them function. Proteins are built on structures called *ribosomes*, so all cells have ribosomes.

The following sections describe these four items.

Customs: Plasma membrane

The *plasma membrane,* shown in Figure 2-1, separates the cell from its environment and is *selectively permeable,* which means it chooses what enters and exits the cell. You can think of the plasma membrane as an international boundary where customs officers inspect the traffic and determine what is allowed to cross back and forth. The molecules that act like customs officers are proteins. Proteins called *receptors* detect signals from the environment of the cell, and *transport proteins* help some molecules get across the membrane. (For more details on how molecules cross the plasma membrane, see Chapter 9.)

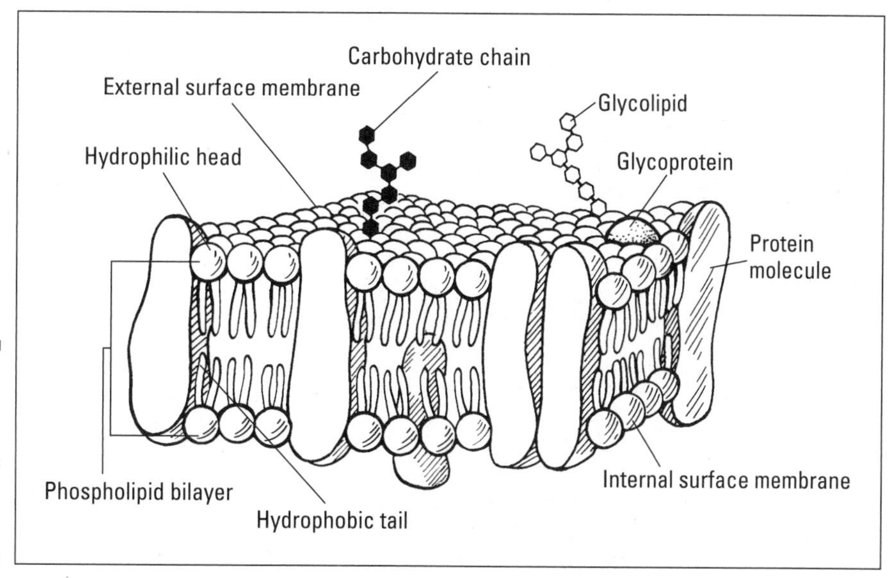

Figure 2-1: The fluid-mosaic model of plasma membranes.

The plasma membrane is made up of two layers of phospholipids along with proteins, sterols, and carbohydrates:

- **Phospholipids** are molecules that are similar in structure to fat molecules. Like fat molecules, part of the phospholipids — the *hydrophobic tails* — doesn't mix well with water. Phospholipids also have a *hydrophilic head* that is attracted to water. Phospholipids make up almost 50 percent of the plasma membrane.
- **Proteins** are stuck in the membrane and associated with the edges of the membrane. Proteins make up almost 50 percent of the plasma membrane.
- **Sterols** are also embedded in plasma membranes. The type of sterol depends on the type of cell. Animal cells have cholesterol in their plasma membranes. Sterols are present in small amounts in the plasma membrane.
- **Carbohydrates** are attached to receptors on the outside of the plasma membrane. They're present in small amounts in the plasma membrane.

The components of the plasma membrane are organized into a *phospholipid bilayer.* Because the hydrophobic tails of the phospholipids don't mix well in water, the two layers of phospholipids make a "fat sandwich" with their two rows of hydrophilic heads pointed toward the water and the hydrophobic tails sandwiched between them and away from the water. Transport proteins and sterols are embedded within the phospholipid bilayer, and carbohydrates are attached to receptors on the outside of the cell.

The structure and behavior of the plasma membrane are described by a theory called the *fluid mosaic model* of the plasma membrane, which basically says that membranes are made of several components and that these components can move within the membrane.

The phospholipids and proteins move back and forth within the plasma membrane, making the plasma membrane a fluid structure. Thus, the plasma membrane is flexible and able to fuse with other membranes. For example, small membranes may carry proteins up to the surface of the cell so that the protein can leave the cell. The membranes carrying the proteins simply melt into the plasma membrane, just like two soap bubbles merge with each other.

A happenin' place: The cytoplasm

The fluid-filled interior of the cell is called the cytoplasm. The cytoplasm is filled with molecules, structures, and activity, like a crowded party packed with people and conversations. Many of the chemical reactions that make up the *metabolism* of the cell happen in the cytoplasm, including important reactions that build proteins. Molecules and cellular components, including organelles, are constantly moving around in cells, being transported from one place to another or just moving randomly around due to their own *kinetic energy* (energy of motion) and attraction to other molecules.

The library: DNA-containing region

All cells contain DNA as their genetic material. (For more on DNA, see Chapter 7.) However, the location of the DNA is different in the two structural types of cells:

- In eukaryotic cells, the DNA is separated from the cytoplasm by membranes inside the cell, forming a structure called the nucleus. (For more information on the nucleus, see the section "Home office: The nucleus," later in this chapter.)
- In prokaryotic cells, the DNA is located within the cytoplasm in a region of the cell called the *nucleoid.*

Workbenches: Ribosomes

All cells need to be able to make proteins because proteins are the main worker molecules of the cell. Proteins are made on structures called ribosomes. All ribosomes have certain things in common:

- Ribosomes are made of two types of molecules: *ribosomal RNA* (rRNA) and proteins. (For more on rRNA, see Chapter 7.)
- The molecules that make up ribosomes are twisted together to form two components: the large subunit and the small subunit. These subunits are built separately from each other and come together to form a completed ribosome when protein synthesis begins.

The ribosomes of prokaryotic cells are different from those of eukaryotic cells. Although both types of ribosomes are made of rRNA and protein, the exact composition of those molecules is different.

Ribosome size is measured in *Svedberg units* (S), a unit that describes how fast particles fall out of solution during centrifugation. As a centrifuge spins things around really fast, larger, more dense particles fall to the bottom of the centrifuge tube ("spin out" of solution) faster than smaller, less dense particles. So, centrifugation, and Svedberg units, can tell you the relative size of particles, such as prokaryotic and eukaryotic ribosomes:

- **Prokaryotic ribosomes are smaller than eukaryotic ribosomes.** They're called *70S* ribosomes because complete ribosomal spin out at 70S. If you spin the ribosomal subunits separately, the large subunit spins out at 50S, and the small subunit spins out at 30S. Only in biology does 50+30=70! This answer is because when the two subunits join together to make a completed ribosome, they pack together into a tight package.
- **Eukaryotic ribosomes are larger than prokaryotic ribosomes** and spin out at 80S. The large subunit alone spins out at 60S, and the small subunit at 40S. More strange biological math: 60+40=80! Again, the two subunits pack together to form the complete ribosome.

In eukaryotic cells, ribosomes that are located in different places in the cell have slightly different functions:

- **Free ribosomes** are located in the cytoplasm of the cell. They make proteins that will function in the cytoplasm of the cell.
- **Membrane-bound ribosomes** attach themselves to the membrane of the *rough endoplasmic reticulum,* which is located inside cells. Membrane-bound ribosomes produce proteins that will either be part of membranes or that will be released from the cell.

Your Body, Your Cells: Eukaryotic Cells

The eukaryotic cells of animals, plants, fungi, and microscopic creatures called *protists* have many similarities in structure and function. They have the structures common to all cells: a plasma membrane, cytoplasm, and ribosomes.

All eukaryotic cells have a nucleus, organelles, and many internal membranes.

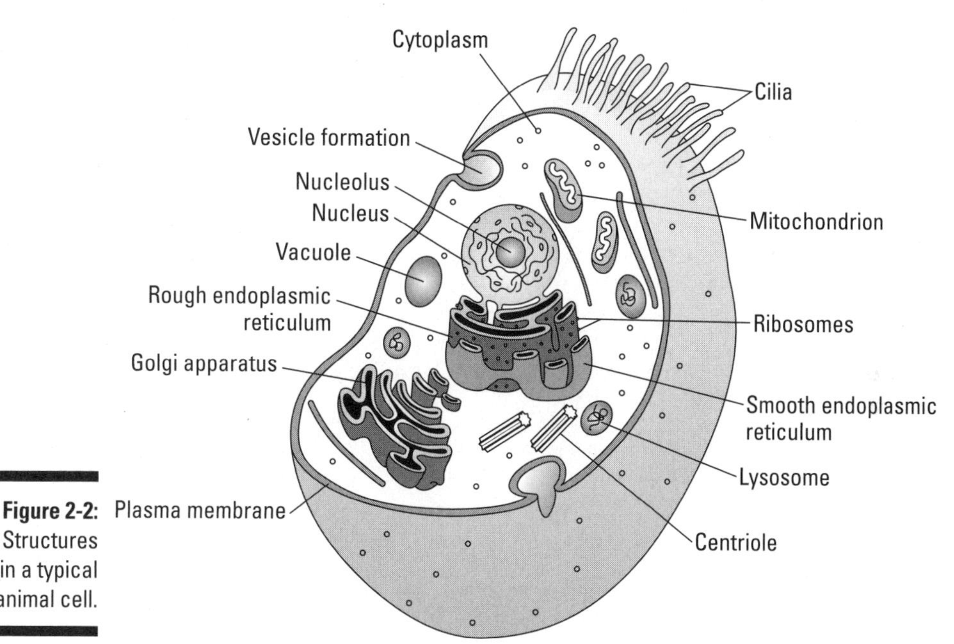

Figure 2-2: Structures in a typical animal cell.

With all the wonderful diversity of life on Earth, however, you're probably not surprised to discover that eukaryotic cells have differences, too. By comparing the structure of a typical animal cell, shown in Figure 2-2, with that of a typical plant cell, shown in Figure 2-3, you can see some of the differences among eukaryotic cells.

- *Cell walls,* additional reinforcing layers outside the plasma membrane, are present in the cells of plants, fungi, and some protists, but not in animal cells.
- *Chloroplasts,* which are needed for photosynthesis, are found in the cells of plants and algae, but not animals.
- *Large, central vacuoles,* which contain fluid and are separated from the cytoplasm with a membrane, are found in the cells of plants and algae, but not animals.
- *Centrioles,* small protein structures that appear during cell division, are found in the cells of animals, but not plants.

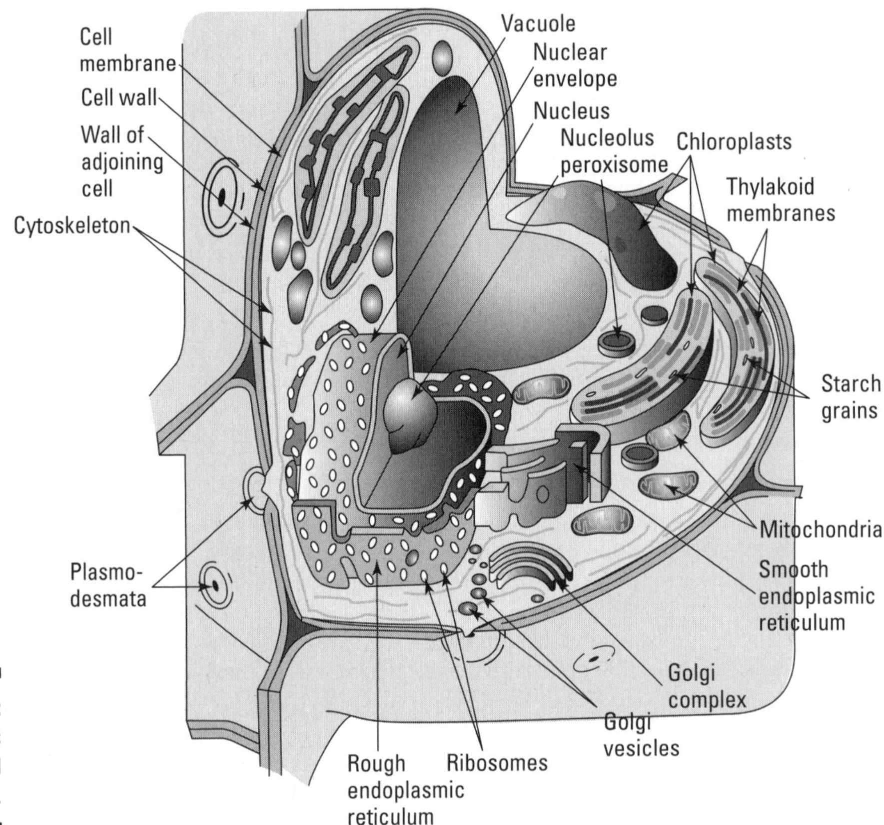

Figure 2-3: Structures in a typical plant cell.

Home office: The nucleus

The nucleus, shown in Figure 2-4, houses and protects the cell's DNA, which contains all of the instructions necessary for the cell to function. The DNA is like a set of blueprints for the cell, so you can think of the nucleus as the office where the blueprints are kept. If information from the blueprints is required, the information is copied into RNA molecules and moved out of the nucleus. The DNA plans stay safely locked away.

The boundary of the nucleus is the *nuclear envelope,* which is made of two phospholipid bilayers similar to those that make up the plasma membrane.

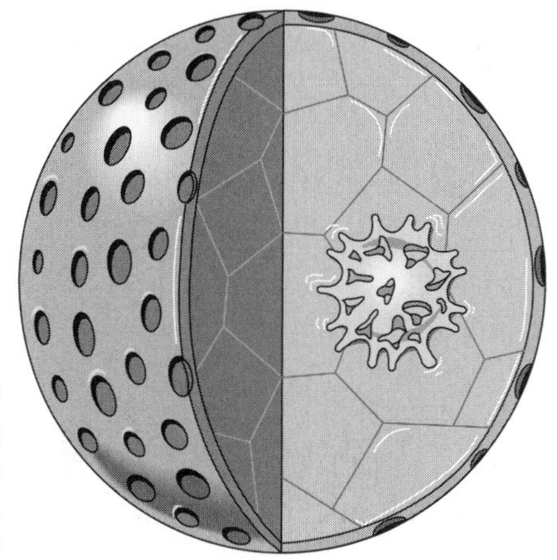

Figure 2-4: The nucleus.

The phospholipids bilayers of the nuclear envelope are supported by a scaffold of protein cables, called the *nuclear lamina,* on the inner surface of the nucleus. The nuclear envelope separates the contents of the nucleus from the cytoplasm. The structures within the nucleus are

- **DNA in the form of chromosomes or chromatin:** When a cell is about to divide to make a copy of itself, it copies its DNA and bundles the DNA up tightly so that the cell can move the DNA around more easily. The tightly bundled DNA molecules are visible through a microscope as little structures in the nucleus called *chromosomes.* Most of the time, however, when a cell is just functioning and not about to divide, the DNA is very loose within the nucleus, like a bunch of long spaghetti noodles. When the DNA is in this form, it is called *chromatin.*
- **Nucleoli where ribosomal subunits are made:** Information in the DNA needs to be read in order to make the small and large subunits needed to build ribosomes. The cell builds the ribosomal subunits in areas of the nucleus called nucleoli. Then, the cell ships the subunits out of the nucleus to the cytoplasm, where they join together for protein synthesis. When you stain cells and look at them under the microscope, nucleoli look like large spots within the nucleus.

The DNA plans for the cell are kept in the nucleus, but most of the activity of the cell occurs in the cytoplasm. Because the DNA is separate from the rest of the cell, a lot of traffic crosses back and forth between the nucleus and the cytoplasm. Molecules enter and exit the nucleus through small holes, called nuclear pores, that pass through the nuclear membrane. Nuclear pores are made by groups of proteins organized into little rings that penetrate through the nuclear envelope. The traffic in and out of the nuclear pores include the following:

- **RNA molecules** and **ribosomal subunits** that are made in the nucleus must exit to the cytoplasm.
- **Proteins** that are made in the cytoplasm but needed for certain processes, such as copying the DNA, must cross into the nucleus.
- **Nucleotides**, building blocks for DNA and RNA, must cross into the nucleus so that the cell can make new DNA and RNA molecules.

Traffic through the nuclear pores is controlled by proteins called *importins* and *exportins.* Proteins that are to be moved into or out of the nucleus have specific chemical tags on them that act like zip codes, telling the importins and exportins which way to move the protein with the tag. The movement of molecules into and out of the cell requires the input of energy from the cell in the form of adenosine triphosphate (ATP) — see Chapter 10.

Post office: The endomembrane system

The *endomembrane system,* shown in Figure 2-5, of the eukaryotic cell constructs proteins and lipids and then ships them where they need to go. Because this system is like a large package-shipping company, you can think of it as the post office of the cell.

The endomembrane system has several components:

- The *endoplasmic reticulum* is a set of folded membranes that begins at the nucleus and extends into the cytoplasm. It begins with the outer membrane of the nuclear envelope and then twists back and forth like switchbacks on a steep mountain trail. The endoplasmic reticulum comes in two types:
  - *Rough endoplasmic reticulum* (RER) is called rough because it's studded with ribosomes. Ribosomes that begin to make a protein that has a special destination, such as a particular organelle or membrane, will attach themselves to the rough endoplasmic reticulum while they make the protein. As the protein is made, it's pushed into the middle of the rough ER, which is called the *lumen.* Once inside the lumen, the protein is folded and tagged with carbohydrates. It will then travel to the Golgi apparatus for further processing.
  - *Smooth endoplasmic reticulum* (SER) does not have any attached ribosomes. It makes lipids — for example, phospholipids for cell membranes.

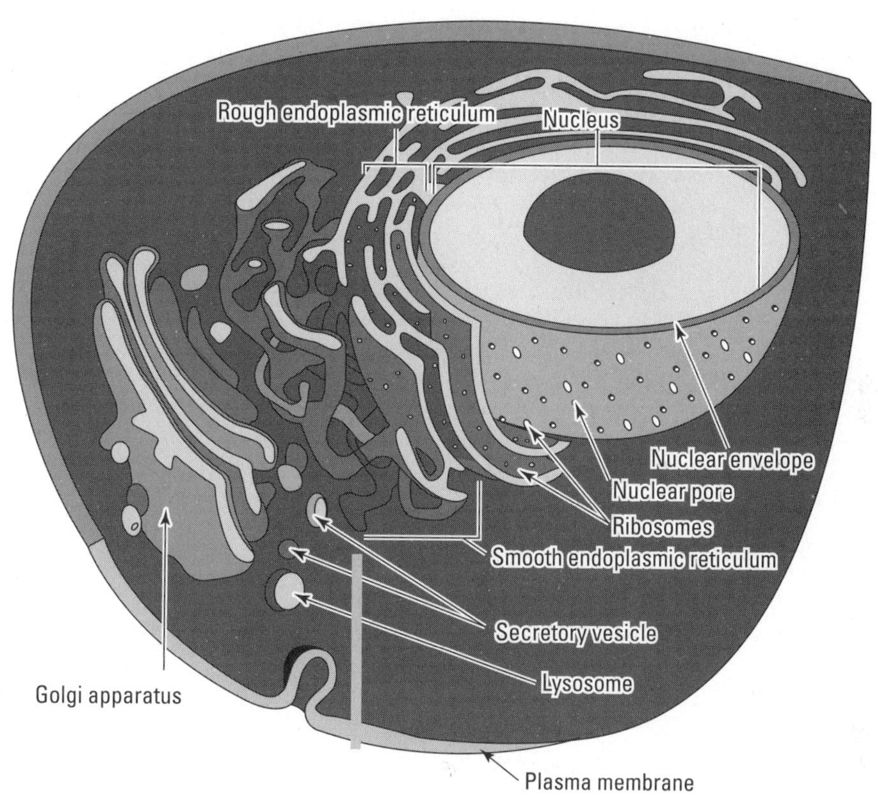

Figure 2-5: The endo-membrane system.

- The *Golgi apparatus* is a stack of flattened membrane sacs that looks a little bit like a stack of pancakes. The side of the stack closest to the nucleus is called the *cis* face of the Golgi, while the stack farthest from the nucleus is called the *trans* face. When proteins arrive at the cis face of the Golgi, they're pushed into the middle of the lumen. Inside the lumen, the proteins are modified — for example, by the addition of sugars — marked with chemical tags, and sorted so that they'll get shipped to their proper destination.
- *Vesicles* are little bubbles of membrane in the cell and come in several types:
  - *Transport vesicles* carry molecules around the cell. They're like the large envelopes that you put your letters in. Transport vesicles travel from the ER to the Golgi and then to the plasma membrane to bring molecules where they need to go. They travel by gliding along protein cables that are part of the cytoskeleton. (For more information on the cytoskeleton, see the section "Scaffolding and railroad tracks: The cytoskeleton," later in this chapter.)

- *Lysosomes* are the garbage disposals of the cell. They contain digestive enzymes that can break down molecules, organelles, and even bacterial cells.
- *Secretory vesicles* bring materials to the plasma membrane so that the cell can release, or secrete, the materials.

- *Peroxisomes* are small organelles encircled by a single membrane. Often, they help break down lipids, such as fatty acids. Also, depending on the type of cell they are in, peroxisomes may be specialists in breaking down particular molecules. For example, peroxisomes in liver cells break down toxins, such as the ethanol from alcoholic beverages. In plants cells, *glyoxisomes,* a special kind of peroxisome, help convert stored oils into molecules that plants can easily use for energy.

All together, the endomembrane system works as a sophisticated manufacturing, processing, and shipping plant. This system is particularly important in specialized cells that make lots of a particular protein and then ship them out to other cells. These types of cells actually have more endoplasmic reticulum than other cells so that they can efficiently produce and export large amounts of protein.

As an example of how the endomembrane system functions, follow the pathway of synthesis and transport for an exported protein:

1. **A ribosome begins to build a protein, such as insulin, that will be exported from the cell.**

   At the beginning of the protein is a recognizable marker that causes the ribosome to dock at the surface of the rough endoplasmic reticulum.

2. **The ribosome continues to make the protein, and the protein is pushed into the lumen of the RER.**

   Inside the lumen, the protein folds up, and carbohydrates are attached to it.

3. **The protein is pushed into the membrane of the RER, which pinches around and seals to form a vesicle, and the vesicle carries the protein from the RER to the Golgi.**

4. **The vesicle fuses with the cis face of the Golgi apparatus, and the protein is delivered to the lumen of the Golgi, where the protein is modified.**

5. **The protein eventually leaves in a vesicle formed at the trans face, which** travels to the plasma membrane, fuses with the membrane, and releases the protein to the outside of the cell.

The fireplace: Mitochondria

The *mitochondrion* (see Figure 2-6) is the organelle where eukaryotes burn their food by cellular respiration (see Chapter 11).

Mitochondria are like the power plants of the cell because they transfer energy from food to ATP. ATP is an easy form of energy for cells to use, so mitochondria help cells get energy.

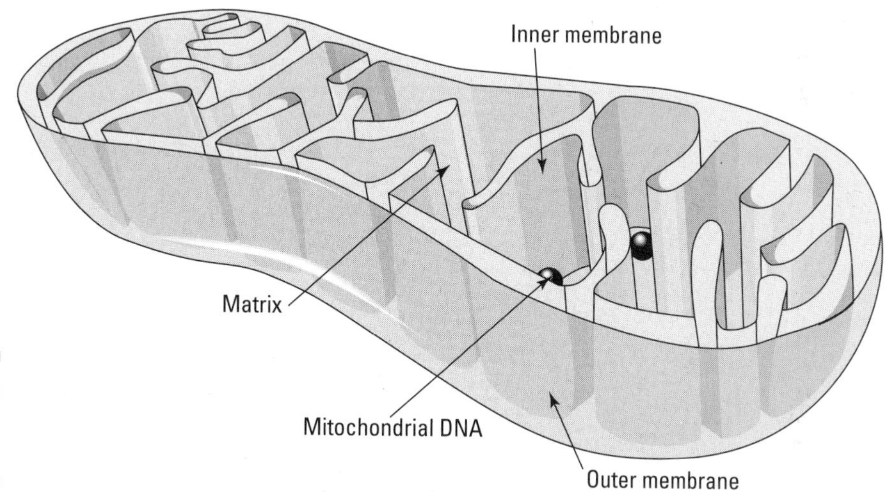

Figure 2-6: The mitochondrion.

Part of the process that extracts the energy from food requires a membrane, so mitochondria have lots of internal folded membrane to give them more area to run this process. Mitochondria actually have two membranes, the *outer membrane* and the *inner membrane*. The inner membrane is the one that is folded back and forth to create more area for energy extraction; the folds of this membrane are called *cristae*. The outer membrane separates the interior of the mitochondrion from the cytoplasm of the cell.

The two membranes of the mitochondrion create different compartments within the mitochondrion:

- The space between the two membranes of the mitochondrion is the *intermembrane space.*
- The inside of the mitochondrion is the *matrix.*

Mitochondria also contain ribosomes for protein synthesis and a small, circular piece of DNA that contains the code for some mitochondrial proteins.

In the kitchen: Chloroplasts

Chloroplasts, shown in Figure 2-7, are the place where eukaryotes make food molecules by the process of photosynthesis (see Chapter 12). Chloroplasts are found in the cells of plants and algae.

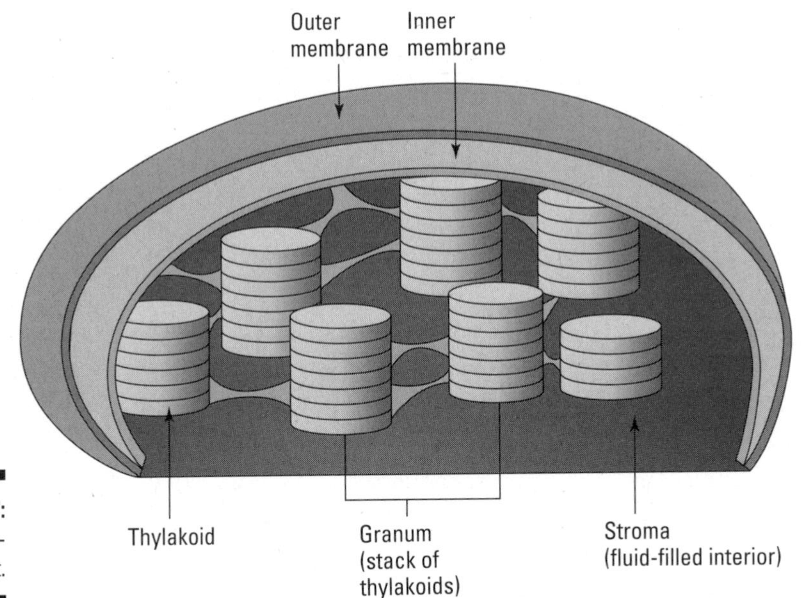

Figure 2-7: The chloroplast.

Like mitochondria, choloroplasts have two membranes, an inner membrane and an outer membrane. In addition, they have little sacs of membranes called *thylakoids* stacked up in towers called *grana*.

The multiple membranes of the chloroplast divide it into several different spaces:

- The *intermembrane space* is between the inner and outer membranes.
- The central, fluid-filled part of the chloroplast is called the *stroma.*
- The interior of the thylakoid is another fluid-filled space.

Where do little organelles come from?

Almost since people first peered at eukaryotic cells under a microscope, they've been wondering about the little structures they saw inside the cells. What were they? Where did they come from? Some people thought these structures looked like little bacteria living inside of the eukaryotic cells. They wondered if cells living inside of other cells, which is a type of *symbiosis,* had led to the formation of these organelles.

In the early 1900s, a French scientist named Portier proposed that mitochondria were the descendants of symbiotic bacteria. Around the same time, a Russian scientist named Mereschovsky proposed that chloroplasts had a similar origin. Both of these scientists based their ideas on a comparison of the physical appearance of the organelles to that of bacteria. Their ideas fell by the scientific wayside until the 1960s, when an American scientist named Lynn Margulis brought them back and unified them into the *serial endosymbiotic theory* of the origin of the eukaryotic cell. She proposed that symbiosis was very important in the origin of the eukaryotic cell and that it had led to the development of cell structures, including the mitochondrion and chloroplast.

Dr. Margulis's ideas were considered radical, and many scientists disagreed. However, as the science of molecular and cell biology advanced and new tools were developed, Dr. Margulis had the last word! The presence of ribosomes and DNA within the mitochondria and chloroplasts suggested that they may have once been free-living cells. And when the ribosomes of chloroplasts and mitochondria were examined, they were found to be prokaryotic ribosomes like those of bacteria rather than eukaryotic ribosomes like those in the cytoplasm of the cell. Likewise, when DNA sequences from chloroplasts and mitochondria were read and compared to those of bacteria and the DNA from the eukaryotic nucleus, the organelles again proved to be closer cousins to the bacteria than they were to the nuclei of the cells in which they lived. These data have led biologists today to agree that chloroplasts and mitochondria actually belong on the bacterial family tree!

In other words, a long time ago, bacteria lived inside the ancestor of the eukaryotic cell. Over time, these bacteria changed slightly and became the mitochondria and chloroplasts scientists know today.

Like mitochondria, chloroplasts contain their own ribosomes for protein synthesis and a small, circular piece of DNA that contains the code for some chloroplast proteins.

Scaffolding and railroad tracks: The cytoskeleton

The structure and function of cells are supported by a network of protein cables called the *cytoskeleton,* shown in Figure 2-8. These proteins underlie membranes, giving them shape and support, much like scaffolding can support a building. Cytoskeletal proteins run like tracks through cells, enabling the movement of vesicles and organelles like trains on a railroad track. When cells swim by flicking whip-like extensions called *cilia* and *eukaryotic flagella,* they're using cytoskeletal proteins. In fact, you use cytoskeletal proteins literally every time you move a muscle.

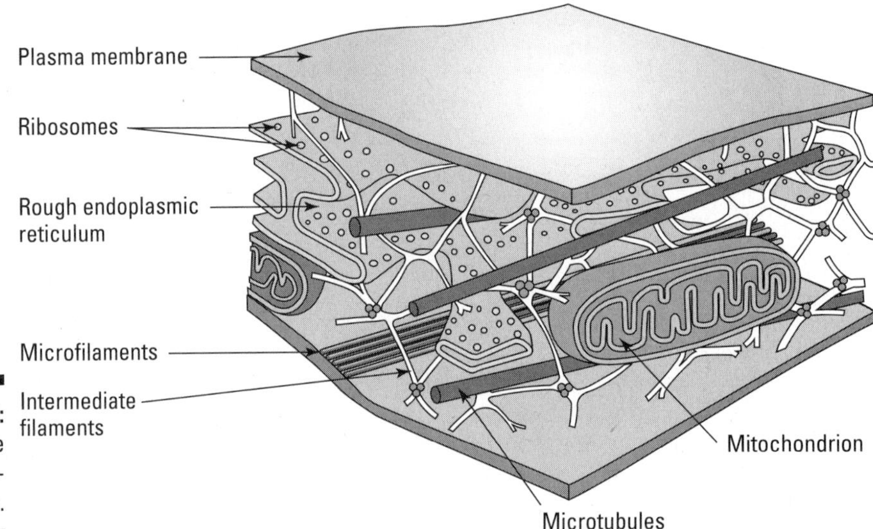

Figure 2-8: The cytoskeleton.

Cytoskeletal proteins come in three main types, with each one playing a different role in cells:

- *Microfilaments* are made of the protein *actin.* Microfilaments are the proteins that make muscle cells contract, help pinch animal cells in two during cell division, allow cells like amoebae to crawl, and act as railroad tracks for organelles in some types of cells.
- *Microtubules* are made of the protein *tubulin.* Microtubules are the proteins inside of *cilia* and *flagella.* They move chromosomes during cell division and act as railroad tracks for the movement of vesicles and some organelles.
- *Intermediate filaments* are made of various proteins. They often act as reinforcing proteins. For example, the protein lamin that strengthens the nuclear membrane is an intermediate filament. Likewise, the keratin that strengthens your skin cells and makes them resistant to damage is an intermediate filament.

You can easily mix up the words "microtubules" and "microfilaments." Remember that "microtubules" are made of "tubul-in," and they're found in the "tube-shaped" cilia and flagella. (Okay, I'm stretching it on that last bit, but if it helps to remember it. . . .)

Motor proteins

Actin microfilaments and microtubules are long, cable-like proteins. They partner with motor proteins, proteins that use ATP to "walk" along the cables by repeatedly binding, changing shape, and releasing. Thus, the motor proteins use chemical energy to do cellular work in the form of movement. Several motor proteins work with microfilaments and microtubules:

- *Myosin* often acts as a partner to actin. For example, when myosin walks along actin microfilaments in muscle cells, it causes the actin microfilament to slide. The sliding of actin microfilaments is what causes muscle contraction. Myosin also attaches to cellular components, such as chloroplasts in plant cells, and then walks along microfilaments. The movement of the motor proteins causes the cellular components to flow around the cell in a process called *cytoplasmic streaming.*
- *Dynein* partners with microtubules inside of cilia and eukaryotic flagella. When dynein walks along microtubules on one side of a cilium or flagellum, it causes the microtubules to bend. The bending of different parts of cilia and flagella makes them flick back and forth like little whips.
- *Kinesin* is another partner with microtubules. One end of the kinesin molecule attaches to vesicles, while the other end walks along the microtubules. The movement of kinesin causes the vesicles to slide along the microtubules like freight cars on a railroad track.

Cilia and flagella

Cilia and flagella are essentially the same structure, but cilia are typically shorter and more numerous on the surface of the cell whereas flagella are typically longer in length and fewer in number. Cilia are found on cells that make up the surfaces of tissues, such as cells in the respiratory and genital tracts of humans, where the cilia beat to move fluid and materials along the surface. For example, in the human respiratory tract, the beating of cilia moves mucus upward where you can cough it out of the body. Some cells, such as microscopic protists and sperm cells, swim using cilia and flagella.

The internal structure of cilia and flagella is distinctive. If you cut a cilium or a flagellum crosswise and look at the circular end with an electron microscope, you'll see the same pattern of microtubules in in both cilia and flagella. The microtubules are grouped in pairs, called *doublets,* that are similar to two drinking straws laid tightly together side by side.

Your cilia are smoked, dude

Smoking cigarettes paralyzes the cilia on the cells that make up the surface of your respiratory tract. The job of those cilia is to beat like little whips, moving mucus upward where you can cough it out. Mucus is the sticky, gooey stuff that traps dust, pollen, and germs when you breathe them in. So, when a smoker paralyzes their cilia, they stop the *ciliary escalator* that was moving all that junk out of their body. The junk stays in their lungs and makes them more susceptible to colds and other diseases. Also, poisons from the cigarette smoke stay in their lungs and combine to form tar. When smokers sleep (and thus aren't smoking), some of their cilia recover. So, first thing in the morning, those cilia try to do their job again. The result is the deep hack of a smoker's cough as they try to clear the junk out of their lungs.

The microtubules appear in a *9+2 arrangement,* where nine pairs of microtubules (nine doublets) are arranged around the outside of the circle, while one pair of microtubules is in the center of the circle.

Rebar and concrete: Cell walls and extracellular matrices

The plasma membrane is the selective boundary for all cells that chooses what enters and exits the cell. However, most cells have additional layers outside of the plasma membrane. These extracellular layers provide additional strength to cells and may attach cells to neighboring cells in multicellular organisms. Typically, these layers are composed of long cables of carbohydrates or proteins embedded in a sticky matrix. The long, cable-like molecules work like rebar in concrete to create a strong substance. Two main types of extracellular layers support eukaryotic cells:

- Cell walls are extra reinforcing layers that help protect the cell from bursting. Among eukaryotes, cell walls appear around the cells of plants, fungi, and many protists.
    - The cell walls of plants and algae are made of cellulose. If the plant is a woody plant, lignin is also present.
    - Fungal cell walls are made of chitin.
- The layer around animal cells is the *extracellular matrix* (ECM), shown in Figure 2-9. This layer is made of long proteins, such as *collagen,* embedded in a polysaccharide gel. The ECM supports animal cells and helps

bind them together. Animal cells actually attach themselves to the ECM via proteins, called *integrins,* that are embedded in the plasma membrane. The integrins bind to the actin microfilaments inside the cell and to ECM proteins called *fibronectins* that are outside the cell.

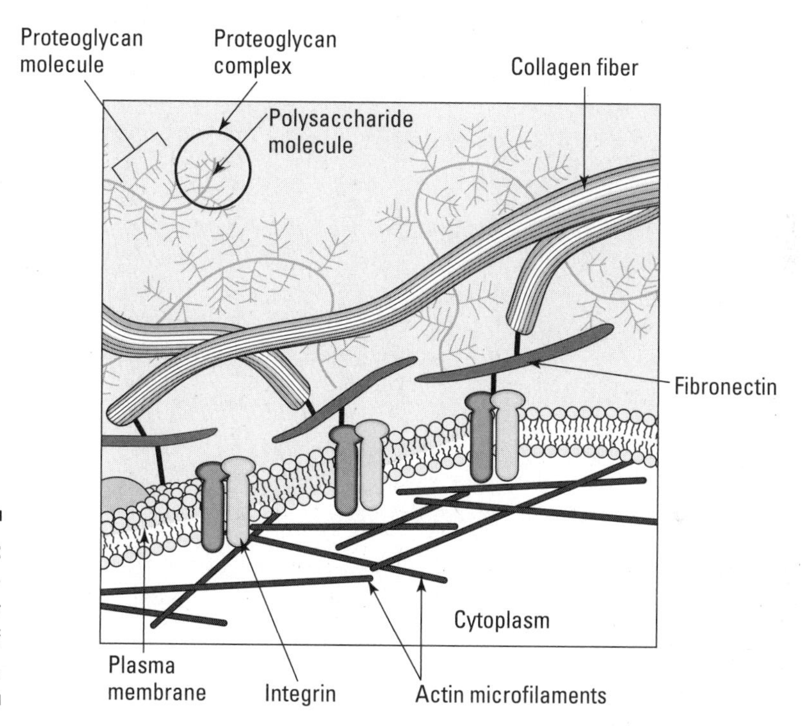

Figure 2-9: The extracellular matrix of animal cells.

Tiny but Mighty: Prokaryotic Cells

Prokaryotic cells are smaller than eukaryotic cells, and, at first glance at least, they appear simpler in structure. However, prokaryotic cells have many structures in common with eukaryotic cells, including a plasma membrane and DNA-containing region, as well as unique structures that perform similar functions as those in eukaryotic cells.

As a group, prokaryotes are very successful — they live in every environment on Earth, from the saltiest seas to the hottest hot springs to dark caves without any light, to the intestinal tracts of animals, including yours. Most of what people know regarding prokaryotic cell structure (see Figure 2-10) was learned by studying bacteria, so that is what I emphasize in this section.

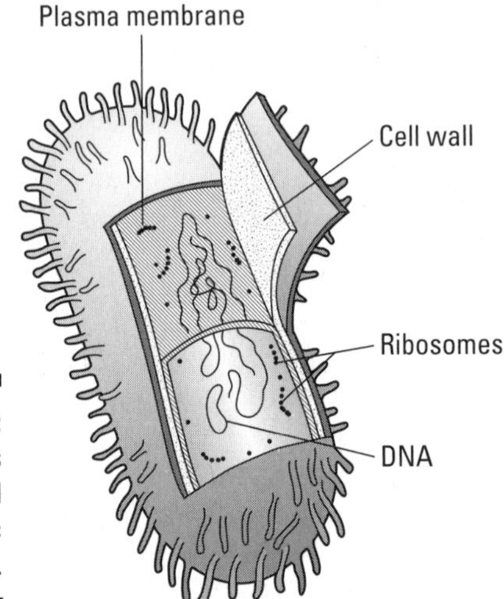

Figure 2-10: Structures in a typical prokaryotic cell.

Castle walls: The cell wall

Almost all prokaryotic cells have a cell wall outside the plasma membrane. As in eukaryotes, the cell wall reinforces the cell and protects it from bursting. Several types of cell walls support prokaryotic cells, but the biggest difference is between the walls of bacteria and archaeans:

- In bacteria, the cell wall is made of the carbohydrate-protein complex *peptidoglycan.*
- Some archaeans have cell walls, and some do not. Archaeans don't have peptidoglycan in their cell walls.

Taking aim on bacterial infections

You have a raging bacterial infection, so your doctor prescribes some antibiotics to help you fight it off. Antibiotics are chemicals that kill bacteria. But wait — bacteria are cells, right? And your body is made of cells, too, right? So, why is it safe to take an antibiotic? The secret lies in the target of the antibiotic — the drug has to target something that bacterial cells have, but your cells don't. That way, you kill them without hurting yourself.

Peptidoglycan is one such target. It's present in the cell walls of bacterial cells, but your cells don't make it. (Your cells don't even have cell walls.) All the drugs that end in *-cillin,* such as penicillin, amoxicillin, and methicillin, target peptidoglycan and thus cause the death of bacterial cells. Your immune system gets a helping hand, and you feel better!

Ooze, slime, and grappling hooks: Capsules, pili, and fimbriae

Many bacteria produce an additional layer called a *capsule* outside the cell wall. This sticky layer is composed of proteins and carbohydrates that help bacterial cells attach to surfaces and to other cells. The capsule also helps bacterial cells avoid capture by certain cells of the human immune system.

In addition, some bacteria attach to surfaces and to other cells by slender protein threads called *pili* and *fimbriae.* For bacteria that cause disease, attachment to the human body is the first step in disease. So, for some bacteria, pili and fimbriae are essential to the disease-causing process.

Outboard motors: Bacterial flagella

Many bacteria swim via *bacterial flagella.* Bacterial flagella are filaments of a protein called *flagellin*. They're attached to a wheel-like structure in the cell wall of bacterial cells. When the wheel rotates, the bacterial flagella spin and cause the bacterial cell to move forward, much like an outboard motor causes a boat to move. Bacterial cells can have anywhere from one to many flagella.

Bacteria and eukaryotes both have structures called flagella, but they are different in structure. Eukaryotic flagella contain microtubules in a 9+2 arrangement and they move by flexing and beating like little whips. Bacterial flagella are made of a protein called flagellin and they move by rotating like outboard motors.

Chapter 3

Dead or Alive: Viruses

In This Chapter

- Designing viruses
- Multiplying viruses
- Comparing viruses and bacteria
- Attacking eukaryotic cells

Viruses are microscopic particles of nucleic acid and protein that attack cells and turn them into factories for producing more viral particles. Viruses have different shapes and patterns of multiplication within cells, and the ones that attack humans cause many significant diseases, including measles, polio, influenza and AIDS. In this chapter, I explore the structure of viruses and compare the steps of viral multiplication in bacteria and eukaryotic cells.

Viruses: Hijackers of the Cellular World

Viruses and bacteria make people sick and are both invisible to our eyes. As a result, they're often confused with each other. However, they aren't the same thing at all. Bacteria are made of cells (see Chapter 2) and can live freely on their own, obtaining energy from food, reproducing, and responding to their environment. In contrast, viruses are not cells — they're tiny, cellular hijackers that are stripped down to the most essential parts. They attach themselves to cells, slip in, and take over, using the materials and structures of the cell to make more viruses. Without a cellular victim, viruses can't do anything. Because they can't live and reproduce on their own, many scientists don't consider viruses to be truly alive. Some viruses have the ability to become dormant inside of a host cell. The genetic material of dormant viruses may remain in the host cell for long periods of time and is copied as the cell reproduces.

Just the basics: The structure of viruses

The simplest viruses have just two components: a nucleic acid *core* and protein *capsid*. The nucleic acid core, which may be DNA or RNA, contains the instructions for taking over cells and making more *virions*, or viral particles. The nucleic acid is surrounded by the capsid, a protective protein coat shown in Figure 3-1. Each individual protein that makes up the capsid is called a *capsomere*.

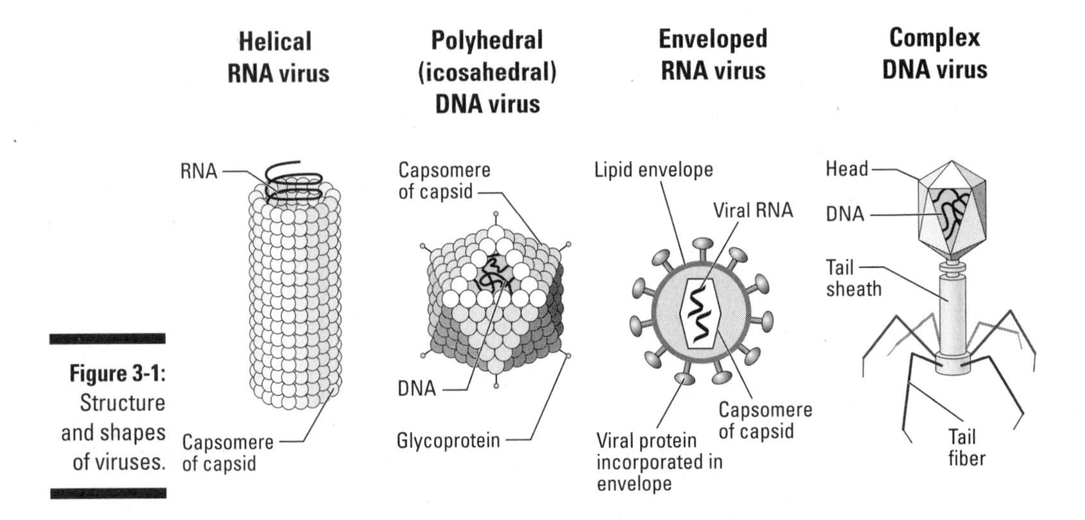

Figure 3-1: Structure and shapes of viruses.

All viruses have at least a capsid and a nucleic acid core. The core consists of one of four types of nucleic acid:

- Double-stranded DNA
- Single-stranded DNA
- Double-stranded RNA
- Single-stranded RNA

One difference between cells and viruses is that cells contain DNA and RNA. However, a single viral particle contains only DNA or RNA. Also, single-stranded DNA and double-stranded RNA are commonly found in viruses, but not in cells.

In addition to the capsid and the core, some viruses have an outer membrane layer called an *envelope*. It's no coincidence that the envelope of a virus is similar to the plasma membrane of a cell (see Chapter 2) — viruses that have envelopes steal them from their cellular victims as they leave the cell! Viral envelopes aren't exactly the same as plasma membranes because they've been changed to suit the needs of the virus by the addition of viral proteins. Once modified and adopted, the envelope helps the virus enter and exit from host cells.

Viruses may also have proteins that stick out of the envelope or off the surface of the capsid. These proteins, called *spikes,* help the virus attach to host cells.

Viruses come in three common shapes:

- *Helical viruses* have a capsid that forms a twisting helix around the nucleic acid core.
- *Polyhedral viruses* have a regular geometric shape. The most complex polyhedral viruses are *icosahedrons* with 20 faces.
- *Complex viruses* have separate patches of proteins, often forming unique structures or extensions on the virus.

Under the microscope, enveloped viruses appear irregular in shape. However, a helical or polyhedral capsid may be located underneath the envelope.

Chemical warfare

You're sick and miserable, so you go see your doctor. Your doctor examines you, tells you that you have a cold caused by a virus, and advises you to go home, get some rest, and drink plenty of fluids. The doctor doesn't give you a prescription because antibiotics don't work on viruses. In fact, for most viruses, scientists don't have any effective antiviral drugs. To be safe and effective, antibiotics and antiviral drugs must attack whatever is making you sick, without harming your healthy cells.

When bacteria make you sick, they present plenty of targets for chemical attack because they have many things your cells don't — cell walls, different ribosomes, and different enzymes. Viruses, however, are a completely different story. They're minimalists — stripped-down attack machines of protein and nucleic acid. That's not much to target! Viruses reproduce themselves by using the parts and materials from your cells, and you certainly don't want to target your own cells. So, in order to develop antiviral drugs, scientists need to spend a great deal of time researching individual viruses in order to discover the few unique molecules they bring into your cells. So far, scientists have managed to develop antiviral drugs for only a few highly studied viruses, including the HIV virus that causes AIDS, herpes viruses, and the influenza virus.

Knock, knock, virus calling: How viruses get into cells

Viruses attach to cells when viral proteins successfully bind to receptors on the host cell. If the viral protein has the right shape, it will tuck into the corresponding shape on the host cell receptor. You can think of viral attachment as a virus having the right key to fit into the lock on the host cell. After the virus is attached, it may force itself into the cell by digging a hole through a cell wall (see Figure 3-2), slip in by fusing its envelope with the membrane of the host cell, or trick the cell into bringing it inside.

The ability of a virus to infect a host cell depends on a match between proteins on the surface of the virus and receptors on the surface of the host cell.

The type of cells a particular virus can infect is called the *host range* of the virus. Because viruses can infect only cells that they can attach to with their proteins, each virus has a very *specific* range of hosts it can infect. In other words, each virus can infect only the host cells for which it has keys. Some viruses have a key that works in the lock on many types of cells. These viruses have a *broad host range.* For example, the rabies virus can infect humans and many other mammals. On the other hand, some viruses have a key that fits into the lock on only a few cells. These viruses have a *narrow host range.* The HIV virus, which infects only certain cells of the human immune system, is a good example of a virus with a very narrow host range.

War on a Microcosmic Scale: Viruses of Bacteria

Viruses and bacteria can both be the enemy of people, but did you know that they can be enemies of each other too? Viruses attack bacterial cells in a way that is very similar to the way they attack your cells. Viruses that attack bacteria are called *bacteriophage* or just *phage* for short. ("Phage" means "eat," so bacteriophage are bacteria eaters.) Bacteriophage attach to bacterial cells, enter the cells, and hijack them, using the materials and structures in the cell to make more phage. This series of events that occurs as viruses hijack cells is called the *multiplication cycle* of the virus. (It may also be called the replication cycle, or more rarely, the life cycle of the virus.)

Bacteriophage reproduce via two different multiplication cycles (Figure 3-2):

- In the *lytic cycle,* bacteriophage immediately use the host cell, making more phage and destroying the host cell in the process.
- In the *lysogenic cycle,* bacteriophage enter a dormant, or resting phase, where their genetic material is incorporated into that of the bacterium. As the bacterial cell reproduces, it copies the phage DNA along with its own. This process creates a population of bacterial cells that are all infected with the bacteriophage.

Some bacteriophage attack only in the lytic mode. Other bacteriophage can switch between lytic and lysogenic modes and are called *temperate* phage.

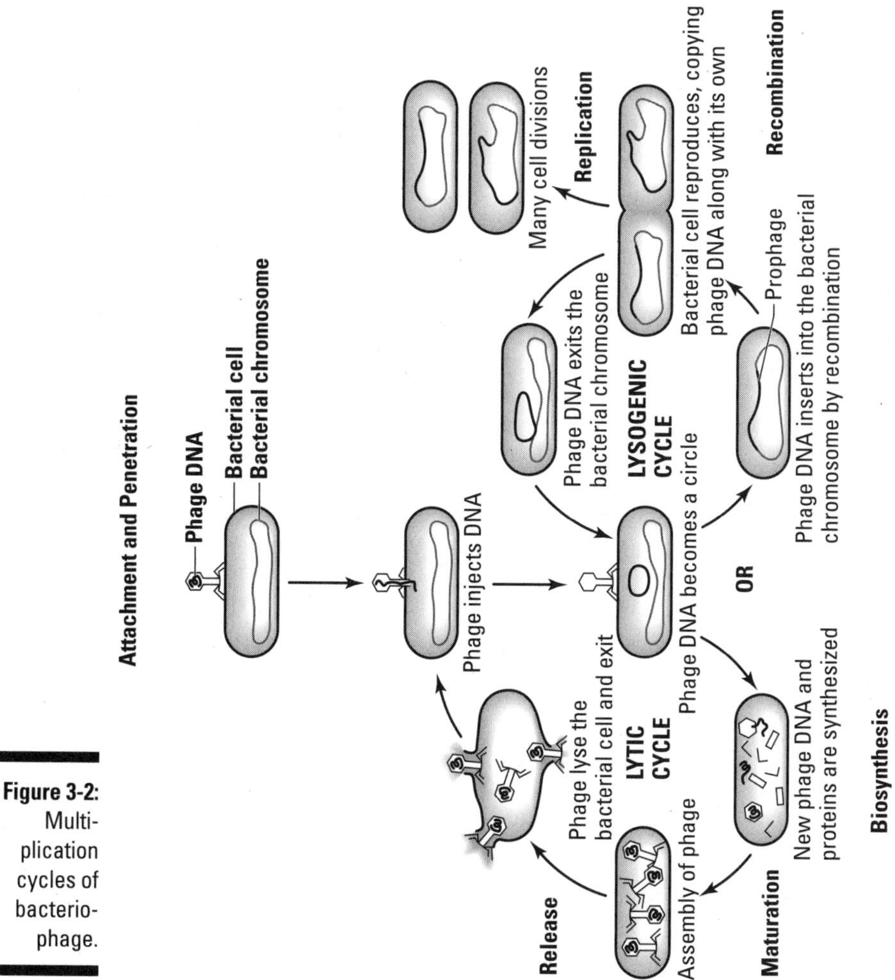

Figure 3-2: Multiplication cycles of bacteriophage.

Seek and destroy: The lytic cycle

Bacteriophage that attack by the lytic cycle immediately convert the bacterial cell to a virus-producing factory. The steps of the lytic cycle, which are shown in Figure 3-2, are as follows:

1. **Attachment.**

   Proteins on the surface of the virus attach to receptors on the surface of the host cell.

2. **Penetration.**

   The genetic material of the virus enters the host cell.

3. **Biosynthesis.**

   The virus uses building blocks from the host cell such as nucleotides and amino acids to make viral proteins and nucleic acids. ATP from the host cell supplies the energy the virus needs to reproduce.

4. **Maturation.**

   The parts of the viral particles assemble into new bacteriophage.

5. **Release.**

   The new bacteriophage exit the host cell, *lysing,* or destroying, the cell as they leave.

I think I'll take a little nap: The lysogenic cycle

Bacteriophage that enter the lysogenic cycle remain inactive within the bacterial cell for a period of time. During their dormant period, their DNA is integrated into the DNA of the bacterial cell. A phage that is integrated into bacterial DNA is called a *prophage.* The steps of the lysogenic cycle are as follows (see Figure 3-2):

1. **Attachment.**

   Proteins on the surface of the virus attach to receptors on the surface of the host cell.

2. **Penetration.**

   The genetic material of the virus enters the host cell.

The enemy of my enemy is my friend

Viruses that attack bacteria may prove to be your friend in the search for new ways to fight antibiotic-resistant bacteria. Bacteriophage attack bacterial cells, not human cells, so they have the potential to be very specific killers of infecting bacteria. Scientists around the world are testing this idea by introducing bacteriophage into animals that have been infected with antibiotic-resistant bacteria. And in Eastern Europe, bacteriophage therapy is already being used to treat human infections. Some day in the not too distant future, your "prescription" for a bacterial infection may be an injection of virus!

3. **Recombination.**

   The phage DNA inserts into the bacterial DNA by recombination.

4. **Replication.**

   The bacterial cell multiplies, copying the viral DNA along with its own DNA.

Environmental signals may trigger the phage to return to the lytic mode, at which time they'll remove themselves from the bacterial DNA and become active.

I've Got a Cold: Viruses of Eukaryotes

For all of recorded history, humans have done a deadly dance with viruses. Measles, smallpox, polio, and influenza viruses changed the course of human history: Measles and smallpox killed hundreds of thousands of native Americans; polio killed and crippled people, including American president Franklin Delano Roosevelt; and the 1918 influenza epidemic killed more people than were killed during all of World War I.

For most viruses that attack humans, your only defenses are prevention and your own immune systems. Antibiotics don't kill viruses, and scientists haven't discovered many effective antiviral drugs.

Vaccines are little pieces of bacteria or viruses injected into the body to give the immune system an education. They work by ramping up your own defensive system so that you're ready to fight the bacteria or virus upon first contact, without becoming sick first. However, for some viral diseases no vaccines exist, and the only option is to wait uncomfortably for your immune system to win the battle.

An ounce of prevention is worth a pound of cure

Most people living in the United States don't worry about measles anymore. Before the measles vaccine became available in 1963, however, people — especially parents — had much more cause for alarm. Before 1963, approximately 8 million deaths per year occurred worldwide due to measles. When measles is contracted by young children, particularly before age 2, it can lead to serious complications and a nasty death. The development of a vaccine for measles saved countless children from an early death. The dramatic beneficial effect of measles immunization is seen in this figure that shows the effect of three major campaigns to get kids in the United States vaccinated against measles.

Today, however, some kids in the United States are in danger again. Flawed scientific studies reported a possible link between the MMR vaccine and autism. This information, combined with confusion about vaccines, has made some parents afraid to get their kids vaccinated. This fear persists despite five extensive scientific studies that failed to find a link between the MMR vaccine and autism. Most people don't remember the dangers of measles, so the fear of autism looms larger than their fear of measles. In a way, vaccines are a victim of their own success — they prevent disease, so people lose their fear of the disease and stop vaccinating. The consequences of this choice can be severe — notice the rise in measles cases in the 1980s when vaccination rates fell.

All vaccines have real risks of side effects that occur in a small percentage of people who get vaccines. In order for a vaccine to be approved, however, the risks due to side effects must be smaller than the risks of the disease itself. When making decisions about yourself and your loved ones, you need good information from a reliable source so that you can clearly separate real risks from rumors. Good information on vaccines and their safety is available from the Vaccine Education Center at the Children's Hospital of Philadelphia. Check it out at `www.chop.edu/consumer/jsp/division/generic.jsp?id=75697`.

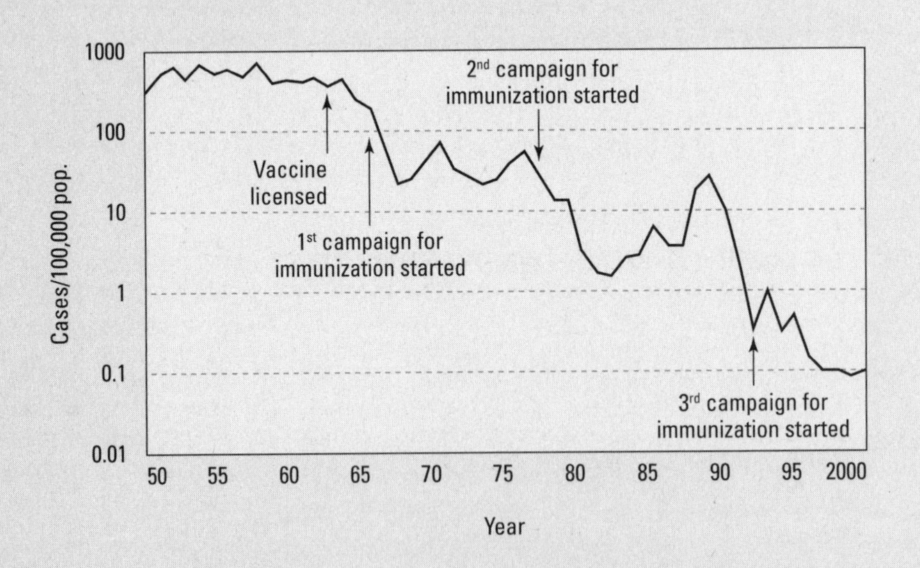

Both antibiotics and vaccines help you fight against disease. However, they work in entirely different ways. Antibiotics are chemicals that poison bacteria by attacking structures found in bacterial cells but not the cells of humans. Antibiotics don't work against viruses because viruses don't have the same structures as bacteria. Vaccines work by educating your immune system and preparing you in advance to fight a particular bacterium or virus. Vaccines work against both bacteria and viruses.

Same story, different players

Viruses that attack eukaryotic cells follow the same basic principles as viruses that attack bacterial cells, but some of the terms used are different:

- **Latent instead of lysogenic:** Eukaryotic viruses, such as herpes, that can have a dormant period within the host cell are called *latent* viruses.
- **Provirus instead of prophage:** A virus that is dormant inside the eukaryotic cell is called a *provirus*.
- **Acute disease instead of lytic cycle:** For viruses that attack humans, a virus that strikes quickly, reproduces rapidly, and causes a rapid but relatively short illness is said to cause an *acute illness*. A good example of viruses that causes an acute illness are the viruses that cause the common cold.

Despite some differences in terminology, the basic events in the multiplication cycle are the same for a virus that attack eukaryotic cells as they are for a virus that attacks prokaryotic cells: attachment, penetration, biosynthesis, maturation, and release. However, viruses of eukaryotes have to navigate around the eukaryotic cell, which is structurally more complex than the bacterial cell (see Chapter 2). Also, the genetic material of eukaryotic viruses is more varied than that of bacterial viruses. As a result, the multiplication cycles of viruses that attack eukaryotes are generally more complex than those of viruses that attack bacteria.

Come in and take your coat off

Viral attack of a eukaryotic cell begins with the attachment of viral proteins to receptors on the host cell (see Figure 3-3). After the virus binds to the host cell, it crosses the plasma membrane of the host. The methods by which viruses enter the cell are best understood for viruses that have an envelope. (See the section "Just the basics: The structure of viruses," earlier in this chapter, for more details on envelopes.)

After an enveloped virus attaches to a host cell, it may enter the cell in one of two ways:

- *Fusion* occurs when the envelope of the virus melts into the plasma membrane of the animal cell. This fusion of membranes is just like two soap bubbles merging into one. Once the membranes are fused, the viral capsid is inside the host cell.
- *Receptor-mediated endocytosis* occurs when binding of the virus to the host cell causes the cell to bring the virus inside. The plasma membrane of the host cell reaches out, wraps around the virus to form a vesicle, and brings the virus inside, envelope and all.

When a virus enters a host cell by fusion or receptor-mediated endocytosis, the viral capsid enters the cell. Before the virus can begin to make more of itself, it has to free its genetic material from the capsid. In other words, it has to take its coat off — which is why the process of removing the capsid is called *uncoating*. Usually, uncoating happens when enzymes from the host cell break down the viral capsid, releasing the genetic material of the virus.

1. **Free Virus.**
2. **Binding and Fusion:** Virus binds to CD4 and coreceptor on host cell and then fuses with the cell.
3. **Penetration:** The viral capsid enters the cell and releases its contents into the cytoplasm.
4. **Reverse Transcription:** The enzyme reverse transcriptase converts the single-stranded viral RNA molecules into double-stranded DNA.
5. **Recombination:** The enzyme integrase combines the viral DNA into the host cell DNA.
6. **Transcription:** Viral DNA is transcribed to produce long chains of viral protein.
7. **Assembly:** Sets of viral proteins come together.
8. **Budding:** Release of immature virus occurs as viral proteins push out of the host cell, wrapping themselves in a new envelope. The viral enzyme protease begins cutting the viral proteins.
9. **Release:** Immature virus breaks free of the host cell.
10. **Maturation:** The viral enzyme protease finishes cutting the viral proteins, and the proteins combine to complete the formation of the virus.

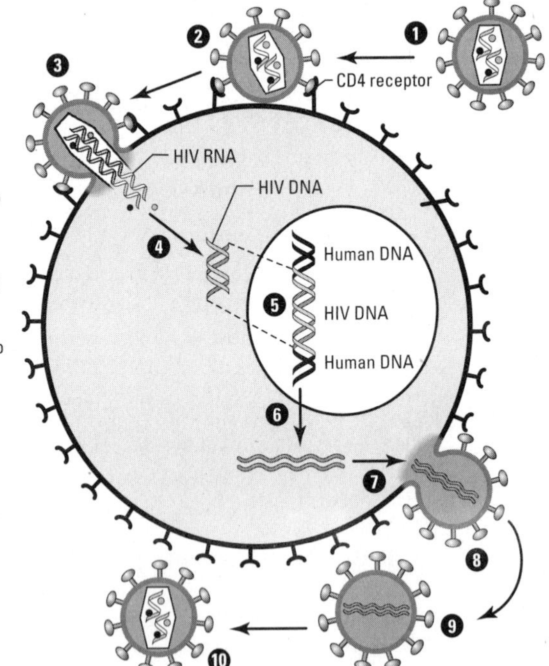

Figure 3-3: The multiplication cycle of the human immunodeficiency virus (HIV).

There's more than one way to copy a virus

In order to reproduce, viruses need to copy their genetic material and make more viral proteins. In eukaryotic cells, production of nucleic acids (DNA and RNA) occurs in the nucleus, while synthesis of proteins occurs in the cytoplasm. So, many viral invaders of eukaryotic cells must visit both of these locations to complete their multiplication cycle.

The main groups of eukaryotic cell viruses and their multiplication strategies are as follows:

- **Double-stranded DNA viruses,** such as smallpox, usually insert their genetic material into the host nucleus. The genetic material is copied as the host cell copies its own genetic material as it prepares to divide.
- **Retroviruses,** such as HIV, contain single-stranded RNA for their genetic material. Once this RNA enters the cytoplasm of the cell, the virus uses a unique enzyme called *reverse transcriptase* to make a DNA copy of the RNA. The double-stranded DNA enters the nucleus, where it inserts into the host DNA by recombination. HIV can be latent, or it can reproduce actively by making new RNA molecules from the viral DNA. During active replication, the RNA molecules move back out into the cytoplasm where they're used to make viral proteins.
- **Double-stranded RNA viruses,** such as many viruses that attack important crop plants, introduce their RNA molecules into the cytoplasm of the cell. The double-stranded RNA molecules of the genetic material are used to make the single-stranded RNA molecules that are necessary for the production of viral proteins. The entire multiplication cycle of these viruses occurs in the cytoplasm of the cell.
- **Single-stranded RNA viruses** are divided into two groups:
    - **Negative-sense single-stranded RNA viruses,** such as influenza and measles, insert their single-stranded RNA molecules into the cytoplasm of the cell. These viruses are called *negative-sense viruses* because their RNA molecules can't be read directly to create proteins. First, their negative-sense RNA molecules must be used to make *complementary,* or "mirror image" RNA molecules. Then, these new complementary RNA molecules make viral proteins. It's as if the genetic material of these viruses is written backwards and must first be turned around before it can be read.
    - **Positive-sense single-stranded RNA viruses,** such as mosaic viruses that attack crop plants, insert their single-stranded RNA molecules into the cytoplasm of the cell. These RNA molecules can be read directly for the synthesis of viral proteins.

The genetic material of viruses can come in many forms, so it's not surprising that several different types of enzymes are involved in copying this genetic material. Some viruses use host enzymes to copy their genetic material; others use viral enzymes. In general, the viral enzymes make more mistakes than do host enzymes. These mistakes lead to genetic variation in viruses and make them harder to defeat with your immune system, vaccines, or anti-viral drugs.

Leaving it all behind

After the viral genetic material is copied, the viral particles assemble and prepare to leave the host cell. Some viruses destroy the host cell as they exit, causing it to burst, while others leave gradually, allowing the host cell to remain intact for longer periods of time. Enveloped viruses exit the cell by *budding*. The capsids of the virus push up against the plasma membrane, which is already studded with viral proteins, and get wrapped up in the membrane as they leave the cell.

Putting it all together

The basic steps of the multiplication cycles of eukaryotic viruses are the same as those for bacterial viruses. However, the multiplication cycles of eukaryotic viruses are typically more complex. For example, when an enveloped virus attacks a eukaryotic cell, the stages of the multiplication cycle would be as follows:

1. **Attachment.**

   Proteins on the surface of the virus bind to receptors on the surface of the host cell.

2. **Penetration.**

   The virus enters the host cell by *fusion* or *receptor-mediated endocytosis,* bringing its genetic material into the cell.

3. **Uncoating.**

   The viral capsid is removed, releasing the genetic material of the virus into the cell.

4. **Biosynthesis.**

   Host or viral enzymes copy the viral genetic material. Energy and materials from the host cell are used to build viral proteins.

5. **Maturation.**

   The viral particles assemble. This step may occur in the cytoplasm or nucleus of the cell, depending on the particular virus.

6. Release.

The virus exits the host cell by *budding.* As the viral particles leave the host cell, they wrap themselves in pieces of the host plasma membrane that were modified by the addition of viral proteins.

HIV and AIDS: Viruses in the real world

Acquired immunodeficiency syndrome (AIDS), a disease in which the entire immune system collapses, affects millions of people worldwide. The human immunodeficiency virus (HIV) causes AIDS.

HIV destroys the immune system because it infects a type of white blood cell, called a *T-cell,* which is critical to human immunity against disease.

T-cells are basically the generals of the immune system army — they give the chemical commands that activate other cells who seek out and destroy viruses. As HIV hijacks and takes over T-cells (refer to Figure 3-3), the T-cells stop functioning, messages are no longer sent to other immune system cells, and the entire immune system stops working. When destruction of T-cells by HIV reaches a certain critical point, the immune system collapses, and the infected person becomes susceptible to any and all infections, such as pneumonia. People with AIDS usually die not from the HIV virus itself, but rather due to complications from these other infections.

The multiplication cycle of HIV is a good example of the features of viruses that attack eukaryotic cells. HIV is an enveloped virus that enters the cell by fusion and exits by budding. It can be latent or active and completes parts of its multiplication cycle in both the nucleus and the cytoplasm. It uses viral enzymes with a high error rate to copy its genetic material and so is highly variable.

In some ways, HIV represents the perfect storm in the form of a virus. Not only does it attack immune system cells, making it hard for bodies to fight the virus, but it has other features that make it hard to target with vaccines or antiviral drugs. HIV is hard to fight with vaccines and drugs because

- The virus can be latent. When the virus is dormant in cells, the immune systems or antiviral drugs have nothing to recognize. The virus is just a sleeping piece of DNA. Antiviral drugs can block all of the active virus in the body, but they can't clear the latent virus from the cells. The virus can reactivate at a later time.
- The virus is highly variable. Reverse transcriptase makes lots of mistakes, which result in a constantly changing virus that is hard to target.

The variability of the HIV virus helps it to eventually develop resistance to antiviral drugs and makes it a difficult target for vaccine development, although research on a vaccine against the virus continues.

With no cure or vaccine, the best defense against the virus is to block transmission. Because transmission requires exchange of bodily fluids, such as semen or blood, people can utilize protective measures to prevent this transmission. Public education campaigns have been very successful in some countries in bringing the rates of HIV infection under control.

The once and future plague

During the Middle Ages, the Black Death killed approximately one-third of the people living in Europe. This plague was so huge that it had far-reaching effects on agriculture, social class structure, religion, and economics. Even today, the plague remains a terrifying reminder of what disease can do to human populations.

Could such a devastating plague ever happen again? Is it happening now? Over 40 million people today, or about 6 percent of the global population, are estimated to be infected with the HIV virus. The situation is the worst in sub-Saharan Africa, where as many as 20 percent of adults are infected with the HIV virus. As these adults sicken and die, their children, who often have HIV from birth, are left without parents and the fabric of their societies is torn apart, just as the Black Death tore apart the societies of Europe during the Middle Ages.

Fortunately, human knowledge and communication are considerably better today than they were during the Middle Ages. Scientists and other people around the globe are working to stop the spread of HIV and to find a cure for HIV infection. Many countries and organizations, such as UNAIDS, AVERT, and the One campaign, are rallying support for the global effort to fight AIDS through research, education, prevention, and treatment of infected people.

Part II

Molecules: The Stuff of Life

In this part . . .

Molecules form the structures of cells and carry out cellular functions. Chemical bonds within and between biological molecules are important to their structure. Four groups of large molecules — carbohydrates, lipids, proteins, and nucleic acids — are the most important molecules to cells. Carbohydrates provide energy and structure to cells. Lipids form membranes, send signals, and store energy. Proteins are the main worker molecules of cells, providing structure and doing lots of different jobs.

The most famous molecule of life — DNA — is a nucleic acid. The blueprints for life are encoded in the structure of DNA, and copies of this code are passed to future generations. In this part, I explain the basic structure and function of molecules and the importance of the four groups of macromolecules to cells.

Chapter 4

Better Living through Chemistry

In This Chapter

- Discovering the atomic nature of matter
- Understanding how and why bonds form between atoms
- Exploring types of bonds
- Taking a look at acids, bases, and pH
- Making and breaking polymers

Chemistry is fundamental to molecular and cellular biology. Cells are made of molecules and communicate through the language of chemistry. Modern biologists work at the molecular level, isolating and studying chemical components of cells like proteins and DNA. In this chapter, I cover the concepts of chemistry that are most important to your understanding of molecular and cellular biology, including the atomic theory of matter, types of chemical bonds, acids and bases, and the nature of polymers.

Life Really Matters

Everything on Earth — this book, the dirt, water, food, buildings, animals, plants, your body, cells, even the air — is made of *matter.* Matter is the "stuff" that gives everything its substance or *mass*. Basically, if it takes up space and you can weigh it (no matter how big or small a scale you'd need!), it's made of matter. Matter can be a solid like your body, a liquid like water, or a gas like the atmosphere.

Virtually all of the matter on Earth has been here since the planet was formed more than 4.5 billion years ago. Living things grow, or increase their matter, by eating food and using the matter in the food to build their bodies. When living things die, the matter in their bodies is recycled back to the soil and the atmosphere by the bacteria and fungi that eat the dead. It is like all living things are made up of tiny Lego building blocks. When living things die, bacteria and fungi help take all the blocks apart and return some to the planetary supply, where living things pick up the blocks and use them again to build their bodies. Just think about it — the matter that makes up your body has been around in various forms for 4.5 billion years! Some of the matter that makes up your cells could even have once been part of a *T. rex*!

It's Elemental!: Atoms That Make Up Living Things

Matter can be separated into *elements*. You've probably heard of some elements, such as gold, iron, copper, oxygen, and aluminum, but many more exist that you've probably never heard of. All the elements discovered by scientists have been organized into the Periodic Table of Elements, shown in Figure 4-1.

Elements are pure substances composed of only one type of *atom*.

An atom is the smallest unit of an element that has the properties of that element. If you had a piece of gold and you cut it into smaller and smaller pieces, until you couldn't even see the pieces anymore, the smallest piece that you could cut that would still be gold would be an atom.

When atoms join together, they form larger structures called *molecules*. Atoms and molecules belong to a hierarchy along with cells (Chapter 2) and *organisms,* or living things:

- Organisms are made of cells.
- Cells are made of molecules.
- Molecules are made of atoms.
- Atoms are made of subatomic particles.

1 H Hydrogen 1.01																	2 He Helium 4.00
3 Li Lithium 6.94	4 Be Beryllium 9.01											5 B Boron 10.81	6 C Carbon 12.01	7 N Nitrogen 14.01	8 O Oxygen 16.00	9 F Fluorine 19.00	10 Ne Neon 20.18
11 Na Sodium 22.99	12 Mg Magnesium 24.31											13 Al Aluminum 26.98	14 Si Silicon 28.09	15 P Phosphorus 30.97	16 S Sulfur 32.06	17 Cl Chlorine 35.45	18 Ar Argon 39.95
19 K Potassium 39.10	20 Ca Calcium 40.08	21 Sc Scandium 44.96	22 Ti Titanium 47.90	23 V Vanadium 50.94	24 Cr Chromium 52.00	25 Mn Manganese 54.94	26 Fe Iron 55.85	27 Co Cobalt 58.93	28 Ni Nickel 58.71	29 Cu Copper 63.55	30 Zn Zinc 65.38	31 Ga Gallium 69.72	32 Ge Germanium 72.59	33 As Arsenic 74.92	34 Se Selenium 78.96	35 Br Bromine 79.90	36 Kr Krypton 83.80
37 Rb Rubidium 85.47	38 Sr Strontium 87.62	39 Y Yttrium 88.91	40 Zr Zirconium 91.22	41 Nb Niobium 92.91	42 Mo Molybdenum 95.94	43 Tc Technetium (99)	44 Ru Ruthenium 101.07	45 Rh Rhodium 102.91	46 Pd Palladium 106.42	47 Ag Silver 107.87	48 Cd Cadmium 112.41	49 In Indium 114.82	50 Sn Tin 118.69	51 Sb Antimony 121.75	52 Te Tellurium 127.60	53 I Iodine 126.90	54 Xe Xenon 130.30
55 Cs Cesium 132.91	56 Ba Barium 137.34	57 La Lanthanum 138.91	72 Hf Hafnium 178.49	73 Ta Tantalum 180.95	74 W Tungsten 183.85	75 Re Rhenium 186.21	76 Os Osmium 190.2	77 Ir Iridium 192.22	78 Pt Platinum 195.09	79 Au Gold 196.97	80 Hg Mercury 200.59	81 Tl Thallium 204.37	82 Pb Lead 207.19	83 Bi Bismuth 208.98	84 Po Polonium (210)	85 At Astatine (210)	86 Rn Radon (222)
87 Fr Francium (223)	88 Ra Radium (226)	89 Ac Actinium (227)	104 Rf Rutherfordium (257)	105 Db Dubnium (260)	106 Sg Seaborgium (263)	107 Bh Bohrium (262)	108 Hs Hassium (265)	109 Mt Meitnerium (266)									

Lanthanides	58 Ce Cerium 140.12	59 Pr Praseodymium 140.91	60 Nd Neodymium 144.24	61 Pm Promethium (147)	62 Sm Samarium 150.35	63 Eu Europium 151.96	64 Gd Gadolinium 157.25	65 Tb Terbium 158.93	66 Dy Dysprosium 162.50	67 Ho Holmium 164.93	68 Er Erbium 167.26	69 Tm Thulium 168.93	70 Yb Ytterbium 173.04	71 Lu Lutetium 174.97
Actinides	90 Th Thorium (232)	91 Pa Protactinium (231)	92 U Uranium (238)	93 Np Neptunium (237)	94 Pu Plutonium (242)	95 Am Americium (243)	96 Cm Curium (247)	97 Bk Berkelium (247)	98 Cf Californium (251)	99 Es Einsteinium (254)	100 Fm Fermium (257)	101 Md Mendelevium (258)	102 No Nobelium (259)	103 Lr Lawrencium (260)

Figure 4-1: The Periodic Table of Elements.

Exploring subatomic particles

Atoms are made of smaller components called *subatomic particles*. The most common subatomic particles in ordinary matter are *protons, neutrons,* and *electrons.* Protons, neutrons, and electrons each have distinct properties:

- **Protons** have a positive electrical charge (+1), have mass, and are found in the center, or *nucleus*, of the atom.
- **Neutrons** have no charge (0), or are neutral, have mass, and are found in the nucleus of the atom.
- **Electrons** have a negative electrical charge (–1), have virtually no mass, and orbit the nucleus of the atom.

The arrangement of subatomic particles within the atom is represented in the Bohr model of the atom, shown in Figure 4-2. In Figure 4-2a, protons are drawn as + or *p,* and neutrons are drawn as 0 or *n.* Electrons are shown as – or •. The protons and neutrons are clustered together in the center, or *nucleus,* of the atom.

The electrons of an atom are constantly whirling around the nucleus in paths called *electron orbitals,* which are grouped into layers around the nucleus called *energy levels* or *electron shells.* The Bohr model illustrates these energy levels as circular paths around the nucleus. Even though they're far away from the nucleus of the atom, the negatively charged electrons are attracted to the positively charged protons, keeping them in orbit around the nucleus. (You've heard the phrase, "opposites attract," right?)

The center of an atom is called the nucleus, and so is the central organelle in a eukaryotic cell. Although they have the same name, they're very different things!

A. Bohr's model of an atom: carbon used as an example.

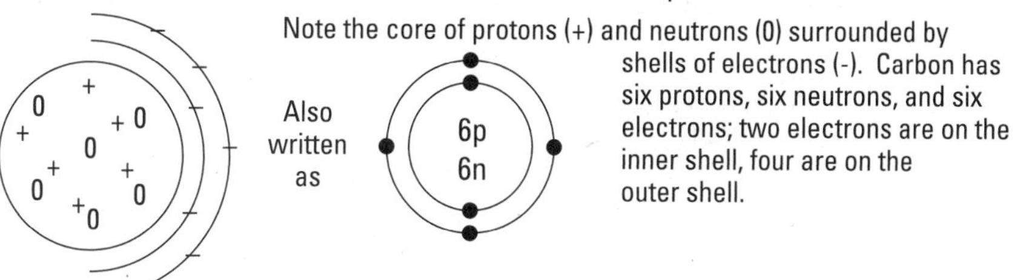

B. Sodium and chloride ions joining to form table salt. The sodium ion has a positive charge because there is one more proton than electrons, so the overall charge is positive. Chloride ion is negative because after it accepts the electron from sodium, it then has one more electron than protons (18 versus 17), so the overall charge is negative. Together, though, NaCl is neutral because the "plus 1" charge is balanced by the "minus 1" charge.

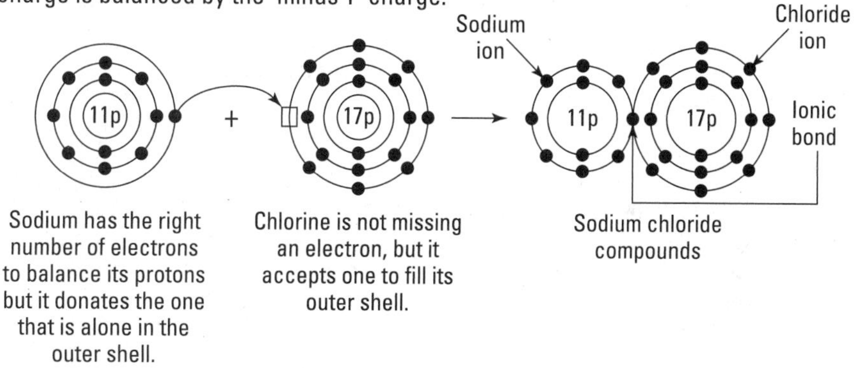

C. Two atoms of oxygen joining to form oxygen gas.

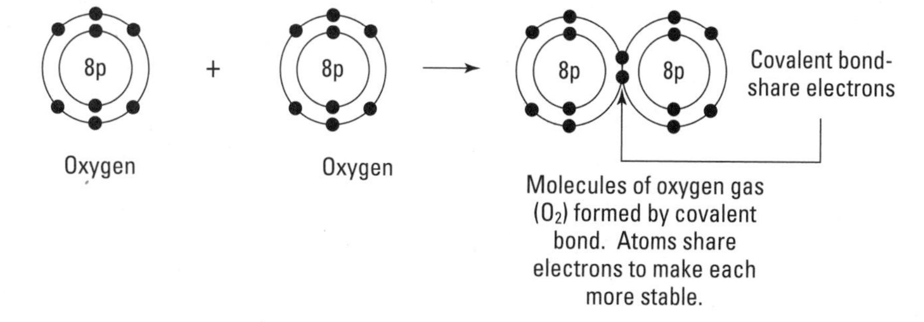

Figure 4-2: Atoms and chemical bonds.

Defining elements

Every atom of a particular element has the same number of protons in its nucleus. That number, called the *atomic number* of the element, is used to organize the elements into the Periodic Table of Elements (refer to Figure 4-1).

The numbers that you see above the letters in the Periodic Table of Elements are the atomic numbers of the elements.

For example, if you look at the second row of the Periodic Table, you can find the letter C, which represents the element carbon. The atomic number 6 is written above the letter C because carbon has six protons. You can also see these six protons in the Bohr model of carbon, shown in Figure 4-2A. In an uncharged atom of an element, the number of electrons is the same as the number of protons. So, in an atom of carbon, there are six protons and six electrons (Figure 4-2A).

Comparing isotopes

The number of protons in the atom of a element never changes. If you have an atom of carbon, it has six protons. However, the number of neutrons that are packed into the nucleus with those protons can vary between different atoms of an element. Most carbon atoms found on planet Earth have six neutrons just like the one shown in Figure 4-2a. However, some carbon atoms found on planet Earth have eight neutrons in the nucleus. These different types of carbon atoms are called *isotopes.*

Isotopes of an element have the same number of protons and electrons as each other, but they don't have the same number of neutrons.

To show how many neutrons a particular isotope has, scientists include the *mass number* of the element along with its name. The mass number is equal to the number of protons plus the number of neutrons. (Remember that only the protons and neutrons in an atom really have any mass to speak of — the mass of electrons is too small to count except in the most careful calculations.) The most common form of carbon atoms on Earth, the ones that have six protons and six neutrons, would have a mass number of 12. You can write this form of carbon in words as carbon-12, or with a symbol as ^{12}C. The heavier isotope of carbon, which has eight neutrons, has a mass number of 14. (All carbon has six protons, so $6+8=14$.)

Want a date with a radioactive isotope?

Radioactive isotopes are atoms of an element that throw subatomic particles out of their nuclei. When they throw out these particles, the isotopes decay into a different type of element. Radioactive isotopes are very useful in science and medicine. For example, scientists can use the radioactive isotope carbon-14 to determine the age of fossils. In living things, the ratio of carbon-14 to carbon-12 is kept the same. But when things die, the carbon-14 that was present at the time of death slowly decays over time, throwing off particles and becoming nitrogen-14. Scientists know how long it takes carbon-14 to decay. So, when scientists find a fossil, they can analyze it to determine the ratio of carbon-14 to carbon-12 to see how much carbon-14 has been lost over time. Because they know how long it takes carbon-14 to decay, they can calculate the approximate age of the fossil.

The Periodic Table of Elements shows the average mass of all the atoms of an element on Earth. That number is written below the letter representing the element (refer to Figure 4-1). If you look at the letter C for carbon in the second row of the table, you can see the *atomic mass* given as 12.01 g/mol of carbon. Most carbon atoms on Earth have a mass number of 12, but a few have a mass number of 14. So, if you take the average of all the carbon atoms on Earth, the average comes out a little bit higher than 12.

As a comparison, imagine that you and three friends want to figure out the average weight of the apples in your lunches. Three of you have apples that weigh exactly 4 ounces, but one of you has an apple that weighs 6 ounces. If you added up the weights of all four apples, then divided by four, you'd come up with an average weight of 4.5 ounces. Because of the effect of the larger apple, the average comes out a little more than the weight of most of the apples. The heavier isotopes of an element have the same effect on the average atomic mass for that element.

If you're given the mass number of an atom, you can always figure out how many neutrons are in the nucleus. Just remember that protons + neutrons = mass number and that the number of protons in an element is the atomic number that is written in the period table. For example, if you wanted to figure out how many neutrons were in an atom of phosphorous-32, you could look up phosphorous in the Periodic Table, see that it has an atomic number of 15, and do the simple math: mass number – protons = neutrons, so 32–15=17 neutrons.

Let's Bond: How Atoms Are Attracted to Each Other

Atoms come together to form larger, more complicated structures called molecules. *Chemical bonds* are the forces, or glue, that hold atoms together. Bonds form when atoms give, take, or share their electrons with each other. Atoms within molecules rearrange to form new molecules when bonds break and reform.

In cells, atoms are attracted to each other and will form bonds with each other because of two main factors:

- The atoms need to share electrons in order to form stable structures.
- There is an attraction between opposite electrical charges (positive and negative) between two atoms.

One clue that reveals what atoms need in order to be stable comes from the atoms in column 18 of the Periodic Table of Elements (refer to Figure 4-1). These elements are called the *inert gases* (or *noble gases*) because they do not tend to react with other atoms easily. They're so stable, in fact, that they're used in MIG and TIG welding as a sort of protective gas blanket between regular atmospheric gases and the electrode that is used during welding. The secret to the stability of the inert gases is in how they arrange their electrons.

Feeling fulfilled by arranging your electrons just right

Whether one atom reacts with another atom to form a bond is all about how their electrons are arranged, which is called their *electron configuration*. The electron configuration of an atom depends on how many electrons they have. As you read from left to right and top to bottom in the Periodic Table of Elements, the elements in the Periodic Table increase in size. Each element has one more proton than the element before it and thus also one more electron. For example, the element carbon is atomic number 6, so it has six protons and six electrons. The next element in the Periodic Table is nitrogen, atomic number 7, so it has seven protons and seven electrons.

Valence electrons

Understanding the exact electron configuration of an atom is actually pretty complicated, but you can take advantage of an excellent shortcut to predict how an atom will bond with another atom: Focus on the *valence electrons.* The valence electrons, which are located in particular orbitals in the outermost energy levels of the atoms, are the most important electrons when it comes to forming bonds. You can think of them as the electron face that atoms show to other atoms. Valence electrons are the electrons that will either be shared or given to another atom during bond formation.

To figure out how an atom will react to become stable, you need to know several things:

- **The Periodic Table of Elements can tell you the number of valence electrons for an element.** In the Periodic Table, each column indicates a *family* of elements that have similar properties. In the elements that are marked with a Roman numeral and an A, the Roman numeral tells you the number of valence electrons found in the atoms within that family. No muss, no fuss, just look it up!
- **Atoms seek to have a full set of valence electrons in their outermost energy level**. If an atom doesn't have a full set of valence electrons, it "feels unfulfilled" and will either give, take, or share electrons to complete its set. The inert gases have a full set of valence electrons, which is why they're so stable.
- **Most atoms follow the octet rule, or Rule of Eight, in order to have a full set of valence electrons.** The Rule of Eight states that the energy level will be full and stable once it contains eight electrons. So, atoms will give, take, or share electrons until they have eight valence electrons in their outermost energy level. This rule doesn't apply to the first, smallest energy shell. The first energy shell is complete with two valence electrons. So, the Rule of Eight doesn't apply to hydrogen and helium, which have two or fewer electrons — they follow their own Rule of Two.

Another way to look at the Rule of Eight is that atoms will give, take, or share electrons until their electron configuration matches that of the closest inert gas. For example, hydrogen is closest to helium, which has two electrons in its outermost energy level, so hydrogen will be stable when it has two electrons. Flourine is closest to neon, which has eight electrons in its outermost energy level, so fluorine will be stable when it has eight electrons.

- **Atoms on the left side of the Periodic Table tend to give up electrons, while atoms on the right side of the Periodic Table tend to take electrons.** The atoms on the left side of the period table, all the way up to aluminum in the column marked IIIA, are metals. Metals have certain properties, including a tendency to give up electrons. The atoms on the right side of the Periodic Table, including the biologically important atoms carbon, nitrogen, oxygen, phosphorous, and sulfur, are nonmetals. Nonmetals tend to share or accept electrons from other atoms. Hydrogen is also considered a nonmetal, even though it's in the upper-left corner of the Periodic Table.

Putting atoms together

Once you know how atoms are likely to react to become stable, you can predict how they'll react with each other. You look at what each atom needs to become stable and whether the two atoms could fulfill each others' needs. Figuring out how atoms can pair up sounds a bit like a dating service for atoms, doesn't it? As an example, look at how the element carbon (Figure 4-1A) would react with the element hydrogen. Follow these steps to figure out how the two atoms are likely to react with each other:

1. **Look the element up in the Periodic Table to find out which family it's in and how many valence electrons it has.**

   a. Carbon is located in family IVA in the Periodic Table, so it has four valence electrons.

   b. Hydrogen is located in family IA, so it has one valence electron.

2. **Look at the position of the element in the Periodic Table to predict whether it will give up or receive electrons to complete its set of valence electrons.**

   a. Carbon is on the right side of the Periodic Table, so it is a nonmetal. It's likely to receive electrons from another atom.

   b. Although hydrogen is on the left side of the Periodic Table, it's also a nonmetal. It's also likely to receive electrons.

3. **Apply the Rule of Eight (if the element is hydrogen or helium, apply the Rule of Two).**

   a. In order for carbon to be stable, it needs eight valence electrons. It has only four, so it needs four more valence electrons.

   b. In order for hydrogen to be stable, it needs two valence electrons. It has only one, so it needs one more to be stable.

After you analyze both atoms, the dating can begin! Carbon and hydrogen are electron receivers, so it's not likely that one would give the other its electrons. If they share, they can both receive. You know that carbon needs four valence electrons, and each hydrogen atom has one available. So, one carbon atom would need to share with four hydrogen atoms to get enough valence electrons to fill its set. If one carbon atom shares electrons with four hydrogen atoms, the carbon atom will fill its set of valence electrons and become stable. Likewise, the carbon atom is sharing its four valence electrons with the four hydrogen atoms, so each hydrogen atom gets to share one and complete its set of two. The hydrogen atoms are also stable. The atoms live happily ever after in the form of gas methane, which has the chemical formula CH_4.

Working without a Periodic Table

The Periodic Table of Elements and the Rule of Eight are useful for predicting the behavior of many elements, but those shortcuts don't work for everything. Plus, you may not always have access to a Periodic Table (on an exam, perhaps?). So, sometimes you have to figure out the electron configuration of an atom on your own in order to determine the number of valence electrons.

To figure out the electron configuration for a particular atom, you need to know how many electrons the atom has and how atoms organize their electrons. Electrons are organized according to the following rules:

- Electrons travel in pairs within orbitals around the nucleus of the atom.
- Electron orbitals are organized into energy levels. Energy levels are identified by their quantum number. In other words, they're numbered beginning with the number 1 for the energy level closest to the nucleus of the atom.
- Each energy level can have as many sublevels as its quantum number. So level 1 can have one sublevel, level 2 can have two sublevels, and so on. Sublevels are identified by letters that represent the shapes of the electron orbitals within that sublevel.
- The first sublevel is called an s sublevel.
- The second sublevel is called a p sublevel.
- The third sublevel is called a d sublevel.
- Each sublevel holds a different number of orbitals and maximum number of electrons. Because electrons travel in pairs within orbitals, each sublevel can hold a maximum of electrons for each orbital.
- s sublevels hold one orbital with up to two electrons.
- p sublevels hold three orbitals with up to six electrons.
- d sublevels hold five orbitals with up to ten electrons.
- The maximum number of electrons that an energy level can hold is represented by the formula $2n^2$. The letter n is the quantum number for the level. So the first level can hold a maximum number of $2(1)^2$, which is two electrons. This maximum number makes sense because the first level has only one s sublevel that holds one pair of electrons. The second level can hold a maximum of $2(2)^2$ which equals eight electrons. These eight electrons represent the two electrons in the s sublevel and the six electrons in the p sublevel. The third level can hold a maximum of 18 electrons, 2 in s, 6 in p, and 10 in d.

(continued)

(continued)

- Energy levels fill with electrons in order. Level 1 would fill first, level 2 would fill second, and so on.

If you put all these rules together, you can figure out the electron configuration for a particular atom. For example, look at the element chlorine. Its atomic number is 17, so it has 17 electrons. If you put those electrons into the energy shells, you'd arrange them as follows:

1. Put two electrons in the first energy level. That leaves 15 electrons remaining.
2. Put eight electrons in the second energy level. That leaves seven electrons remaining.
3. Put seven electrons in the third energy level.

By following these steps, you figured out that chlorine has seven electrons in its outermost energy level. These electrons are its seven valence electrons.

Holding on: Electronegativity

Not all atoms treat electrons equally. Some atoms have a really strong pull for electrons and tend to attract them. Other atoms have a weaker pull for electrons and even give up electrons rather easily.

The tendency of an atom to attract electrons is called its *electronegativity*.

An atom that has a strong pull for electrons is highly electronegative or, as I like to call it, an electron hog (because of its tendency to hog the electrons).

The electronegativity of an atom is important in cellular chemistry for several reasons:

- **Electronegativity affects bonding behavior of atoms.** If an atom that is highly electronegative interacts with an atom that is not, the electronegative atom is likely to take electrons away from the other atom.
- **Electronegativity is important during reactions between molecules.** One of the most important processes in cells, the transfer of energy from food to a usable form for the cell, relies heavily upon the high electronegativity of oxygen (Chapter 11).

Electronegativity increases as you read from left to right in the Periodic Table of Elements. The exception to this rule is the last column on the right of the Periodic Table, which contains the inert gases. (These elements have a full set of valence electrons and have very low electronegativity.)

Give and take: Oxidation and reduction

Atoms react with one another in order to have a full set of valence electrons. Atoms that are highly electronegative will attract electrons from other atoms. The result of an interaction between two atoms is going to depend on how electronegative they are relative to each other. You can think of this relationship as a tug of war for electrons. A highly electronegative atom, the electron hog, is like a big guy with lots of muscle that can pull hard on the rope. Atoms that aren't very electronegative can't pull as hard. If a highly electronegative atom goes toe to toe with an electron wimp, the big guy is going to take electrons from the little guy. However, if two equally electronegative atoms pull against each other, it may end up in a tie, and the atoms will share electrons.

If one atom takes an electron from another atom, an *oxidation reduction reaction,* or *redox reaction,* has occurred. The atom that gave up the electron was *oxidized;* the one that received the electron was *reduced.*

An example of oxidation that you're probably familiar with is when something made of iron rusts. Iron gives up electrons to oxygen to form a compound called iron oxide, which is visible as the red rust on the iron object. So, iron is oxidized during the formation of iron oxide, while oxygen is reduced.

Another good example of a redox reaction is the interaction of sodium and chlorine to form sodium chloride, which is the fancy chemical name for table salt. In this reaction, which Figure 4-2B shows in detail, chlorine takes an electron from sodium. Chlorine is highly electronegative and needs only one electron to become stable with eight valence electrons. Sodium is not very electronegative and will become stable (like its nearest inert gas neon) if it gives up its electron. So, sodium gives up its electron and is oxidized. Chlorine accepts the electron and is reduced.

To remember the definitions of oxidation and reduction, memorize this phrase: "Leo the lion goes ger." LEO stands for loss of electrons is oxidation. GER stands for gain of electrons is reduction.

Opposites attract: Ionic bonds

Ionic bonds are electrical attractions between positively and negatively charged atoms, called *ions*. In solids, such as sodium chloride or table salt, ionic bonds are very strong. However, in the watery environment of the cell, they are weak bonds.

The compound sodium chloride is the classic example of an ionic bond. When sodium gives an electron to chlorine, as in Figure 4-2B, sodium no longer has the same number of protons and electrons. Because it has one more proton than electron, it becomes the positively charged sodium ion. Likewise, when chloride accepts an electron, it becomes unbalanced in charge, having one more electron than proton. It becomes the negatively charged chloride ion. (For every extra proton an atom has, it gains a charge of +1; for every extra electron, it gains a charge of –1.) The sodium ion and chloride ion have opposite electrical charges and so are attracted to each other. The electrical attraction between the two ions is an example of an ionic bond.

Sharing is caring: Covalent bonds

When two atoms have about the same pull for electrons — in other words they're equal in electronegativity — they'll probably form a *covalent bond* with each other.

Covalent bonds are bonds based on shared pairs of electrons. In the watery environment of the cell, covalent bonds are the strongest bonds.

Figure 4-2C shows an example of a covalent bond where two oxygen atoms are shown joining together to form oxygen gas. Oxygen is family VIA, so it has six valence electrons and needs two more to become stable. The two oxygen atoms are equal in electronegativity, so they'll share electrons with each other. If each atom shares two electrons with the other, then they'll both have eight valence electrons (6+2=8). The shared electrons "belong" to both atoms and spend time orbiting both atomic nuclei, so both atoms become stable. The four dots located between the two atoms represent the two pairs of shared electrons. Each pair of shared electrons is one covalent bond, so the two pairs of shared electrons are called a *double bond.*

A molecule by any other picture

Scientists draw atoms and molecules in several different ways, depending on what they want to emphasize. For example, the Bohr model of the atom (refer to Figure 4-2) shows the configuration of electrons within each energy level. Other ways of drawing atoms and molecules illustrate different aspects of molecules, such as the bond angles or the shape of the molecule.

You can compare several different ways of drawing molecules by looking at the drawings in Figures 4-3 and 4-4:

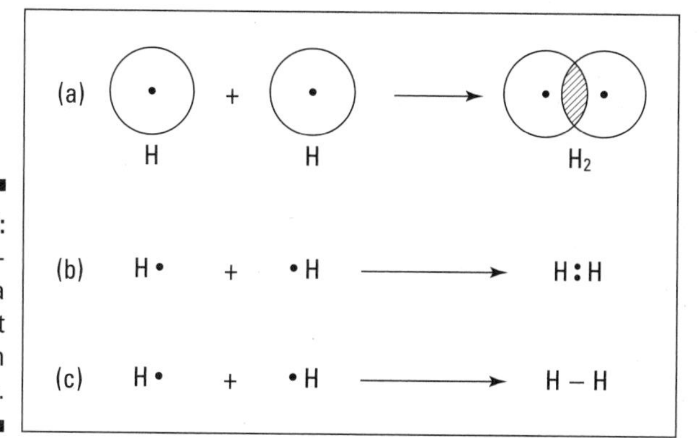

Figure 4-3: The formation of a covalent bond in hydrogen.

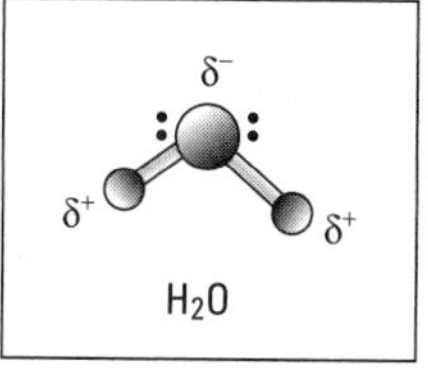

Figure 4-4: Polar covalent bonding in water.

- A version of the Bohr model in Figure 4-3a shows the single proton of the hydrogen atom and the energy shell as a ring around the nucleus. The covalent bond between the two hydrogen atoms is shown by the shading between the two overlapping energy levels.
- An *electron dot model* is shown in Figure 4-3b. The valence electrons are shown as dots around the letter H, which represents hydrogen. The shared pair of electrons between the two hydrogen atoms when they join to form hydrogen gas is shown by the two dots between the letters.
- A *Lewis structural model,* or structural formula, is shown in Figure 4-3c. The valence electrons are again shown as dots around the letter H for hydrogen. However, in this model, the shared pair of electrons in hydrogen gas is shown as a single straight line.
- In a *ball and stick model,* shown in Figure 4-4, the nuclei of the atoms are drawn as balls while the covalent bonds are drawn as sticks. The shape of the drawing and the angles of the bonds represent the actual shape of the molecule.

Don't hog the toys! Polar covalent bonds

Not all covalent bonds are equal. When an atom forms a covalent bond with another atom that is somewhat different in electronegativity, the bond may become unequal, or *polar.*

A *polar covalent bond* is formed when electrons are shared unequally between two atoms.

In a polar covalent bond, the shared electrons spend more time orbiting the nucleus of the more electronegative atom. (You can think of this unequal sharing as the electrons spending more time playing with one of the atoms than the other.) Because electrons have a negative charge, the end of the molecule where the electrons tend to hang out becomes slightly negative. The end of the molecule that gets ignored by the electrons has fewer negative charges than positive charges (its electrons are hanging out somewhere else), so it becomes slightly positive.

An important molecule that has polar covalent bonds is water (Figure 4-4). A water molecule has two hydrogen atoms and one oxygen atom (H_2O). Both hydrogen and oxygen are nonmetals, which are usually electron receivers, so they'll share electrons with each other. Oxygen needs two valence electrons to become stable. Each hydrogen can supply one. So, one oxygen atom shares electrons with two hydrogen atoms. However, oxygen is much more electronegative than hydrogen, so oxygen doesn't share equally. It's an electron hog and keeps the electrons playing at its end of the molecule. The oxygen end of the water molecule develops a slight negative charge (represented by $\delta-$). The hydrogen ends of the water molecule develop a slight positive charge ($\delta+$). Each covalent bond within the water molecule is a polar covalent bond, and the entire molecule itself is also said to be polar.

Molecular Velcro: Hydrogen bonds

After polar covalent bonds are formed, they set up conditions for the formation of another type of bond, the *hydrogen bond.* Hydrogen bonds are weak electrical attractions that form between the ends of polar molecules. As always, opposites attract, so the negatively charged end of one molecule would be attracted to the positively charged end of another.

The term hydrogen bond doesn't refer to a covalent bond with a hydrogen atom. The bond got its name because the hydrogen atoms on polar molecules are often seen sticking to atoms on other polar molecules. For example, in the water molecules shown in Figure 4-5, the hydrogen bonds occur *between* the water molecules, not *within* the water molecules

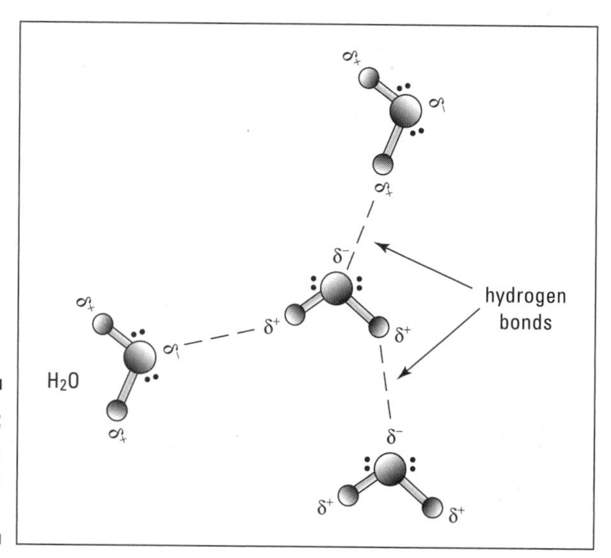

Figure 4-5: Hydrogen bonding in water.

Hydrogen bonds are the molecular equivalent of Velcro. Have you ever looked closely at a piece of Velcro? One side has tiny little hooks. On the other side are soft fuzzy stuff for the hooks to grab. Each little hook on its own forms a very weak attachment to the fuzzy stuff — it would take almost nothing to pull it free again. However, when you have a whole strip of the stuff, you can hold all kinds of things together!

Hydrogen bonds work the same way. Each individual hydrogen bond is very weak. However, many hydrogen bonds together do important things inside cells. Hydrogen bonds hold the two halves of your DNA molecules together so that they form a twisting, ladder-like double helix (see Chapter 7). Hydrogen bonds are also important in holding proteins together in their proper shapes so that they can do their many jobs inside the cell (see Chapter 6). And finally, without hydrogen bonds, water would not have its amazing properties. Hydrogen bonds between the positively and negatively charged ends of the water molecules cause them to stick together (see Figure 4-5), making the water molecules cling to each other.

If you have ever seen a bug walking on the surface of water, you have seen the power of hydrogen bonds in action. The hydrogen bonds between water molecules give water its *surface tension,* which is just enough cling to support the weight of small water bugs.

Molecular cliques: Hydrophobic interactions

Some molecules just don't play nicely with water. Because water is a polar molecule, it tends to stick to itself via hydrogen bonds (see Figure 4-5). Other polar molecules also stick to water molecules and can mix right in, dissolving into the water. However, *nonpolar molecules* have evenly shared covalent bonds and lack the slight negative and positive charges of polar molecules. Because they're uncharged, nonpolar molecules don't mix well with water.

Nonpolar molecules are also called *hydrophobic molecules* because "hydro" means water and "phobic" means to fear.

When nonpolar molecules are placed in a watery environment, the polar molecules will all stick to each other and push the nonpolar molecules away. You can think of the scenario as if the polar molecules all belong to a clique that refuses to hang out with the nonpolar molecules. (The name of this clique, by the way, is the *hydrophilic molecules*.) Because the nonpolar molecules all get pushed together, they become associated with each other.

The interaction between nonpolar molecules is called a *hydrophobic interaction.*

You can easily demonstrate a hydrophobic interaction to yourself. Just go into your kitchen, put some water in a cup, and then add a little oil. Even if you stir the mixture vigorously to mix the oil into the water, as soon as you stop stirring, all the oil will gather together on top of the water. The water molecules all stick to each other and push the oil molecules away. Hence the saying, "They get along like oil and water!"

Blue Planet: The Ocean Inside Your Cells

Water is essential to life on Earth; in fact, life on Earth probably began in the oceans. When living things moved onto land, they took the ocean with them inside their bodies. Your body, for example, is probably about 60 to 75 percent water. Water is essential to the proper functioning of your body and your cells for several reasons:

- **Water is a solvent.** Ions and polar molecules dissolve in water. Water surrounds these molecules, preventing ions from crystallizing into solids and facilitating the function of polar molecules. All the chemical reactions that happen inside cells happen in a watery environment.

- **Water helps move things across membranes.** As water moves across membranes, it helps dissolved ions and small molecules to pass into and out of cells.
- **Water causes hydrophobic interactions.** When water sticks to itself, it pushes nonpolar molecules away. Hydrophobic interactions are important to the structure of plasma membranes and other membranes in the cell (see Chapter 2).
- **Water helps maintain the structure of important molecules.** Water clings to polar surfaces in proteins and nucleic acids, helping maintain their structure.

Splitting water

Water molecules fall apart easily and get back together easily, shifting back and forth between being water molecules and ions. When water molecules fall apart, oxygen (the little electron hog) keeps an electron from one of the hydrogens and then lets the rest of the hydrogen go. The electron-less hydrogen is now a positively charged *hydrogen ion,* which is really just a proton. The oxygen keeps the other hydrogen with it, forming a negatively charged *hydroxide ion*. The reaction for the reversible ionization of water molecule into ions is

$$H_2O \longleftrightarrow H^+ + OH^-$$

In pure water, the balance between hydrogen ions and hydroxide ions is exactly equal. In cells, the balance between hydrogen and hydroxide ions shifts slightly as conditions within the cell change.

Measuring pH

The concentration of hydrogen ions within a solution, including the solutions inside and surrounding cells, is important to living things. Hydrogen ion concentration is represented as $[H^+]$ and is measured by the *pH scale,* shown in Figure 4-6. The pH scale is useful because hydrogen ion concentration in cells and other solutions is small. For example, in pure water, the concentration of hydrogen ions and the concentration of hydroxide ions is exactly balanced, and the concentration of each is 1.0×10^{-7} M (M stands for *molarity,* which represents the concentration in number of moles of solute per liter). To convert hydrogen ion concentration to a pH value, you take the negative log of the concentration:

$$pH = -\log[H^+]$$

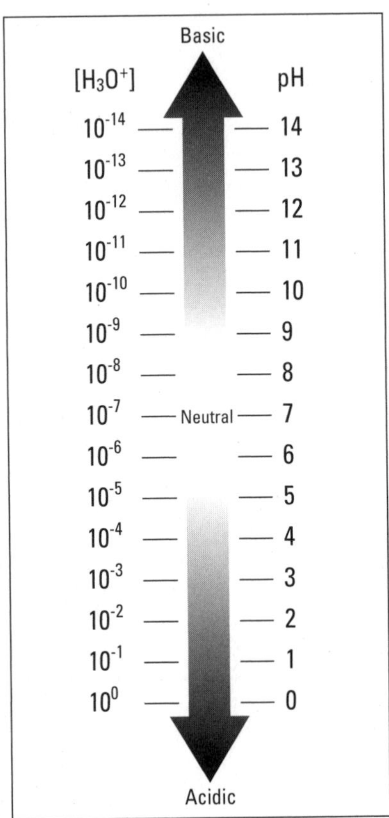

Figure 4-6: The pH scale.

By taking the negative log of the concentration, you convert the $[H^+]$ in pure water from 1.0×10^{-7} M to a pH value of 7.0. By converting all $[H^+]$ values to pH, scientists created a pH scale with values from 0 to 14. I'm sure you'll agree that the numbers 0 to 14 are easier to deal with than the tiny numbers of the unconverted hydrogen ion concentrations!

Cells can survive only within a narrow range of pH. The pH of pure water is 7.0, which is considered *neutral*. Solutions that have pH values lower than 7.0 are *acidic*. Solutions with pH values higher than 7.0 are *basic*. Most cells prefer environments with pH values very close to neutral. As their pH values move to either end of the scale, solutions typically become increasingly damaging to cells. There are some exceptions, such as bacteria that survive in acidic rivers or *alkaline* (basic) lakes.

Changing pH

Some chemicals can change the pH of a solution when they're added to the solution. Chemicals that make solutions more acidic are called *acids;* chemicals that make solutions more basic are called *bases.* Acids are chemicals that release hydrogen ions into the solution, increasing its acidity and thus lowering its pH. Bases, on the other hand, remove hydrogen ions from a solution, lowering its acidity and increasing pH.

I like to think of bases as little sponges for hydrogen ions. When you add a base to a solution, it will suck up the hydrogen ions, decreasing their concentration in the solution.

Maintaining pH

Because cells can survive only within a narrow range of pH, maintaining pH is hugely important to cells. To maintain their pH, cells employ *buffers.* A buffer solution resists changes in pH from the effects of acids or bases. Buffer solutions maintain pH because they're switch hitters — they can either release hydrogen ions into a solution or remove them. Usually, buffer solutions contain a matched pair of a weak acid and base. That way, no matter which way the pH starts to change, a chemical is on hand to counteract the change.

A good example of a buffer system is cruising through your veins right now. Your blood contains the acid/base pair of carbonic acid and bicarbonate to help maintain your blood pH at just the right level. When the concentration of hydrogen ions in the blood increases, which happens when you exercise, the bicarbonate in the blood snatches them up so that your blood doesn't become dangerously acidic. When the concentration of hydrogen ions drops too low, carbonic acid releases more into the blood before your blood can become dangerously alkaline. The reactions for this buffer system look like this:

$$H^+ + HCO_3^- \longleftrightarrow H_2CO_3$$

This reaction proceeds in either direction, depending on the conditions in the blood. If the blood becomes acidic, bicarbonate combines with the hydrogen ions, producing carbonic acid. If the blood becomes alkaline, carbonic acid releases hydrogen ions, correcting blood pH.

When bicarbonate picks up hydrogen ions, it's converted into carbonic acid. Alternatively, when carbonic acid releases hydrogen ions, it's converted into bicarbonate. Together, the two compounds keep blood pH very close to a healthy level of 7.4.

Chain, Chain, Chain: Building and Breaking Polymers

As cells use and recycle matter, they constantly take molecules apart and rearrange the combinations of atoms to make new molecules needed by the cell. Molecules are like the building blocks that matter, including living things, are made of. Each molecule is made of even smaller building blocks, which represent the atoms. If you imagine that each element from the Periodic Table is represented by a different color block, then molecules would be built in many different shapes and color combinations. Cells take these large, colorful molecules, break them back down into smaller groupings of blocks, and then recombine the smaller groupings into new structures needed by the cell.

Identifying the parts and the whole

Macromolecules are just what they sound like — big molecules. Four groups of macromolecules are vitally important to cell structure and function:

- Carbohydrates (see Chapter 5)
- Proteins (see Chapter 6)
- Nucleic acids (DNA and RNA — see Chapter 7)
- Lipids (fats and related molecules — see Chapter 8)

Three of the four types of macromolecules — carbohydrates, proteins, and nucleic acids — are very long molecules made of the same repeating type of building block. A long repetitive molecule is called a *polymer*. The building block a polymer is made out of is called a *monomer*.

Getting together and breaking up again

Polymers are made by stringing together monomers to form a long chain. As each monomer is added to the chain, two hydrogen atoms and an oxygen atom are removed from the new monomer and growing chain (Figure 4-7A). The hydrogen and oxygen atoms are combined to form water during the process of adding the monomer. Because a water molecule is formed when a monomer is added, the reaction is called *condensation* or *dehydration synthesis*. (Condensation forms water and so does adding monomers to the growing polymer.)

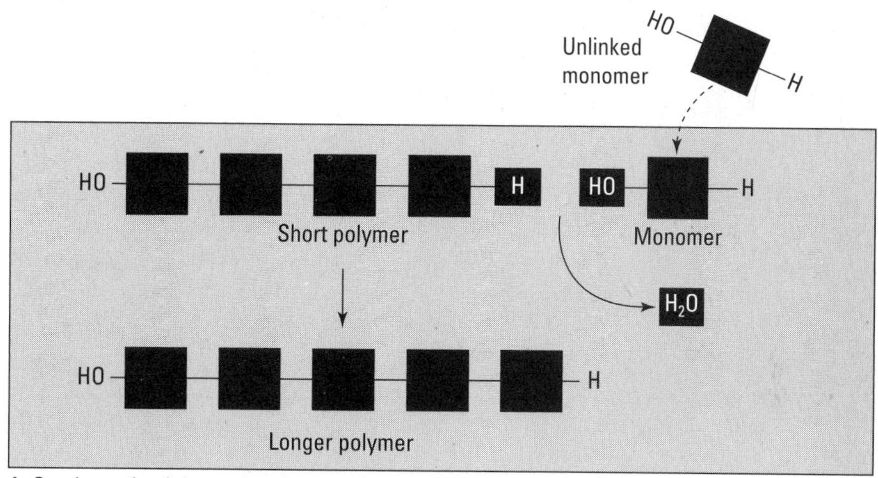

A. Condensation joins molecules together.

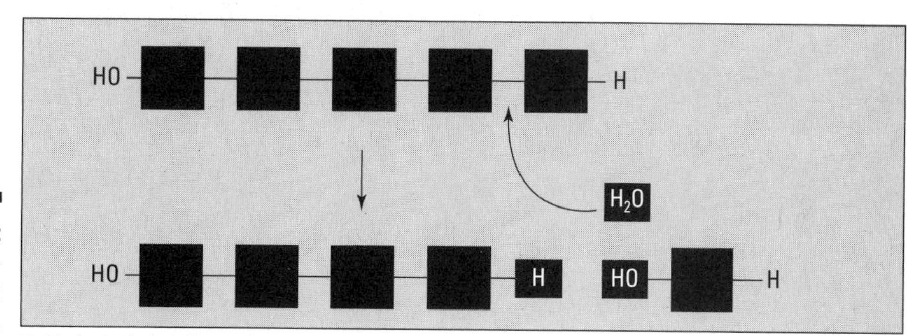

B. Hydrolysis breaks molecules apart.

Figure 4-7: Condensation and hydrolysis.

Polymers can be broken back down into individual monomers by *hydrolysis*. During hydrolysis, a water molecule is inserted into the bond between two monomers (Figure 4-7B). The bond is broken and the monomer is freed from the chain.

Hydrolysis is easy to remember if you know what the word means. *Hydro* means water, and *lysis* means breaking. So, hydrolysis is breaking something with water.

Playing the name game with polymers

When molecules of the same type are joined together, they form repetitive molecules of various lengths. The name of the chain depends on the number of repeating units in the chain. One molecule by itself is a *monomer,* because *mono* means one. Attach two monomers together, and you have a *dimer,* because *di* means two. Stringing several monomers into a short chain forms an *oligomer,* because *oligo* means few. Chains of many monomers are called *polymers* because poly means many. These prefixes remain the same no matter what type of polymer you're talking about, while the suffix changes depending on the type of monomer. So, a chain of two sugars is a disaccharide, a short chain of nucleotides is an oligonucleotide, and a long chain of amino acids is a polypeptide.

Chapter 5

Carbohydrates: How Sweet They Are

In This Chapter

- Examining monosaccharides
- Building polysaccharides
- Exploring carbohydrate function

Carbohydrates are made of carbon, hydrogen, and oxygen atoms. Many different monosaccharides, or simple sugars, can combine into polysaccharides, or complex carbohydrates. Even though they have a bad reputation among some diet plans, carbohydrates perform many essential functions for cells. In this chapter, I present the basic structure of carbohydrates and explain their importance to cells.

CH_2O: Structure of Carbohydrates

In recent years, due to the comeback of the low-carb diet, carbohydrates have gotten a bad rap. Some people have started thinking that proteins are good, and carbohydrates are bad. However, the idea that carbohydrates aren't good for you is overly simplified. After all, carbohydrates are an essential component of your cells. What can make a difference is the type of carbohydrates you eat. *Carbohydrates* are organic molecules composed of carbon, hydrogen, and oxygen. The two main types of carbohydrates are as follows:

- **Monosaccharides** are also called simple sugars. (Most diets recommend that you avoid eating too much of this type of carbohydrate.) Glucose is a monosaccharide that is usually available to your cells.
- **Polysaccharides** are also called complex carbohydrates. (Fiber is an example of a complex carbohydrate that is a recommended part of your daily nutrition.)

Finding your way around

When scientists want to discuss the structure of molecules, they need a way to make sure that they're all looking at the same part of the molecule. To mark the molecule, they number the carbon atoms in the structure.

For carbohydrates, the rule is that the carbon atoms are numbered consecutively, beginning with the end of the carbohydrate that is nearest the carbonyl group (double-bonded oxygen). This numbering is pretty easy to do when you're looking at a linear structure of a carbohydrate as in Figure 5-1B, but when carbohydrates are in their ring form, it can be more challenging.

Typically, you begin numbering with the carbon on the right side of the oxygen in the ring and then number consecutively away from the oxygen. So for the glucose molecule in Figure 5-1B, the #1 carbon is represented by the angle on the right side of the ring. Numbering away from that carbon brings you to the #6 carbon that sticks up off the ring.

Keeping it simple: Monosaccharides

Monosaccharides, or simple sugars, are single sugars. ("Mono" means "one" and "sacchar" means sugar.) Many monosaccharides have the generic formula CH_2O: For every one carbon atom they have, they have two hydrogen atoms and one oxygen atom. Two monosaccharides that may be familiar to you are glucose (see Figure 5-1A) and *fructose* (a sugar found in fruit and also in high-fructose corn syrup).

All monosaccharides have certain features in common:

- **A backbone of 3, 4, 5, 6, or 7 carbons**. Sugars are categorized based on the number of carbons: In order of the numbers, they are called trioses, tetroses, pentoses, hexoses, and heptoses. For example, glucose is a hexose, or 6-carbon sugar.
- **Hydroxyl groups (–OH) attached to every carbon but one.** The hydroxyl groups make sugars polar, which is why they dissolve easily in water.
- **One double-bonded oxygen attached to the carbon backbone**. An oxygen double-bonded to a carbon is called a *carbonyl group*. If the carbonyl group is located at the end of a monosaccharide, the sugar is an *aldose*. If the carbonyl group is located within the carbon backbone, the sugar is a *ketose*. Glucose is an aldose because its carbonyl group is at the end of the carbon backbone.

Of the four groups of macromolecules (carbohydrates, proteins, nucleic acids, and lipids), carbohydrates have the greatest number of hydroxyl groups (–OH) attached to their carbon atoms. When you're trying to distinguish between the four types of macromolecules, a structure with hydroxyl groups attached to almost every carbon is probably a carbohydrate.

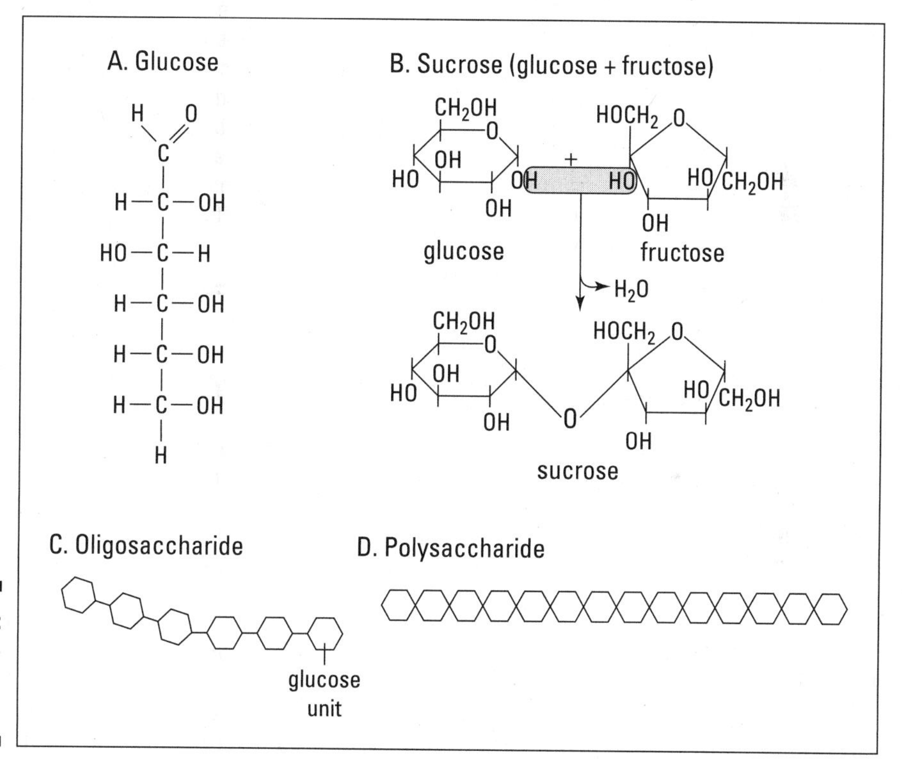

Figure 5-1: Carbohydrate molecules.

Two monosaccharides can have the same numbers of carbon, hydrogen, and oxygen atoms and yet have very different properties. When two monosaccharides have the same atoms, but those atoms are arranged differently, the sugars are *isomers* of each other ("iso" means same). For example, if the hydroxyl group (–OH) and hydrogen atom (–H) attached to the fourth carbon from the top in glucose (see Figure 5-1A) were swapped with each other, the sugar would be converted to galactose. Glucose and galactose are almost identical, except for that one change in the arrangement of the atoms, and yet they behave very differently in cells.

Got lactose intolerance?

Lactose intolerance is the inability to break down the disaccharide lactose, or milk sugar. People who don't produce the enzyme lactase can't break down lactose as it passes through their small intestine. If a person who is lactose intolerant consumes dairy products, the lactose will pass through their small intestines undigested.

The arrival of undigested lactose is happy news for the bacteria who live there and are just waiting for their next meal. Unfortunately, for the person, one of the side effects of lactose breakdown by the bacteria is gas production. Too much gas in the intestines can lead to discomfort and even diarrhea.

Most people begin to produce less lactase after they reach age 2 (when milk becomes less important as a source of nutrition) and gradually become more intolerant as they age. One exception is people of Northern European descent, in whom lactose intolerance is least common. Scientists speculate that, in the past, Northern Europeans were heavily dependent on dairy products for their nutrition and that a mutation occurred that allowed people to produce lactase further into their lifespan. This mutation would have given people who had it an advantage because they could survive easily on dairy products, and so this gene spread through the population.

Today, people who suffer from lactose intolerance can still consume dairy products if they want to. They just need to purchase some lactase at their local drug store and mix it with the dairy products before they eat them. The lactase will break down the lactose before it gets into the small intestine, and the bacteria won't be able to produce the uncomfortable side effects.

The way the atoms are bonded together is very important in the structure and function of sugars. Isomers are made from exactly the same atoms, but their atoms are arranged differently.

In the watery environment of the cell, monosaccharides convert into ring-shaped structures. A bond forms between two atoms in the backbone of the sugar, causing the sugar to bend around to form the ring. As an example, compare the linear structure of glucose shown in Figure 5-1A with the ring structure shown in Figure 5-1B.

Making it complex: Polysaccharides

Polysaccharides, or complex carbohydrates, are polymers (see Chapter 4) of monosaccharides. ("Poly" means many, and "sacchar" means sugar, so a polysaccharide is "many sugars" strung together.) To make polysaccharides, monosaccharides are joined together by condensation reactions (see Chapter 4). During condensation, a water molecule is removed as a bond is formed between an atom in the growing polysaccharide chain and an atom in the monosaccharide that is being added to the chain (see Figure 5-1B). The bonds between monosaccharides are called *glycosidic linkages.*

Polysaccharides are classified based on the number of monosaccharides in the chain:

- **Disaccharides are chains of two monosaccharides.** *Sucrose* (see Figure 5-1B), or table sugar, is a disaccharide that is probably very familiar to you. Another disaccharide you probably know about is *lactose,* the sugar found in milk.
- **Oligosaccharides are short chains of monosaccharides** (see Figure 5-1C). Oligosaccharides are part of receptors in the plasma membranes of your cells.
- **Polysaccharides are long chains of monosaccharides** (see Figure 5-1D). Starch and cellulose, both shown in Figure 5-2, are two polysaccharides that you probably eat every day. Starch is found in bread, potatoes, rice, and pasta; cellulose is referred to as fiber in your diet.

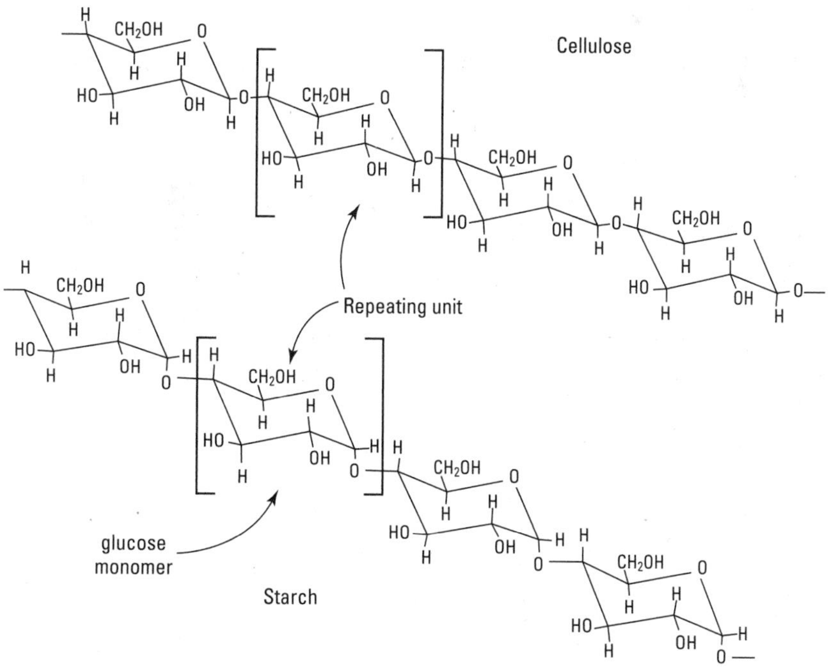

Cellulose

Repeating unit

glucose monomer

Starch

Figure 5-2: Cellulose and starch.

Many cell types produce polysaccharides. Starch and cellulose, which are made by plants, are both polymers of glucose. *Glycogen,* made by animal cells, is also a polymer of glucose. *Chitin,* found in the shells of crustaceans and insects, is a polymer of a nitrogen-containing monosaccharide called N-acetylglucosamine. *Peptidoglycan*, the polysaccharide found in bacterial cell walls (see Chapter 2), is a polymer of two alternating monosaccharides, N-acetylglucosamine and N-acetylmuramic acid.

Polysaccharides can also be different based on how their monosaccharides are strung together. Starch, cellulose, and glycogen are all made entirely of glucose, yet they behave very differently in the body. Starch and glycogen are easily broken down in the human digestive system. Cellulose, or fiber, can't be broken down at all by humans. Instead, it passes right through your digestive system and exits as part of your wastes.

The difference between starch, cellulose, and glycogen isn't what they're made of, but rather in the bonds between the glucose molecules:

- The glucose molecules in starch are joined with a bond called a α–1,4–glycosidic linkage.
- The glucose molecules in cellulose are joined with a β–1,4–glycosidic linkage.
- At approximately every tenth glucose molecule, a branch is joined to the main backbone of glycogen by an α–1,6–glycosidic linkage. Thus, glycogen molecules are highly branched.

The reason humans can digest starch and glycogen, but not cellulose, is that human enzymes can break down some glycosidic linkages, but not others. Human enzymes break down α –1,4–glycosidic linkages and α –1,6–glycosidic linkages, but not β –1,4–glycosidic linkages. Together, starch, cellulose, and glycogen demonstrate how important different types of glycosidic linkages can be to polysaccharide structure and function.

The type of glycosidic linkage between monosaccharides is very important in determining structure and function of polysaccharides.

Sticky and Sweet: Functions of Carbohydrates

Carbohydrates are probably most famous for their role in providing energy to bodies (and, of course, cells), but they perform many other important functions for cells as well:

- **Carbohydrates are an important energy source for cells.** The monosaccharide glucose is a rapidly used energy source for almost all cells on planet Earth. In addition, many cell types store matter and energy for later use in the form of polysaccharides. Plants, algae, and bacteria store energy in starch, and animals and bacteria store energy in glycogen.
- **Carbohydrates are important structural molecules for cells.** Polysaccharides are the major components of the cell walls of plants, algae, fungi, and bacteria. The cell walls of plants and algae contain cellulose, the cell walls of fungi contain chitin, and the cell walls of bacteria contain peptidoglycan.
- **Carbohydrates are important markers of cellular identity**. The surfaces of cells are marked with *glycoproteins,* molecules of protein that have an attached sugar. Different cells have different glycoproteins on their surface, marking the cells with their identity. In your body, liver cells are marked as liver cells, heart cells are marked as heart cells, nerve cells are marked as nerve cells, and so on.
- **Carbohydrates are important extracellular molecules**. Polysaccharides are a major component of the sticky matrix that surrounds cells. They help bacteria stick to surfaces and are also important in the attachment of animal cells to each other.

Splendid substitutes

Humans have taste receptors that specifically recognize sweet tastes. The ability to recognize sweets was probably an advantage in the distant past, when the ability to recognize a good source of energy may have meant the difference between life and death. In many developed nations today where refined foods are plentiful, however, the craving for sweets seems more of a curse than a blessing. So, humans search for a way to satisfy cravings without packing on the pounds.

A recent entry into the artificial sweetener game is *sucralose.* Sucralose is produced from the modification of sucrose. Three hydroxyl groups (–OH) on sucrose are replaced with chlorine atoms. The result is a molecule that really triggers sweet receptors — sucralose tastes 600 times sweeter than sucrose! The chemical modification also changes the molecule so that enzymes can't break it down for energy, which means that sucralose passes through bodies undigested and has no calories.

Chapter 6

Proteins: Workers in the Cellular Factory

In This Chapter

- Examining the four levels of protein structure
- Exploring protein function
- Focusing on enzymes
- Describing membrane proteins
- Introducing DNA-binding proteins

Most jobs in the cell are done by proteins. To do their jobs, proteins must interact in a specific way with other molecules. The specific relationship between proteins and other molecules is based on the shape of the protein. Protein shape is complex and is essential to protein function. In this chapter, I present the basic structure of proteins and introduce their many functions, with a special emphasis on enzymes, membrane proteins, and DNA-binding proteins.

Get into Shape: Levels of Protein Structure

Proteins are the "movers and shakers" of the cell — whatever the job, they get it done. Proteins control metabolism, transport, communication, structure, division, and many other aspects of cell function. One of the things that makes proteins so successful is their ability to specifically target a particular molecule or process. Protein specificity depends upon protein structure.

Proteins begin as polymers of *amino acids,* shown in Figure 6-1. A polymer of amino acids is called a *polypeptide*. Once a polypeptide is folded and becomes functional, the polypeptide is called a *protein*. Some proteins are made of more than one folded polypeptide chain.

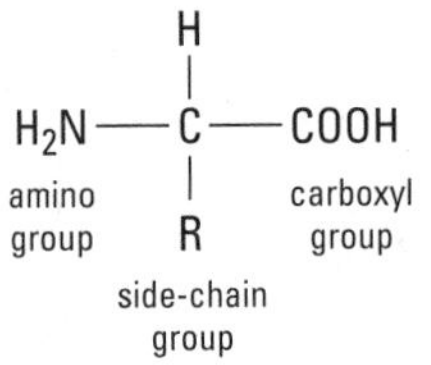

$$H_2N - \underset{R}{\overset{H}{C}} - COOH$$

amino group

carboxyl group

side-chain group

The central carbon atom is flanked by an amino group and a carboxyl group. The name of the amino acid depends on which one of the 20 side-chain groups is at R. For example, if $CH_2-C(=O)O^-$ was at R, the amino acid would be aspartic acid. Proteins are amino acids joined together by peptide bonds. Specific proteins are created based on the order of amino acids connected together. The order of amino acids is determined by the genetic code, which is discussed in Chapter 14.

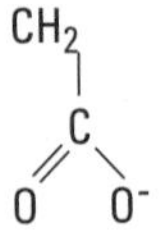

Figure 6-1: Amino acid structure.

Each polypeptide is carefully folded, creating unique sections that are tailored for their particular job. In fact, protein shape is so important to their ability to do their job that if a protein unfolds, or *denatures,* it will no longer function. Protein structure, or *conformation,* is fairly complex and is organized into four categories, as shown in Figure 6-2:

- **Primary structure** is the sequence of amino acids in the protein.
- **Secondary structure** are small areas that are folded into alpha helices and pleated sheets.
- **Tertiary structure** is the final 3D shape of one folded amino acid chain.
- **Quarternary structure** is found in proteins that consist of more than one folded amino acid chain.

Protein denaturation with Julia Child

Actually, you can denature proteins all on your own without help from a famous chef. All you have to do is crack an egg into a hot pan. Egg white is made of the protein albumin. When albumin is in the raw egg, it's coiled up in its normal three-dimensional structure. The heat from the pan, however, will cause the protein to unfold, or denature. As it denatures, its properties change. You can see denaturation happening as the egg white turns from its translucent form in the raw egg to the opaque white form of the cooked egg.

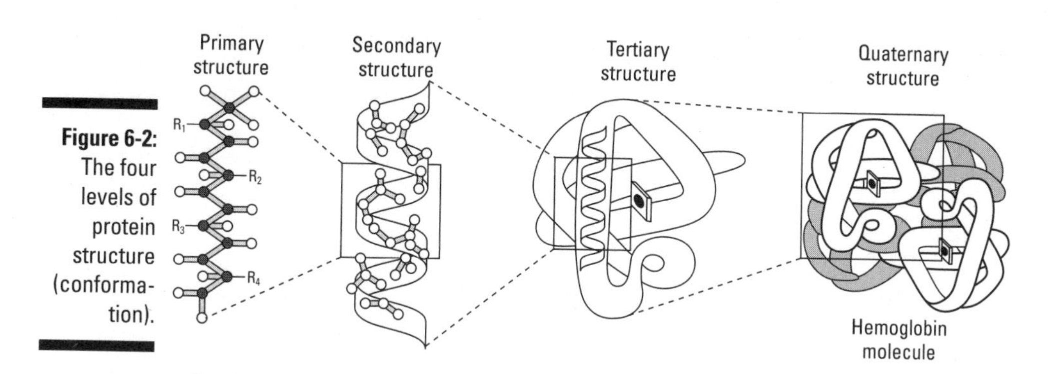

Figure 6-2: The four levels of protein structure (conformation).

Get in line: Primary structure

The primary structure of a protein is the sequence of amino acids in the polypeptide chain. Twenty amino acids exist in cells, so the choice and ordering of amino acids makes the primary structure of one polypeptide different from that of another. (The sequence of amino acids in each polypeptide is determined by the DNA code of the cell; see Chapter 18.)

Each of the 20 amino acids found in cells has the same basic structure. A central carbon atom in the amino acid is bonded to four different chemical groups:

- A nitrogen-containing amino group ($-NH_2$ or $-NH_3^+$)
- A carboxyl group ($-COOH$ or $-COO^-$)
- A hydrogen atom (–H)
- A side chain (–R) that varies between the 20 amino acids

The letter *R* or *X* represents the variable side chain of the twenty amino acids. The side chain can be as simple as a hydrogen atom (–H), as in the amino acid glycine (shown as the amino acid on the left in Figure 6-3) or more complicated as in the amino acid aspartic acid (refer to Figure 6-1).

When you're trying to distinguish between the structures of the four groups of macromolecules (carbohydrates, proteins, nucleic acids, and lipids), look for the characteristic N-C-C backbone of the amino acids. Polypeptide chains can get very complicated looking, particularly if the R groups attached to the amino acids are complex, but if you see a repeating pattern of N-C-C, it's a good bet you're looking at a protein.

Amino acids are attached to each other by condensation: As a covalent bond is formed between the two amino acids, a water molecule is removed. (See Chapter 4 for more on the condensation process.) The bond between amino acids is called a *peptide bond*. Figure 6-3 shows peptide bond formation between the amino acid glycine (on the left) and the amino acid alanine (on the right).

$$^{+}H_3N-\underset{\displaystyle H}{\underset{|}{CH}}-\overset{\displaystyle O}{\overset{\|}{C}}-O \qquad ^{+}H_3N-\underset{\displaystyle CH_3}{\underset{|}{CH}}-\overset{\displaystyle O}{\overset{\|}{C}}-O^{-}$$

$$^{+}H_2O \;\rightleftharpoons\; -H_2O$$

$$^{+}H_3N-\underset{\displaystyle H}{\underset{|}{CH}}-\overset{\displaystyle O}{\overset{\|}{C}}-\underset{\displaystyle H}{\underset{|}{N}}-\underset{\displaystyle CH_3}{\underset{|}{CH}}-\overset{\displaystyle O}{\overset{\|}{C}}-O^{-}$$

peptide bond

Figure 6-3: The formation of a peptide bond.

Because peptide bonds are covalent bonds, they're among the strongest bonds in cells. Extreme environmental conditions, such as increases in temperature or changes in pH, can damage proteins, causing them to denature (unfold). Although denatured proteins can't function, they don't lose their primary structure: The covalent bonds are strong enough to withstand these types of changes in the environment of the cell.

The long and winding road: Secondary structure

As polypeptide chains fold up into their final structure, some areas of the chain form very regular folded patterns. These folded areas represent the *secondary structure* of the protein. Two patterns of folds are part of the secondary structure:

- **The *alpha helix* twists around like a spiral staircase.** To help imagine an alpha helix, take a string or piece of ribbon and wind it around your finger. The twisting spiral pattern of the string or ribbon is the same as that of an alpha helix.

- ***Pleated sheets* wind back and forth.** If you've ever hiked on a mountain trail or down into the Grand Canyon, you know what a switchback is like. You go in one direction for a short span and then turn sharply and go back the other way. The same pattern of twists runs through a pleated sheet in a polypeptide.

The number and type of secondary structures found in a protein depend on the protein. Some proteins have multiple alpha helices but no pleated sheets. Other proteins have both structures or only pleated sheets. The structure depends on the protein and that protein's function. The unique shape of each protein depends on what it binds to as part of its job.

Alpha helices and pleated sheets are held together with hydrogen bonds that form between the atoms in the backbone of the polypeptide chain. Polar covalent bonds within the amino acid create polar groups, allowing the formation of hydrogen bonds:

- **The nitrogen and hydrogen atoms in the amino groups of the amino acids are joined together with polar covalent bonds.** Nitrogen is more electronegative than hydrogen, so it "hogs" the electrons and becomes slightly negatively charged. Hydrogen, on the other hand, develops a slight positive charge.
- **The bond between the carbon and oxygen of the carbonyl group is also a polar covalent bond.** In this case, oxygen is the electron hog that develops a slight negative charge, while the carbon atom develops a slight positive charge.

The slight positive and negative charges due to the polar covalent bonds develop all along the backbone of the polypeptide chain. When an alpha helix or pleated sheet forms, the positively charged hydrogen atoms from the amino groups are attracted to the negatively charged oxygen atoms of the carboxyl group. These weak electrical attractions act like molecular Velcro and hold the alpha helices and pleated sheets in their shapes.

Strong as steel, soft as silk

Spider silk is one of the strongest known natural products. The spider proteins owe their strength to their pleated sheets, which are held together with clusters of hydrogen bonds. Even though hydrogen bonds are 100 to 1,000 times weaker on the molecular level than the metallic bonds in steel, the arrangement of these bonds around the pleated sheets of spider silk gives the silk, pound for pound, the same strength as steel. Scientists are trying to discover ways to produce large amounts of spider silk because of its potential use in medicine and the defense industry. If they're successful, soldiers may one day be able to wear bulletproof shirts that are as lightweight as a T-shirt!

3D: Tertiary structure

The tertiary level of protein structure is the final three-dimensional shape of the polypeptide chain. The final shape of any polypeptide chain is unique and will have specific areas that are necessary for the function of the protein. Overall, two categories of proteins are based on overall three-dimensional shape:

- **Globular proteins** have an overall rounded or irregular shape. Many enzymes are globular proteins.
- **Fibrous proteins** are long and cable-like. Proteins important to the structure of the cell, like cytoskeletal proteins, are usually fibrous proteins.

The tertiary structure of a polypeptide chain is held together by various types of bonds between the atoms in the R groups of the amino acids. The 20 different amino acids found in proteins have 20 different R groups. The different structures of the R groups give them different properties. For example, some R groups are hydrophobic and don't mix well with water, while others are hydrophilic and do mix with water. As the polypeptide chain twists and folds upon itself, the R groups that stick off of the backbone come into contact with each other. Depending on the structure of those two R groups, a bond may form between them. The types of bonds that form between R groups, shown in Figure 6-4, are

- **Covalent bonds:** The amino acid cysteine has a sulfhydryl group (–SH) in its R group. When the R groups from two cysteines come near each other in a folded polypeptide chain, they form a covalent bond called a *disulfide bridge*. "Di" means two, and disulfide bridges contain two sulfur atoms (–S–S–). Because they're covalent bonds, disulfide bridges are strong and aren't lost when a protein denatures.
- **Ionic bonds:** Some R groups ionize in the watery environment of the cell. When ionized R groups come near each other in a folded polypeptide chain, an ionic bond forms between them. In the watery environment of the cell, these ionic bonds are weak. They'll be lost when a protein denatures.
- **Hydrogen bonds:** Some R groups contain polar covalent bonds, creating slight differences in positive and negative charges in the atoms. When atoms with a slight positive charge come near atoms with a slight negative charge, hydrogen bonds form between them. These weak bonds are lost when a protein denatures.
- **Hydrophobic interactions:** Some R groups are hydrophobic. In the watery environment of the cell, these R groups can get pushed together in little pockets inside the folded polypeptide chain, forming a hydrophobic interaction. These weak bonds are lost when a protein denatures.

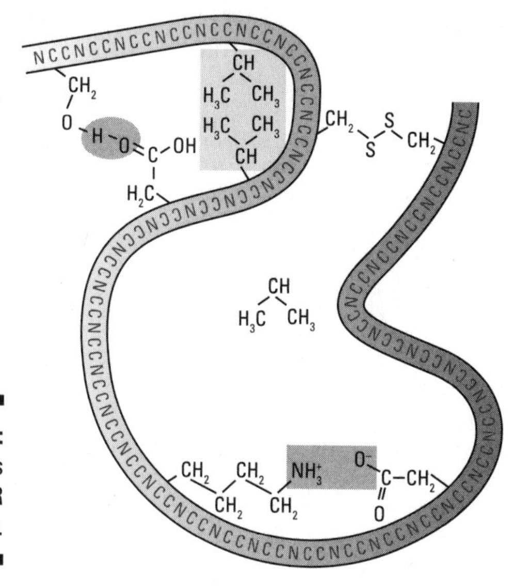

Figure 6-4: Interactions between R groups.

Some polypeptide chains are functional proteins once they have folded into their three-dimensional structure, and they'll begin to do their job for the cell. The protein lysozyme, which fights bacteria in your body, is functional as a single, folded polypeptide chain. Other polypeptide chains require an additional level of structure — quarternary structure — in order to be functional proteins.

A group of proteins called *chaperone proteins* help other proteins fold correctly. The cytoplasm is full of molecules, and it would be very easy for a newly built protein to stick to some of these other molecules and not fold into its correct shape. Chaperone proteins help prevent this mistake by temporarily binding to new proteins and stopping them from misfolding. Chaperone proteins may also give new proteins a relatively clear space in which to fold so that other molecules don't get in the way.

Sometimes one is not enough: Quarternary structure

Some proteins are large and complex, consisting of many polypeptide chains. These proteins have quarternary structure. The hemoglobin that is carrying oxygen around in your blood right now is a protein with quarternary structure. In order to do its job for your body, each hemoglobin protein must be made of four joined polypeptide chains, two of one type (alpha chains) and two of another (beta chains). The collagen that supports your skin is made of three identical polypeptide chains.

The bonds that hold multiple polypeptide chains together to make a protein are the same types of bonds that hold together the tertiary structure of proteins. When polypeptide chains are brought together to build a protein with quarternary structure, the R groups that stick off the backbone of the different polypeptide chains come into contact with each other. The bond that forms depends on the structure of the R groups that come into contact. The possibilities are the same as they are for the bonds between R groups that hold tertiary structures together.

Jacks of All Trades: The Many Functions of Proteins

Proteins are important to the structure of the cell and carry out many jobs within the cell. Here are some of the important jobs that proteins do:

- **Proteins are enzymes.** Enzymes make chemical reactions happen faster (see Chapter 10).
- **Proteins reinforce structures.** Proteins are part of the structure of plasma membranes, and cytoskeletal proteins reinforce the internal structure of cells. Proteins in the extracellular matrix (see Chapter 2) support cells and proteins, such as collagen.
- **Proteins transport materials into and out of the cell.** Proteins in the plasma membrane help molecules enter and exit the cell (see Chapters 2 and 9).
- **Proteins are involved in cellular identity.** Glycoproteins on cell surfaces act as a marker that identifies the cell type.
- **Proteins help cells move.** Cytoskeletal proteins power the movement of flagella and allow cells to crawl like amoebae (see Chapter 2).
- **Proteins help cells communicate.** Proteins are receptors for signals sent to the cell (see Chapter 9). Also, some signaling molecules, like the hormone insulin, are proteins.
- **Proteins organize molecules within the cell.** Chaperone proteins assist folding of new proteins and guide proteins to their proper locations within the cell.

- **Proteins help defend the body against bacteria and viruses.** Proteins called antibodies are key players in the immune system, helping target bacteria and viruses for destruction.
- **Proteins regulate how DNA is used by the cell.** DNA-binding proteins control which sections of DNA are used by the cell and which are silent (see Chapter 19).

Get 'Er Done: Enzymes Make Things Happen

Enzymes are *catalysts,* molecules that speed up chemical reactions (see Figure 6-5). Without enzymes, the reactions within the cell wouldn't happen fast enough to keep cells alive. Because enzymes are necessary for the reactions of the cell to happen, cells can regulate their chemistry by regulating their enzymes.

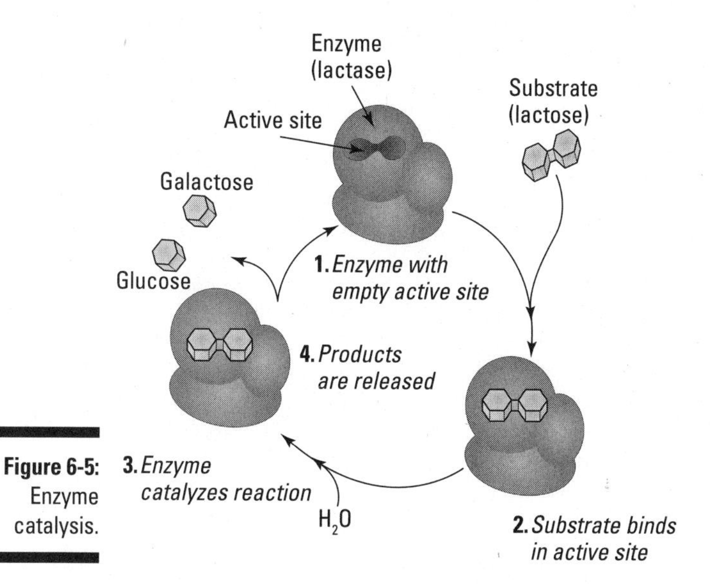

Figure 6-5: Enzyme catalysis.

Made for each other: Enzymes and substrates

Enzymes are very specific — each one catalyzes a particular chemical reaction — because the correct folding of the enzyme creates a pocket called an *active site* that fits only certain molecules. The molecules that fit into the active site of an enzyme are its *substrates*. You can think of the enzyme as a lock and the substrate as the key. Only the correct key can fit into the lock, so the enzyme-substrate relationship is very specific.

The relationship between enzyme and substrate is actually not quite like that of a lock and key. Locks don't shift or change in response to their key, but enzymes do change slightly in response to their substrate. When the substrate first binds into the active site, the fit between enzyme and substrate is not perfect. But when weak bonds form between the substrate and the R groups of the enzyme, the enzyme changes it shape slightly, strengthening the attractions between the substrate and the active site. This view of enzyme-substrate interaction is called the *induced fit* model because the enzyme is induced to fit more closely around the substrate. The enzyme with its induced fit allows the chemical reaction of the substrate(s) to occur by lowering the energy needed (activation energy) for the reaction.

Listening to others: Inhibiting enzymes

Cells can determine whether a particular metabolic pathway is turned on or off by controlling the enzymes that catalyze the reactions in the pathway. You can think of the enzymes in a metabolic pathway as a bunch of assembly line workers in a factory. As the product comes down the assembly line, each worker must make an adjustment to it in order for the product to be ready for the next worker. If you were to tell one of the workers in the line to take a break, production would stop at that point in the line. Enzyme regulation works exactly the same way — if an enzyme is prevented from functioning, the entire metabolic pathway shuts down as a result.

Enzymes are controlled by regulatory molecules that bind to the substrate. The two main types of enzyme regulation (see Figure 6-6) are as follows:

- **Competitive inhibition** occurs when a molecule that is similar in shape to the substrate enters the active site and gets in the way of catalysis. The regulatory molecule, called a *competitive inhibitor,* isn't the right shape for the reaction to actually occur; it just prevents the real substrate from entering. In other words, the competitive inhibitor competes with the substrate for the active site.

- **Noncompetitive or allosteric inhibition** occurs when a regulatory molecule binds to a site other than the active site of the enzyme. When enzymes fold up into their three-dimensional shape, they can have more than one pocket into which molecules can bind. Binding sites other than the active site are called *allosteric sites*. Molecules that bind to allosteric sites to regulate enzymes are called *noncompetitive inhibitors*. When noncompetitive inhibitors bind to allosteric sites, they cause slight shape changes in the enzyme. These shape changes affect the active site as well, which means the substrate can no longer bind to the enzyme.

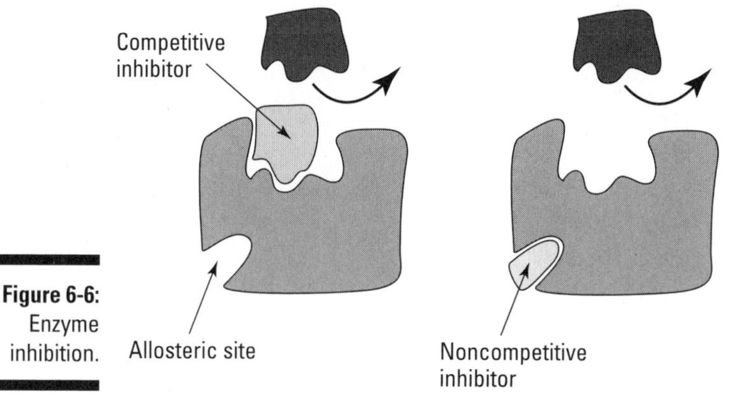

Figure 6-6: Enzyme inhibition.

The most common way that cells regulate their metabolic pathways is through *feedback inhibition* or *end product inhibition* (see Figure 6-7). During feedback inhibition, an end product of a pathway acts as the noncompetitive inhibitor of an enzyme in that pathway. Think about it: If you were managing a production plant, how would you know when it was time to slow production? Your biggest clue would probably be if you had too much of your product on your hands. If that were the case, you'd probably slow production until demand caught up with supply, right? This same principle drives feedback inhibition. If a cell has made enough of a product, then lots of those molecules will be floating around in the cell. The simplest way to enact a slowdown in production is for that product itself to bind the allosteric site on an enzyme and shut it down. When the cell uses up the product, there won't be any excess to bind to the enzyme, and production will start up again.

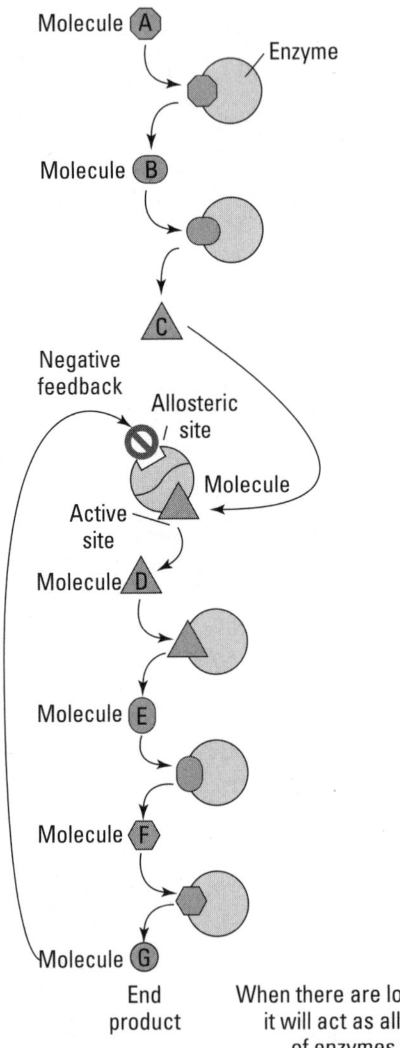

Figure 6-7: Feedback inhibition.

Gatekeepers: Membrane Proteins

The plasma membrane is the barrier between the cell and its environment. Like the border between two countries, the plasma membrane is a busy place. Signals arrive at the membrane with messages for the cell, molecules seek to cross in and out, and defensive proteins from the immune system wander by to make sure that everything is okay inside the cell.

The proteins that make up almost 50 percent of the plasma membrane are responsible for coordinating all activity at the plasma membrane. Some of the proteins span the entire membrane; others are associated with one side or the other. Membrane proteins, shown in Figure 6-8, perform all the following jobs:

- **Transport proteins** help molecules cross the membrane. Some proteins provide open channels for small molecules to pass through, while others bind specifically to certain molecules and move them across (see Chapter 9).
- **Receptor proteins** receive signals on the outside of the cell and relay the message to the inside of the cell. Chemical signals that bind to receptors are called *ligands.*
- **Adhesion proteins** form connections between the outside and the inside of the cell. On the outside of the cell, they bind to molecules in the extracellular matrix (see Chapter 9), while on the inside of the cell they bind to cytoskeletal proteins (see Chapter 2).
- **Identity proteins** are glycoproteins that stick out of the cell and label the cell with its function and identity.
- **Enzymes** are also associated with membranes. They may catalyze reactions that are part of processes occurring within the membrane or they be involved in passing signals from receptor proteins to the inside of the cell.

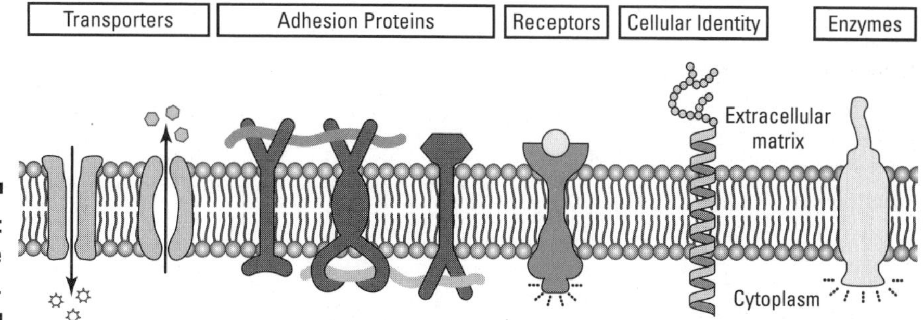

Figure 6-8: Membrane proteins.

Finding your perfect match

You probably know that if a person needs an organ transplant, doctors must find a "close match" between the donor and the recipient. But what exactly are the doctors trying to match?

Well, it turns out that a group of membrane proteins called the Major Histocompatibility Complex (MHC) marks your cells as your cells. If a person receives a transplant of someone else's cells as part of an organ and the MHC proteins don't match, their body will reject the organ as foreign. So, doctors must look for a donor whose MHC proteins are as close as possible in structure to those of the recipient. And even then, the little differences in the MHC proteins are still recognized by the body, so the recipient needs to remain on immunosuppressant drugs for the rest of their life.

Some day, scientists hope to be able to grow organs in the lab from a person's own stem cells. That way, the proteins on the surface would be a perfect match! (By the way, I know Major Histocompatibility Complex seems like a mouthful. But it really means what is says. "Histo" = "tissue." So, Major Histocompatibility Complex means "the most important complex of proteins for determining whether two tissues will be compatible.")

I'm in Charge: DNA-Binding Proteins

The genetic code that contains the instructions for all the structures and functions of the cell is in the form of DNA. So, it would seem that DNA is the most powerful molecule in the cell, right? In some ways, DNA is the most powerful, but it's not as simple as that. Certain proteins, called *DNA-binding proteins* (see Figure 6-9), can actually control which parts of the DNA are being read and which are silenced.

Just as enzymes have active sites that are shaped to fit their substrates, DNA-binding proteins have binding sites, called *DNA-binding domains,* that are shaped to fit DNA. The alpha helices and pleated sheets within the DNA-binding domains fit into the grooves of the DNA molecules. Like other proteins, DNA-binding proteins are very specific, so DNA-binding domains must be able to recognize a specific target sequence.

Several types of DNA binding proteins are known. Each type has a slightly different structure in its DNA-binding region. These different structures have been found again and again as new proteins are discovered, appearing like a repeating pattern, or *motif,* in DNA-binding proteins. The most common motifs found in DNA-binding proteins are listed here:

- **Helix turn helix motifs** have two alpha helices connected by a short stretch of amino acids that forms a turn between the two helices. The helix turn helix motif is common in prokaryotic DNA-binding proteins.
- **Zinc finger motifs** have an alpha helix and two pleated sheets held together with a zinc atom. The zinc finger motif is common in eukaryotic DNA-binding proteins that regulate transcription. (For details on how DNA-binding proteins regulate transcription, see Chapter 19.)
- **Leucine zipper motifs** have two alpha helices that cross over each other. The two helices are held together by the hydrophobic interactions between the R groups on the amino acid leucine. The leucine zipper motif is common in eukaryotic DNA-binding proteins that regulate transcription.

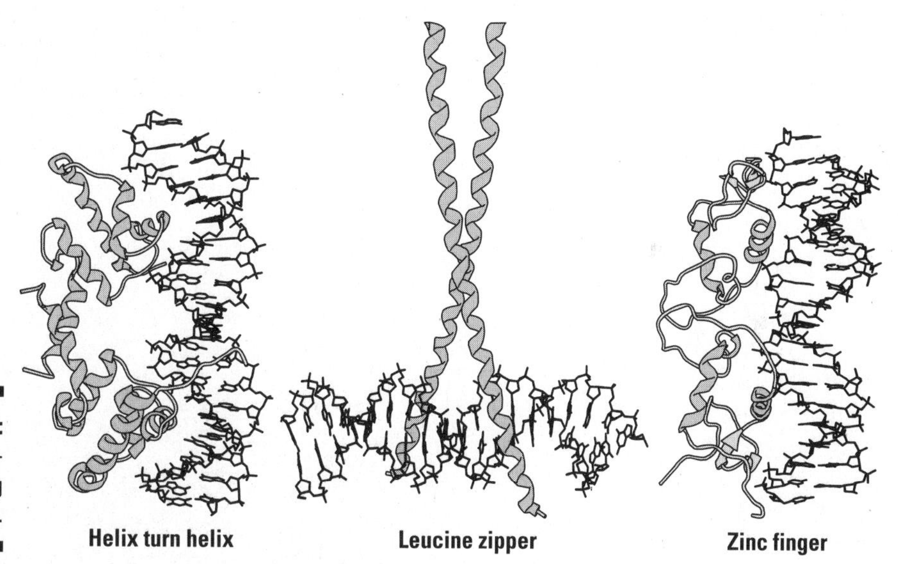

Figure 6-9: DNA-binding domains.

Chapter 7

DNA and RNA: Instructions for Life

In This Chapter

- Building a nucleotide
- Stringing together polynucleotide chains
- Comparing DNA and RNA
- Introducing the functions of nucleic acids

Deoxyribonucleic acid (DNA) makes up the genetic code of all cells, determining their characteristics by specifying the construction of proteins and molecules of ribonucleic acid (RNA). DNA and RNA are both nucleic acids, which are polymers of nucleotides. In this chapter, I present the structure of DNA and RNA, as well as introduce the functions of these important nucleic acids.

It's Puzzling: Structure of Nucleic Acids

Nucleic acids are the information molecules of cells. The most famous nucleic acid is undoubtedly *deoxyribonucleic acid*, or DNA. You probably know that your DNA contains your genetic code, all the messages that determine your traits. Another type of nucleic acid, *ribonucleic acid*, or RNA, is less well known, but also important to cells.

All nucleic acids are polymers, consisting of long chains of repeating monomers like beads on a string. However, nucleic acids look far more complex than either polysaccharides or polypeptides. Nucleic acid structure is complex for several reasons:

- **The monomer of nucleic acids is the nucleotide.** Nucleotides have three different components, so they're pretty complicated even before they get strung into chains.

- **Nucleic acids may be single-stranded or double-stranded.** When nucleotides are strung together, they form *polynucleotide chains*. RNA consists of a single polynucleotide chain. In cells, DNA consists of two polynucleotide chains wound together to form a double helix.

Although nucleotides are the building blocks of nucleic acids, nucleotides themselves are built from three different components:

- **A pentose (5-carbon) sugar:** In DNA, the sugar is deoxyribose. In RNA, the sugar is ribose.
- **A nitrogenous base:** Nitrogenous bases are single- or double-ring structures that contain several nitrogen atoms.
- **A phosphate group:** Phosphate groups consist of a phosphorous atom surrounded by oxygen atoms. They're negatively charged, which gives nucleotides and nucleic acids an overall negative charge as well.

Navigating nucleotides

To help navigate the structure of nucleotides, scientists use the following numbering system for the carbon atoms in the sugar of the nucleotide:

- The carbon atoms in the sugar are numbered sequentially from one to five.
- The numbered carbons are also given the designation of prime, making them 1', 2', 3', 4', and 5'. The prime mark (') distinguishes the carbon atoms in the sugar from the carbon atoms in the nitrogenous base, which are also numbered.
- The first carbon, called the *one prime carbon,* is the carbon atom that is attached to the nitrogenous base (Figure 7-1).
- The rest of the carbon atoms in the sugar are numbered consecutively, moving away from the oxygen in the ring. As an example, the 3' and 5' carbon atoms are labeled in the nucleotide shown in Figure 7-1.

The sugars in the nucleotides that make up DNA and RNA differ slightly. DNA contains the sugar deoxyribose, while RNA contains the sugar ribose. These sugars are both pentose (5-carbon) sugars and are identical except for the presence of one oxygen atom. Ribose has a hydroxyl group (–OH) attached to its 2' carbon, whereas deoxyribose has only a hydrogen atom (–H). The nucleotide shown in Figure 7-1 contains ribose, so it's a *ribonucleotide*, a nucleotide that would be used to build RNA molecules. Nucleotides that build DNA are called *deoxyribonucleotides*.

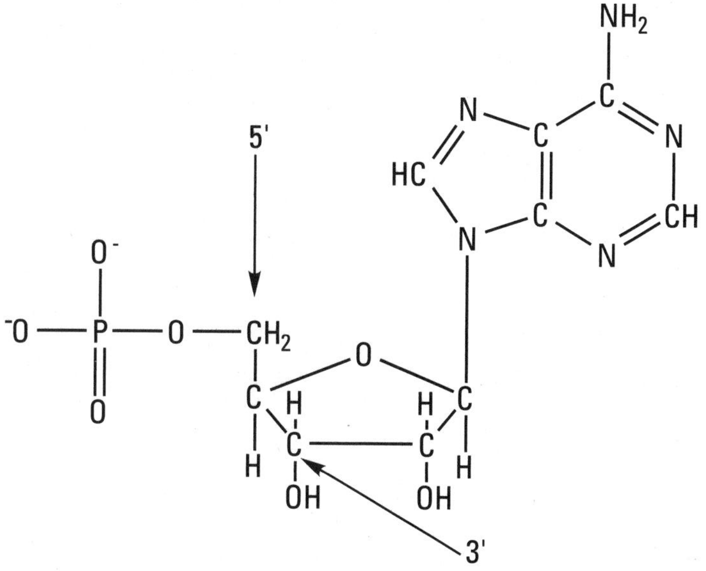

Figure 7-1: Structure of a nucleotide (adenosine monophosphate).

Naming the nucleotide bases

Five different nitrogenous bases are found in nucleotides (see Figure 7-2): adenine (A), guanine (G), cytosine (C), thymine (T), and uracil (U). Each nucleotide is slightly different depending on which nitrogenous base is part of its structure.

As an example, the nucleotide in Figure 7-1 contains the nitrogenous base adenine, which makes the nucleotide adenosine monophosphate. The nucleotides that make up DNA molecules contain adenine, guanine, cytosine, and thymine, whereas the nucleotides that make up RNA molecules contain guanine, cytosine, adenine, and uracil.

Nitrogenous bases fall into two groups, shown in Figure 7-2:

- *Pyrimidines* (cytosine, thymine, and uracil) have a single-ring structure with two nitrogen atoms within the ring.
- *Purines* (adenine and guanine) have two fused rings with two nitrogen atoms within each ring.

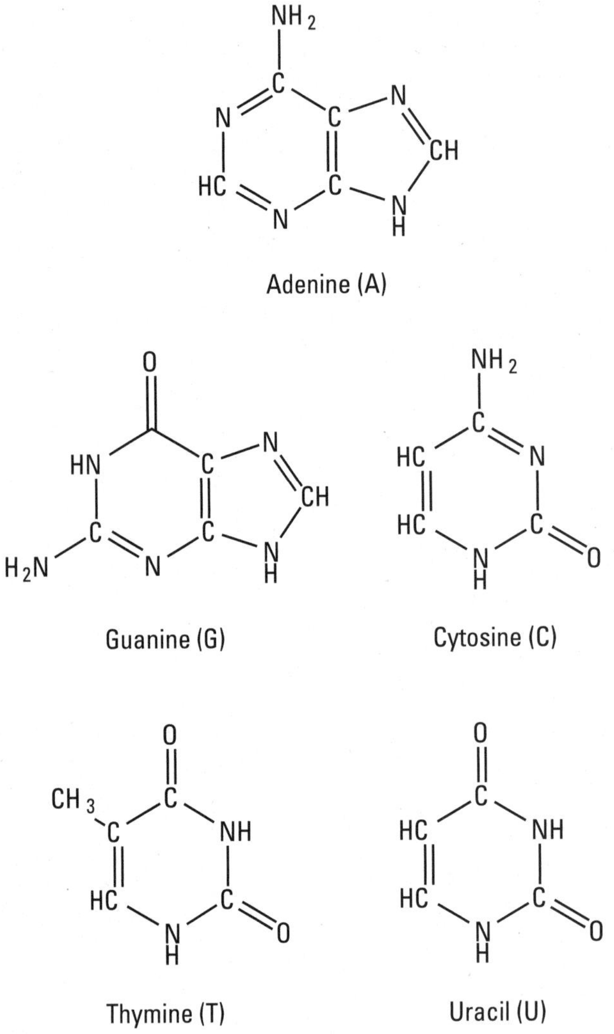

Adenine (A)

Guanine (G)

Cytosine (C)

Thymine (T)

Uracil (U)

Figure 7-2: The nitrogenous bases.

Recognizing nucleotides

Here are some tricks to help you recognize nucleotides and their components:

- To tell the difference between ribose and deoxyribose, look at the 2' carbon. If no oxygen is attached, it's deoxyribose.
- The nitrogenous bases with the longer name (pyrimidines) are actually the smaller molecule (one ring instead of two).

- Within the pyrimidines, only thymine (T) has a methyl group ($-CH_3$). (Remember that there is a *T* in methyl.) Cytosine (C) has an amino group ($-NH_2$) attached to its ring. (Remember that cytosine rhymes with amine.)
- Within the purines, guanine has a double-bonded oxygen attached to one of its rings. Remember that if you close the circle on a *G*, it looks like an *O*.

Making DNA and RNA

Nucleotides are joined together by condensation. (For more on condensation, see Chapter 4.) The phosphate group attached to the 5' end of one nucleotide is joined to the 3' carbon of the nucleotide on the growing chain (see Figure 7-3). A water molecule is removed as the bonds are formed between the two nucleotides. This process is repeated, forming a polynucleotide chain. The order and type of nucleotides in a polynucleotide strand make up its *primary structure*. The order of nitrogenous bases within a polynucleotide chain is also called its *sequence*. All polynucleotide chains have certain key characteristics:

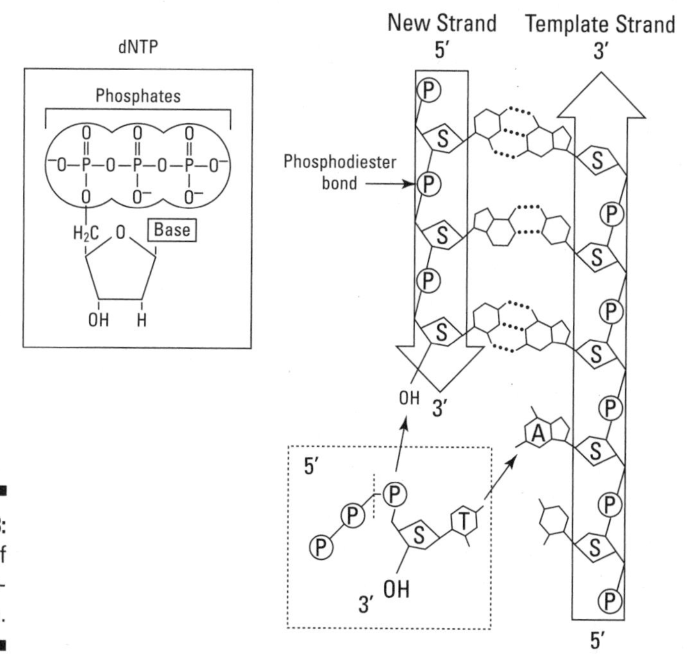

Figure 7-3: Synthesis of a polynucleotide chain.

- A **sugar-phosphate backbone** made from the alternating sugars and phosphate groups in the chain (see Figure 7-3).
- **Nitrogenous bases** that project outward from the sugar-phosphate backbone.
- **5' to 3' polarity** where one end of the chain is very different than the other. At one end of the chain is a phosphate group, attached to the 5' carbon of the topmost sugar. This end is called the 5' end of the strand. At the other end of the chain is a hydroxyl group (–OH) attached to the 3' carbon of the lowermost sugar. This end is called the 3' end of the strand.

Both DNA and RNA are made of polynucleotide chains, but the two types of molecules are unique. Three major differences exist between DNA and RNA:

- DNA contains the nitrogenous base thymine, while RNA contains uracil. (Both contain adenine, guanine, and cytosine.)
- DNA nucleotides have the sugar deoxyribose, while RNA nucleotides have ribose.
- DNA molecules are double-stranded, while RNA molecules are single-stranded.

The double helix of DNA

Polynucleotide chains interact with each other and with themselves to give nucleic acids three-dimensional shape, called their *secondary structure*. The secondary structure of DNA is a double helix. Two polynucleotide chains join together, forming a molecule with the shape of a twisted ladder. The sides of the ladder are made of the sugar phosphate backbones of the two strands (see Figure 7-4). The nitrogenous bases project off of the sugar-phosphate backbone and join together by hydrogen bonds, forming the "rungs" of the ladder.

The two chains of the double helix are *antiparallel* to each other, meaning that they're opposite in polarity. If you hold one of the strands so that its 5' end was up, then its partner strand would have its 5' pointing down. In other words, the two strands are upside down relative to each other. This position is the only way the two strands fit together.

When polynucleotide chains bond together, they're always antiparallel to each other.

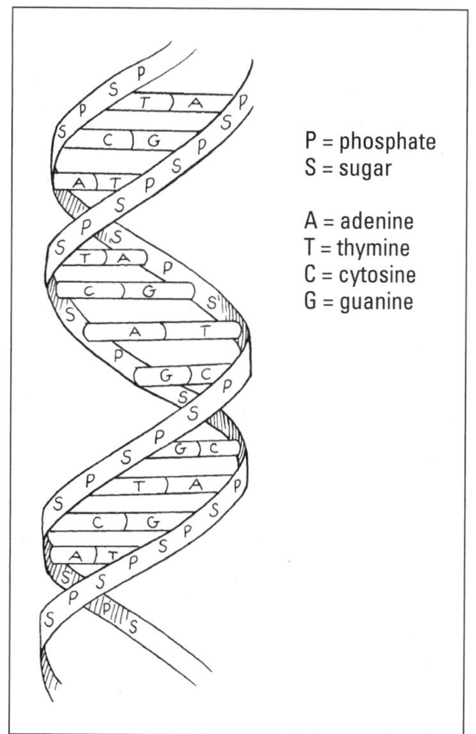

Figure 7-4: The twisted ladder model of the double helix of DNA.

The two antiparallel strands of DNA are held together by hydrogen bonds between their nitrogenous bases. In order for the hydrogen bonds to form, the atoms of the nitrogenous bases must be lined up with each other in just the right way, which happens only when certain bases come together. Adenine (A) forms hydrogen bonds with thymine (T), and cytosine (C) forms hydrogen bonds with guanine (G). You can think of the bases like puzzle pieces that fit together a certain way: In DNA, A fits only with T, and C fits only with G. Scientists say that A *is complementary to* T, and C *is complementary to* G. When these bases match up, they form *complementary base pairs.*The base pairing rules for DNA are A=T and C=G.

Each turn of the double helix is about ten base pairs in length. The distance between each base pair is 0.34 nanometers (nm). As the DNA molecule twists around, the distances between the sugar phosphate backbones vary. When they're farther apart, they make a *major groove.* When they're closer together, they make a *minor groove.* One major and one minor groove alternate with each other in a single turn of the helix. One complete turn of the helix is 3.4 nm long. The diameter of a strand of DNA is 2.0 nm.

Shaping up RNA molecules

RNA molecules are made of single polynucleotide chains, but they can also have three-dimensional secondary structure. Some RNA molecules fold around upon themselves into complex shapes. The shapes are held together by hydrogen bonds between the nitrogenous bases within the same strand. As in DNA, the nitrogenous bases within an RNA molecule can hydrogen bond to each other if they fit together just right. In RNA, cytosine (C) can form hydrogen bonds with guanine (G), and adenine (A) can form hydrogen bonds with uracil (U). Thus, the base-pairing rules for RNA are C=G and A=U.

Breaking the Code: The Function of DNA and RNA

DNA and RNA molecules are involved in information storage and retrieval in cells. DNA stores the information that ultimately determines the characteristics of cells and organisms. The information is written in a chemical code determined by the order of nitrogenous bases within the DNA. You can think of the four bases (A, T, C, and G) like letters of the alphabet being used to write a coded message. When cells decode the DNA message, it provides instructions for important cell structures and functions, including the following:

- **Protein structure** is determined by the sequence of nucleotides in DNA. Through the processes of transcription and translation, the code in DNA is copied and used to specify the sequence of amino acids (primary structure) in polypeptide chains. (For more on transcription and translation, see Chapter 19.)
- **RNA structure** is also determined by the sequence of nucleotides in DNA. Through the process of transcription, the sequence of nucleotides in DNA is used to build RNA molecules.
- **Regulation of DNA** occurs by the interaction between DNA-binding proteins (see Chapter 6) and special regulatory sequences of DNA. These interactions control the organization and retrieval of information from DNA molecules.

Several types of RNA molecules are built from the information in DNA. Each type of RNA has a different function in cells:

- **Messenger RNA (mRNA)** carries the code for protein structure from the DNA to ribosomes where it can be used to produce proteins. (See Chapter 19 for more on this function.)

- **Transfer RNA (tRNA)** decodes the message in mRNA by matching amino acids to the mRNA code. (See Chapter 19 for more on this function.)
- **Ribosomal RNA (rRNA)** is part of the structure of the ribosome.

Jurassic DNA

In the movie *Jurassic Park,* scientists retrieved pieces of dinosaur DNA from mosquitoes preserved in amber and used that DNA to bring dinosaurs back to life. Was this fiction or future fact?

Well, mostly fiction, but maybe a little bit of future fact. Scientists have tried to recover ancient DNA from amber, but so far no one has been successful. (Some scientists have claimed to recover DNA from amber, but no other scientists have been able to repeat this work — and in science, it's a deal breaker when your work can't be repeated.) It turns out that contamination with DNA from living fungi and bacteria is a significant problem when scientists are trying to make copies of the fossil DNA from amber.

So far, scientists have been much more successful in getting ancient DNA that has been frozen in permafrost from places like Antarctica and Greenland. By comparing ancient DNA to the DNA of organisms that are alive today, scientists can figure out what kind of DNA has been isolated.

The oldest DNA extracted so far came from ice sheets on Greenland and is estimated as being between 450,000 and 800,000 years old. Comparisons to existing organisms reveal that Greenland was covered with forests of yew, pine, spruce, and alder trees, very similar to those currently found in Canada. The DNA of beetles, spiders, flies, and butterflies that lived in these forests was also recovered.

Chapter 8

Lipids: Waterproof and Energy Rich

In This Chapter

- Constructing lipids
- Comparing saturated and unsaturated fats
- Exploring the functions of lipids

Lipids are molecules that don't mix well with water. Many types of lipids are made by cells, including fats, phospholipids, waxes, and sterols. In this chapter, I cover the structure of the different groups of lipids and then introduce their functions in cells.

Hydrocarbons: Structure of Lipids

You may not have ever heard the term *lipids,* but you've certainly heard of many of the molecules that fall into this group, including fats, oils, waxes, cholesterol, and steroids. You may even have heard the term hydrocarbons when getting results from an emissions check on your automobile. *Hydrocarbons,* including gasoline and other lipids, are molecules that have long chains of carbon atoms bonded to hydrogen atoms. (Unfortunately, if you have hydrocarbons in your automobile emissions, it means your car isn't burning its fuel efficiently, and you're going to need some repairs.)

All lipid molecules have one thing in common: They're *hydrophobic,* meaning that they don't mix well with water. When you wax your snowboard or your car, you're taking advantage of this characteristic to create a barrier against water. Lipids are hydrophobic because they contain many nonpolar covalent bonds such as those between carbon and hydrogen (for more on nonpolar covalent bonds, see Chapter 4).

As a group, lipids are more diverse in structure than the other three groups of macromolecules (carbohydrates, proteins, and nucleic acids). Also unlike the other macromolecules, lipids aren't polymers. Based on their structures, lipids are organized into several subcategories:

- **Fats and oils** are very similar in structure to each other. They're made of fatty acids attached to a glycerol backbone.
- **Phospholipids** are also made of fatty acids attached to a glycerol backbone. However, they also have a unique hydrophilic head group.
- **Waxes** are made from fatty acids attached to long-chain alcohols.
- **Sterols** contain four fused carbon rings.

Saturating fatty acids

Fatty acids are long chains of carbon and hydrogen that are components of some types of lipids. Fatty acids are described as *saturated* and *unsaturated.* These terms describe the way carbon atoms in fatty acid chains are bonded to each other and with hydrogen:

- In *saturated fats,* the carbon atoms in the fatty acids are bonded to each other with single bonds. Because carbon atoms form four bonds, each carbon atom in a saturated fat can also bond to two hydrogen atoms.
- In *unsaturated fats,* the carbon atoms are bonded to each other with more double bonds. The carbon atoms involved in a double bond can bond to only one hydrogen atom.

As an example of saturated and unsaturated bonds, look at the fatty acids in the typical fat shown in Figure 8-1. The top fatty acid in the fat molecule has saturated bonds: The bonds between the carbon atoms are single bonds, and each carbon atom is bonded to at least two hydrogen atoms. In other words, the carbon atoms are *saturated* with as many hydrogen atoms as they can hold.

The bottom fatty acid in the fat molecule has two unsaturated bonds: The carbon atoms are bonded to each other with double bonds, and each of these double-bonded carbon atoms is bonded to only one hydrogen atom. They aren't saturated with as many hydrogen atoms as they could potentially hold, so they're *unsaturated.*

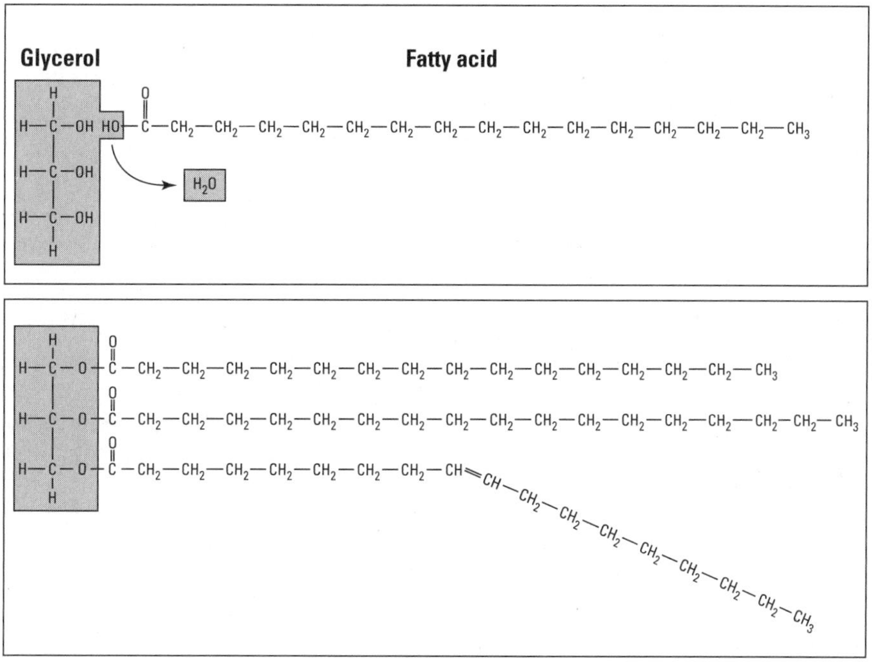

Glycerol Fatty acid

Figure 8-1: Saturated and unsaturated bonds in a typical fat.

Forming fats and oils

Fats and oils form when three fatty acids are joined to a glycerol backbone (see Figure 8-2). Fats and oils are also called *triglycerides* because of the three fatty acids. Glycerol is a three-carbon alcohol, with a hydroxyl radical (–OH) attached to each carbon. Fatty acids are long chains of carbon atoms bonded to hydrogen atoms, making them very hydrophobic.

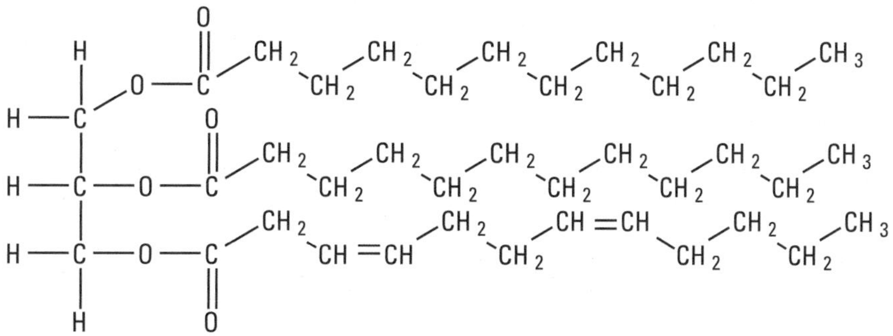

Figure 8-2: The formation of a fat or oil.

Each fatty acid is attached to glycerol in a condensation reaction, leading to the removal of a water molecule as the bond is formed. At one end of a fatty acid molecule is a carboxyl group. When fatty acids are attached to glycerol molecules during the formation of a fat, the carbon atom from the carboxyl group forms a bond with an oxygen in the glycerol molecules.

When you're trying to recognize lipid molecules, remember the word *hydrocarbon.* Lipids have lots of carbon-hydrogen bonds, and few polar groups, such as hydroxyl radicals (–OH) or nitrogen-hydrogen bonds. Also, fat and oil molecules look like a big letter *E.*

Triglycerides that contain high levels of saturated bonds are called fats; those that have more unsaturated bonds are called oils. Fats are solid at room temperature because the saturated fatty acid tails tend to pack together in an organized way, making the fats tighter and more solid. Oils are liquid at room temperature because the unsaturated fatty acid tails have more kinks in them, preventing them from packing together very well. So, oils are looser and more liquid.

In terms of health, you probably already know that you should avoid eating saturated fats as much as possible. Animal fats, such as those in dairy products and meat, are usually saturated. Plant and fish oils, such as those in nuts and fish, are usually unsaturated.

Looking at other types of lipids

Fats and oils are probably the best known lipids, but other types of lipids are also important cellular components. Some lipids are similar in structure to triglycerides, but others are quite different.

Phospholipids, shown in Figure 8-3, are similar in structure to triglycerides and also form by condensation reactions between glycerol and fatty acids. The difference between phospholipids and triglycerides is that phospholipids have only two fatty acid tails. In place of the third fatty acid, they have a head group that contains nitrogen and phosphate.

The negative charges within the phosphate, combined with the polar bonds between nitrogen and hydrogen atoms, make the head group of a phospholipids hydrophilic (Chapter 4). The fatty acid tails, however, have nonpolar bonds between carbon and hydrogen and are hydrophobic. Thus, phospholipids have a split personality — one end of the molecule, the hydrophilic head, is attracted to water, while the other end, the hydrophobic tails, is repelled by water. The word for this split personality, when a molecule has both polar and nonpolar areas, is *amphipathic.*

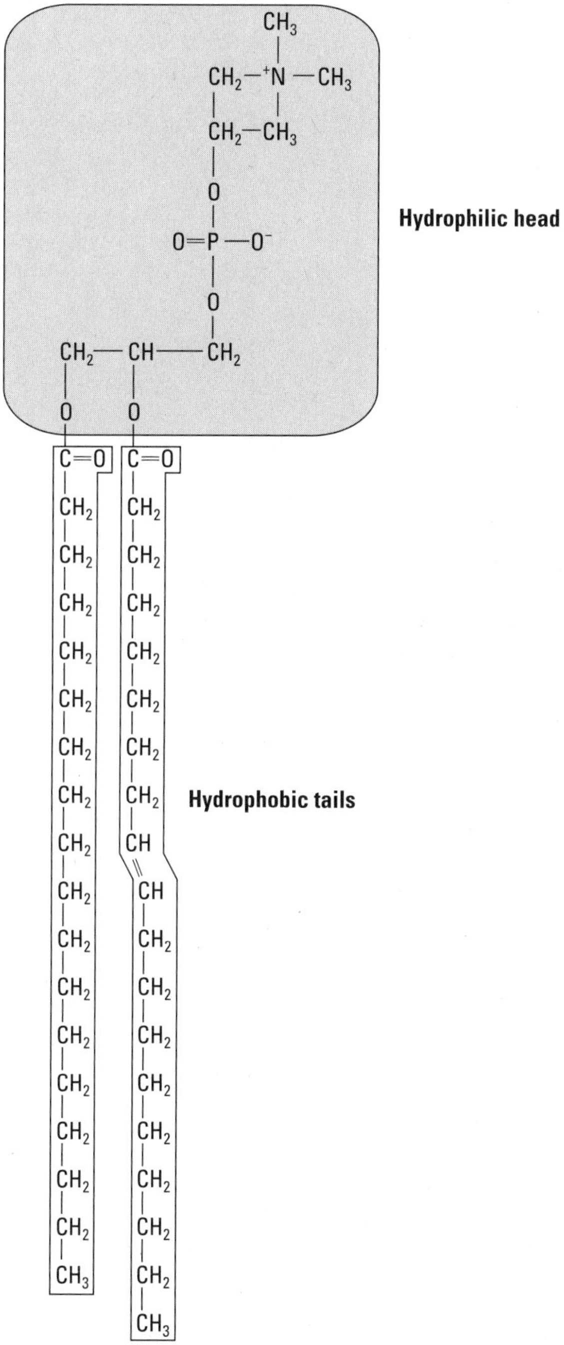

Hydrophilic head

Hydrophobic tails

Figure 8-3: A phospholipid.

The butter wars

Margarine is usually made from corn oil, but margarine is a solid. To turn corn oil into a solid form, guess what you have to do: *Hydrogenate* it — add hydrogen atoms to the fatty acid tails, causing the bonding to change from double bonds into single bonds. In other words, you need to change your corn oil into a saturated fat.

So, if unsaturated fats, like those in corn oil, are healthy, and saturated fats, like those in butter, are unhealthy, where does margarine fit in? Chemically speaking, a saturated fat is a saturated fat no matter how you got there, so really traditional margarine is nutritionally about the same as butter.

This explanation may lead you to wonder why margarine is around at all. The answer goes back to World War II and food rationing. There just wasn't enough butter to go around, so companies started making margarine (or *oleomargarine*, called *oleo* for short). When margarine was first produced, it was white in color like vegetable shortening, but people didn't like its appearance, so the companies started including a packet of yellow dye with the margarine. When you bought your margarine, you had to mix your own dye in to make it look like butter.

When food rationing due to the war ended, the companies that made margarine wanted to stay in business, so they started marketing their product as healthier than butter. The butter companies counterattacked, leading to many years of ad campaigns pitting "natural" butter against "healthier" margarine.

Today, margarine and butter are seen as about equal nutritionally with personal choice based on flavor and price. The dreaded fat today is the *trans fat*, which is a fat that has only been partially hydrogenated. Trans fats are linked to heart disease and should be avoided in your diet. Ironically, these fats used to occur in margarine as a byproduct of the hydrogenation process. (Back when margarine was being marketed as healthier, it probably actually wasn't.) Now, many food production practices have changed to eliminate these fats, and products are labeled as being free from trans fats. Today, diet and nutrition advisors say the best choice is to use "healthy fats" whenever possible. By "healthy fats," the diet and nutrition industry really means "healthy oils" — in other words, oils from fish and plants.

Waxes are formed from fatty acids and long chain alcohols. They're basically very long carbon chains with lots of nonpolar carbon hydrogen bonds. If you imagine two very long fatty acids hooked together with a few oxygen atoms in the middle where they're joined, you'd have a pretty good picture of what a wax looks like.

Sterols are made of four fused carbon rings (see Figure 8-4). As in the other categories of lipids, the rings contain lots of nonpolar carbon hydrogen bonds, making sterols hydrophobic. The sterols include some fairly famous molecules, such as the steroid hormones testosterone and estrogen, as well as cholesterol and vitamin D.

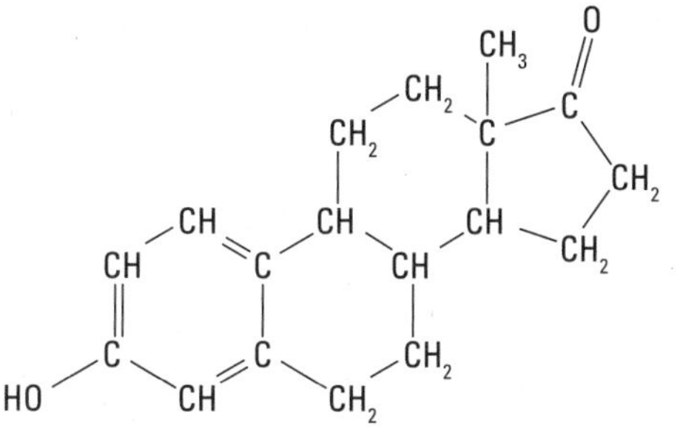

Figure 8-4: The structure of estrogen, a sterol.

You Say Fat Like It's a Bad Thing: Functions of Lipids

Bonds between carbon and hydrogen atoms have a lot of useful energy for cells. Because they have so many carbon-hydrogen bonds, lipids are a great way to store a lot of energy in a relatively small package. For example, you may know that fats have 9 calories per gram, while carbohydrates and proteins have only about 5 calories per gram. (A *calorie* is a measure of energy.) Clearly, fats are the winner among macromolecules in the energy efficiency category. So, energy storage is one of the major functions of lipids in cells. Energy storage is the reason that plants make oils and that people pack on extra pounds of fat when they eat more and exercise less.

Quest for oil

Recent human history demonstrates the value of lipids for energy storage. Since the Industrial Revolution, industrialized countries have sought far and wide for the energy to fuel the lifestyles of their people. One of the most useful forms of this energy is oil.

Unfortunately, the human desire for oil spawned the whaling industry and led to the near extinction of several whale species as they were killed for their oil-rich blubber. More recently, people have fueled their economies with an energy-rich oil legacy left to them by cells of the past. And now, as oil deposits within the Earth begin to run low, people are realizing they need to again seek another source of energy. One part of the solution to this energy crisis may be to find a new source of oil — this time, growing oil-rich crops so that people can create biofuels such as biodiesel.

In addition to energy storage, lipids have other important functions for cells and organisms. The major functions of lipids are as follows:

- **Insulation** by fat is common in many animals. In humans, fat acts as a shock-absorbing layer between the organs and as an insulator around nerve cells (like plastic coating on copper wires). Some animals, such as whales, polar bears, and seals, can survive cold temperatures due to their insulating fat layers.
- The **structure of plasma membranes** is large due to the structure of phospholipids. Plasma membranes are made from two layers of phospholipids that form a bilayer with the hydrophilic heads on the outside of the membrane and the hydrophobic tails pointing inward. Cholesterol is also a component of animal plasma membranes.
- **Water-proofing** by waxes is used by plants to protect their leaves from damage and by water birds to keep their feathers dry.
- **Signaling** by steroid hormones, such as testosterone and estrogen, causes important developmental changes in humans.

Part III
The Working Cell

The 5th Wave — By Rich Tennant

"I'll have the cheese sandwich with the interesting mold on the bread, and the manicotti with the fungal growth, and that really, really old dish of vanilla bread pudding."

In this part . . .

Cells have lots of work to do to keep life happening. Cells take in energy from their environment and transfer it into forms that the cell can use to fuel this work, converting one molecule to another through cellular metabolism. Some cells make the food that all cells rely upon as their source of energy and building materials. Cells grow, move, and communicate with each other.

Cells make new cells, sometimes to repair damaged tissue, sometimes to reproduce themselves. In this part, I explain the working cell, what cells need to function, how they supply those needs, and how they reproduce themselves.

Chapter 9

Hello, Neighbor: How Cells Communicate

In This Chapter

- Crossing plasma membranes
- Forming cell-cell attachments
- Sending signals from cell to cell

Cells live in complex environments surrounded by signals from the environment and from other cells. Cells take in materials from their environment by moving them across their plasma membranes. Cells interact with each other, sending and receiving signals that cause changes in cellular behavior. In this chapter, I present the types of transport that occur across cell membranes, the types of attachments that form between cells, and the ways in which cells send and receive signals.

Shipping and Receiving: Transport Across Membranes

Just as your skin defines your body, the plasma membrane of cells defines the boundary of the cell. Materials can enter and exit the cell only if the plasma membrane allows it; in other words, the plasma membrane is *selectively permeable.* Two major factors determine whether a molecule can cross the plasma membrane:

- **Size:** Smaller molecules cross more easily than larger ones.
- **Attraction to water:** Hydrophobic (nonpolar) molecules cross more easily than hydrophilic (polar) molecules. The hydrophobic molecules can pass through the hydrophobic interior of the phosopholipid bilayer that makes up the membrane. Hydrophilic molecules and ions need help from proteins in order to cross the membrane.

Getting past the bouncer

Transport proteins in the plasma membrane control what is allowed to enter and exit the cell. Although some small, hydrophobic molecules such as carbon dioxide (CO_2) can scoot through the phospholipid bilayer without help, most molecules are either too big or too polar to cross without a transport protein. Even tiny ions like potassium ions (K^+), sodium ions (Na^+), and chloride ions (Cl^-) are too attracted to water to pass through the hydrophobic interior of the membrane on their own. Small molecules such as simple sugars are also too large and too polar to cross without a transport protein. Different types of cells control which materials enter and exit the cell by having different types of transport proteins in their plasma membranes.

Which way should I go?

If more of a molecule is on one side of a membrane than the other, the molecules will move across the membrane until they're evenly distributed across the membrane (see Figure 9-1).

Diffusion is the movement of molecules from an area of higher concentration to an area of lower concentration. Diffusion is a passive process and requires no energy to be input from the cell.

Diffusion happens because a random distribution of molecules is more energetically favorable than an organized distribution (see Chapter 10). When the molecules are randomly distributed, they have reached *equilibrium*.

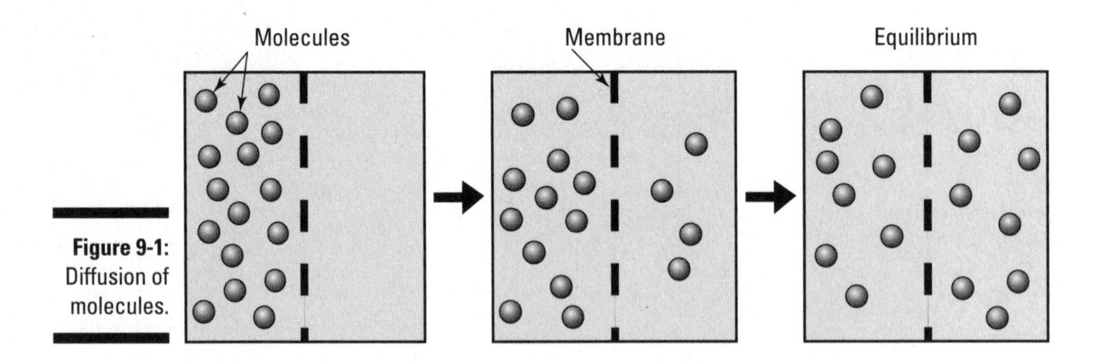

Figure 9-1: Diffusion of molecules.

Crossing the border

If a molecule can cross a plasma membrane, the molecule will diffuse across the membrane until it reaches equilibrium. Diffusion can happen with or without the help of a transport protein (see Figure 9-2). Diffusion is divided into two categories depending on whether a protein is involved:

- **Simple diffusion** occurs when molecules diffuse across membranes without any help from transport proteins.
- **Facilitated diffusion** occurs when molecules diffuse across membranes with the help of transport proteins.

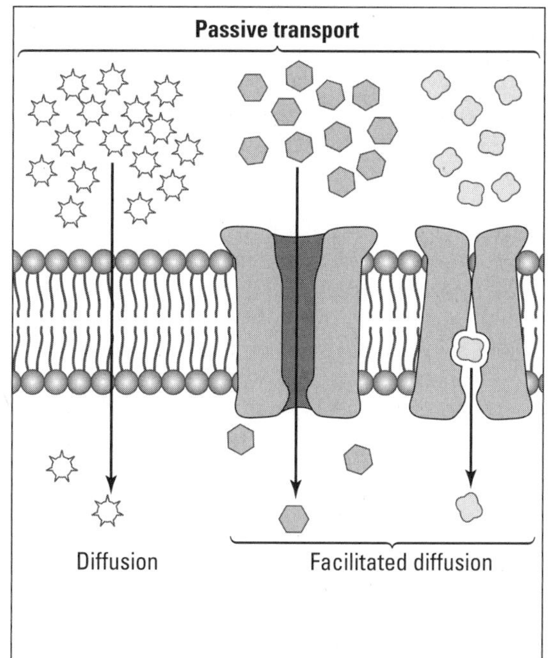

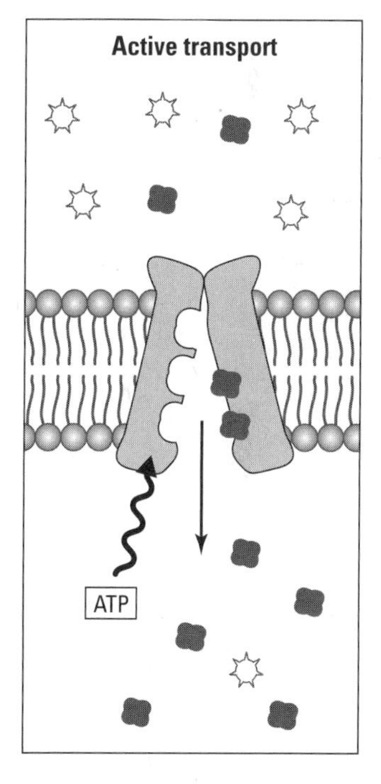

Figure 9-2: Transport across membranes.

Transport proteins facilitate the diffusion of molecules by helping molecules cross the hydrophilic center of the membrane. Two types of transport proteins help with diffusion, and each type helps in a slightly different way:

- **Channel proteins** are proteins that are built like soda straws. The polypeptide chains of these proteins loop around and around to build the walls of the straw, creating an open tunnel down the middle of the protein. When channel proteins are inserted into a membrane, they form tunnels filled with water through the membrane. Small hydrophilic molecules can cross plasma membranes by traveling through the tunnels made by channel proteins. Some channel proteins remain open all the time, while others, known as *gated channels,* open and close in response to signals.

- **Carrier proteins** transport specific molecules across membranes. Each carrier protein has a binding site that is the right shape for the molecule that it can transport. When the right molecule binds to the carrier protein, the protein changes shape and moves the molecule to the other side of the membrane.

Going with the flow

Water also diffuses across membranes. The diffusion of water, called *osmosis,* follows the rules of diffusion: Water moves from an area where water is more concentrated to an area where water is less concentrated. Osmosis is a passive process that requires no energy input from the cell.

Thinking about osmosis in terms of water concentration can be a little hard to wrap your head around: Water is more concentrated when fewer solutes are dissolved in the water. In other words, water is most concentrated where water is most pure. So water moves from where water is most pure to areas where water is less pure.

If thinking about water concentration gives you a headache, you can think of osmosis in terms of solute concentration instead: Water moves to the area of greater solutes. Whichever way you like to think about it, the result is the same: Water molecules become more randomly distributed by osmosis.

The relative solute concentrations on either side of a cellular membrane determine the movement of water molecules by osmosis and can greatly impact the functioning of the cell. Three terms describe the relative concentrations across membranes:

- **Hypertonic** solutions have a greater concentration of solutes.
- **Hypotonic** solutions have a lesser concentration of solutes.
- **Isotonic** solutions have the same concentration of solutes.

If cells are in an environment that has a different solute concentration than the cytoplasm, water will move into or out of the cell by osmosis. For example, for thousands of years people have used salt to preserve food. When food is placed in salt, any bacteria on the surface of the food become surrounded by a hypertonic environment. Water travels by osmosis out of the bacterial cells, and the bacterial cells become shriveled inside their cell walls like deflated balloons. Although the bacteria may not die, they can't grow, multiply, and eat the food, effectively preserving the food for the people who salted it in the first place.

Osmosis also affects people who suffer from diabetes. The blood of diabetics can become full of sugars because their cells don't receive the insulin signal to remove sugar from the blood. When the blood becomes hypertonic to the person's cells, water moves by osmosis into the blood. The water is filtered out of the blood by the kidneys, and the diabetic person produces lots of urine. Because too much water is being filtered out of their bodies, diabetics can become dangerously dehydrated.

In a normally functioning body, cells carefully control the levels of solutes inside and outside of the cell. When cells are in an isotonic environment, water moves equally in and out of the cell.

It's an uphill battle

Cells often need to move molecules from areas where the molecules are less concentrated to areas where they're more concentrated. In other words, cells need to move molecules against their concentration gradient. You can think of this type of transport as cells adding to their stockpile of certain molecules.

Cells move molecules against their concentration gradient by *active transport* (refer to Figure 9-2). Active transport requires the input of energy from the cell, which is usually provided by the energy molecule adenosine triphosphate (ATP).

During active transport, carrier proteins pick up the molecule to be transported on one side of the membrane and, with the help of ATP, change shape in order to move the molecule to the other side of the membrane. For example, a cell may have lots of glucose molecules inside the cell, but it still may need to bring in more because glucose is an important food molecule. Or a nerve cell may need to move a bunch of ions to one side of the membrane to get ready to send and receive electrical signals. Carrier proteins that do active transport are called *pumps*.

One very important active transport protein in animal cells is the *sodium-potassium pump.* Through a series of steps, the sodium-potassium pump actively transports both sodium and potassium to opposite sides of the membrane at the same time:

1. **The sodium-potassium pump picks up three sodium ions (Na+) from the inside of the cell.**
2. **An ATP molecule is split, and one of its phosphate groups is attached to the sodium-potassium pump.**

3. **The sodium-potassium pump changes shape, releasing the three sodium ions (Na+) outside of the cell.**
4. **The sodium-potassium pump picks up two potassium ions from the outside of the cell.**
5. **The phosphate group from ATP is released from the sodium-potassium pump.**
6. **The sodium-potassium pump returns to its original shape, releasing the two potassium ions (K+) to the inside of the cell.**

For every round of action, the sodium-potassium pump moves three sodium ions out of the cell and moves two potassium ions into the cell. Thus, the pump creates a higher concentration of sodium outside the cell, a higher concentration of potassium inside the cell, and a greater positive charge outside the cell. These differences in ion concentration and electrical charge are important in the functioning of nerve and muscle cells in animals.

Chatting through Cellular Connections

Multicellular organisms are made of a community of cells that work together to create the functioning whole organism. Just like the members of any complex organization, these cells must communicate with each other, perform distinct tasks, and respond to each others' requests. Two essential components to the function of a multicellular organism are

- **Attachment** of cells to form tissues. Cells are connected to each other through various types of *cell-cell attachments.*
- **Communication** between cells to coordinate responses to signals. Cells can communicate directly with neighboring cells or send signals over long distances to communicate with cells farther away in the body.

Shaking hands through cell-cell attachments

The cells of animal tissues are connected to each other and to the extracellular matrix that surrounds them. Cells are connected to each other by different types of connections that depend on the function of the tissue. More than one type of connection, described in the following list, can attach a single cell to other cells and the extracellular matrix, as shown in Figure 9-3:

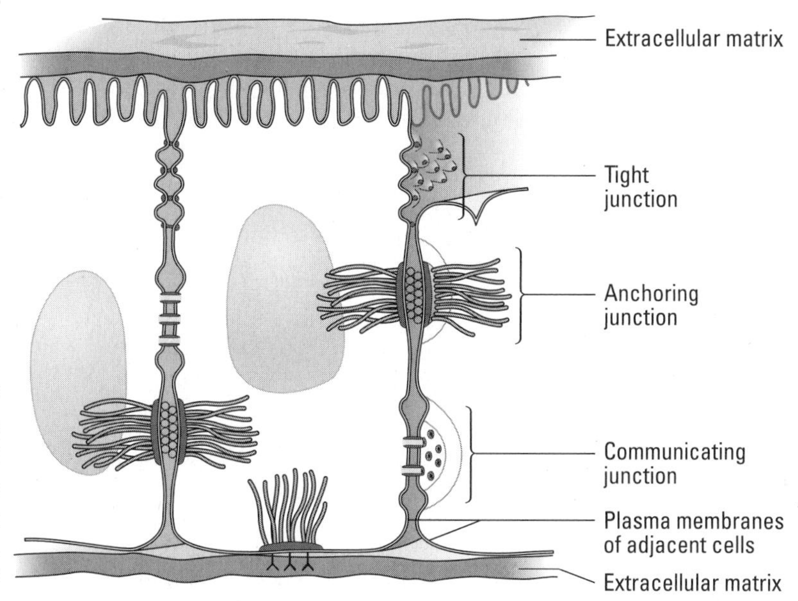

Figure 9-3: Types of junctions between animal cells.

- **Tight junctions** bring cells together so tightly that even water can't pass between the cells. Proteins pass through the membranes of both cells, holding the cells as if they were sewn together. The proteins are arranged in lines so that continuous bands of attachment are formed like seams of stitches through the tissue. Tight junctions are important in surface tissues like skin and mucous membranes where they create an effective barrier to molecules and foreign organisms that would otherwise slip between the cells.
- **Anchoring junctions** hold cells together tightly, but allow materials to move through the intercellular space. They give structure and strength to tissues. Anchoring junctions are very important in tissues that do a lot of work — for example, heart muscle cells. Three types of anchoring junctions use different proteins to make connections between cells, the cytoskeleton, and the extracellular matrix:
  - *Desmosomes* anchor cells to each other by attachments between proteins called *cadherins* in the membranes of both cells. The cadherins also connect with intermediate filaments in the cytoplasm of the cells.
  - *Hemidesmosomes* anchor cells to the extracellular matrix. Proteins called *integrins* in the membranes of cells attach to extracellular matrix proteins toward the outside of the cell and to intermediate filaments in the inside of the cell.

- *Adherens junctions* anchor cells to each other and to the extracellular matrix. Adherens junctions connect cells together by attachment of cadherin proteins in the membranes. The cadherin proteins also attach to actin microfilaments in the cytoplasm of the cells. The actin microfilaments may form bands around the cells that help with movement of the tissue. Adherens junctions also form between cells and the extracellular matrix. In this case, integrins in the plasma membranes of cells form connections to proteins in the extracellular matrix.

Sticking together through thick and thin

Plant cells stick together in tissues because of their cell walls, as shown in Figure 9-4. The first part of the plant cell wall to be made is the *middle lamella,* a layer of sticky molecules that holds adjacent plant cells together. After the middle lamella is formed, long cellulose molecules are embedded in the sticky molecules to add strength and create the *primary wall.* In woody plants, a *secondary wall* containing the tough molecule lignin is also added.

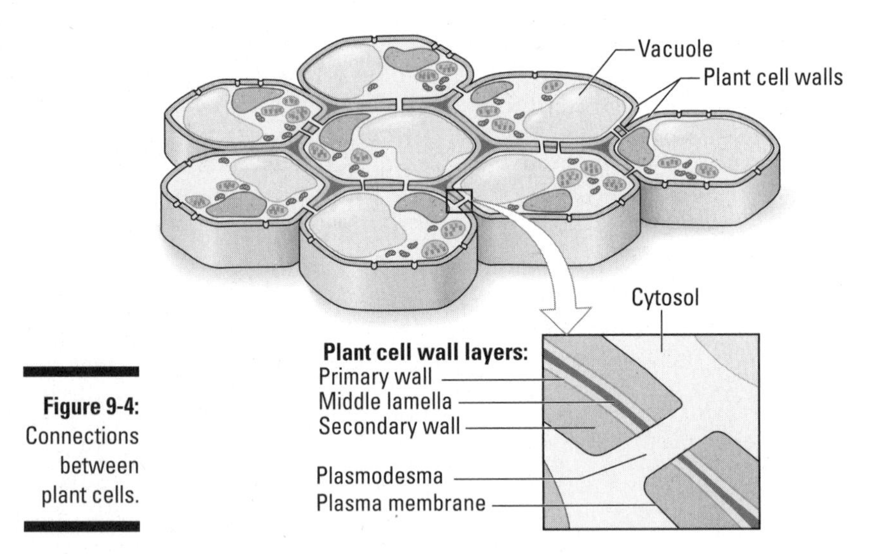

Figure 9-4: Connections between plant cells.

Jumping the cell-cell gap

Some connections between cells allow easy passage of materials between cells. In both plants and animals, tunnel-like structures allow cells to easily exchange ions, signals, and food molecules:

- Between cells in almost all animal tissues, *gap junctions* are formed by proteins, called *connexins,* that form little rings in the membranes of cells. The rings on two adjacent cells line up with each other and connect together so that the open centers of the protein rings, called *pores,* are aligned. In other words, the connected rings of proteins form the walls of a little tunnel that passes between the two cells. Ions and small molecules can travel through these tunnels from one cell to the next. In response to signals, the protein rings can pull together, closing the opening to the tunnel and blocking the movement of materials.
- Between plant cells, tunnels of cytoplasm called *plasmodesmata* function very similarly to gap junctions in animal cells. When plant cell walls form, little gaps in the walls allow the cytoplasm of one cell to touch the cytoplasm of neighboring cells. Plant cells can easily pass materials from one cell to another and have even been observed to exchange small organelles!

Sending and Receiving Signals

In many multicellular organisms, including you, signals produced from one cell cause a response in a cell far away in the body. Signals that travel over a distance to reach their target cells are called *hormones.* For example, the hormone insulin, which is produced by cells in your pancreas, travels around your body through your blood and tells cells to take glucose out of the blood so that the cells can use it as a food source. Likewise, the plant hormone auxin, which is produced at the tips of growing plant stems, travels down the stems and tells side buds on the stem not to grow.

Letting go

Cancer is most dangerous when cancer cells *metastasize,* or spread throughout the body. In order for metastasis to occur, cancer cells must let go of their neighboring cells. A wide variety of cancer cells have defects in their cell junctions, suggesting that part of the pathway that leads to cancer involves changes in the formation of these junctions. Scientists who are studying this link between cell junctions and cancer have found that changes in the proteins that form the junctions can lead to cancer-like abnormalities in cells. For example, mice who can't make normal cadherin proteins develop thick, irregular layers of skin that becomes cancerous very early in the development of the mice. Defects in cadherin also play a role in the development of breast cancer in humans. By studying how defects in cell junctions contribute to the development of cancer, scientists hope to identify new ways of controlling the metastasis of cancer cells.

This stuff is addictive

If receptors for a particular signal are stimulated at high levels over a long period of time, a cell may respond by decreasing the number of receptors for that signal. It's as if the cell gets tired of listening to a loud noise over and over and wants to make things quieter.

For example, when people use drugs that stimulate nerve cells at high levels for long periods of time, the nerve cells may make fewer of the receptors that recognize the drug. Because they have fewer receptors, the drug user will be less sensitive to both the drug and the normal signals usually recognized by the receptor. As a result, the drug user will feel tired and dragged out when they aren't using the drug and will have strong cravings for the drug when they aren't using it. It's a vicious cycle — using the drug leads to decreased receptors, which leads to increased cravings for the drug!

In order for cells to respond to a signal such as these hormones, the cell must recognize the signal and change in response to it. A cell can recognize a signal only if it has a *receptor* for that signal. Cells have receptors in two locations:

- Receptors for signaling molecules that can cross the plasma membrane, such as steroid hormones, are located inside of cells.
- The receptors for signaling molecules that can't cross the plasma membrane are located within the plasma membrane of the cell.

Satellite dishes: Receptors

Receptor proteins have binding sites for the signals they recognize. Many different cell types can respond to the same signal if they all have the same receptor.

When signaling molecules, called *ligands,* bind to receptor proteins, the receptor proteins change shape. The changed receptor protein sets off a series of events that causes the cell to change its behavior in response to the signal.

Relaying the message: Signal transduction

Receptor proteins located in plasma membranes bind to ligands on the outside of the cell and then cause a change in behavior on the inside of the cell. Because the signal is passed across the plasma membrane, this process is called *signal transduction.* Signal transduction involves several steps, shown in Figure 9-5:

1. **The signal is received when the ligand binds to the receptor.**

The ligand is the signaling molecule, such as a hormone. Because the ligand is the original signal to the cell, it's also called the *primary messenger*.

2. **The signal is transduced when the receptor changes shape and becomes ready to cause a change inside the cell.**

   The binding of the ligand to the receptor changes the shape of the receptor protein, causing new binding sites on the receptor to become available.

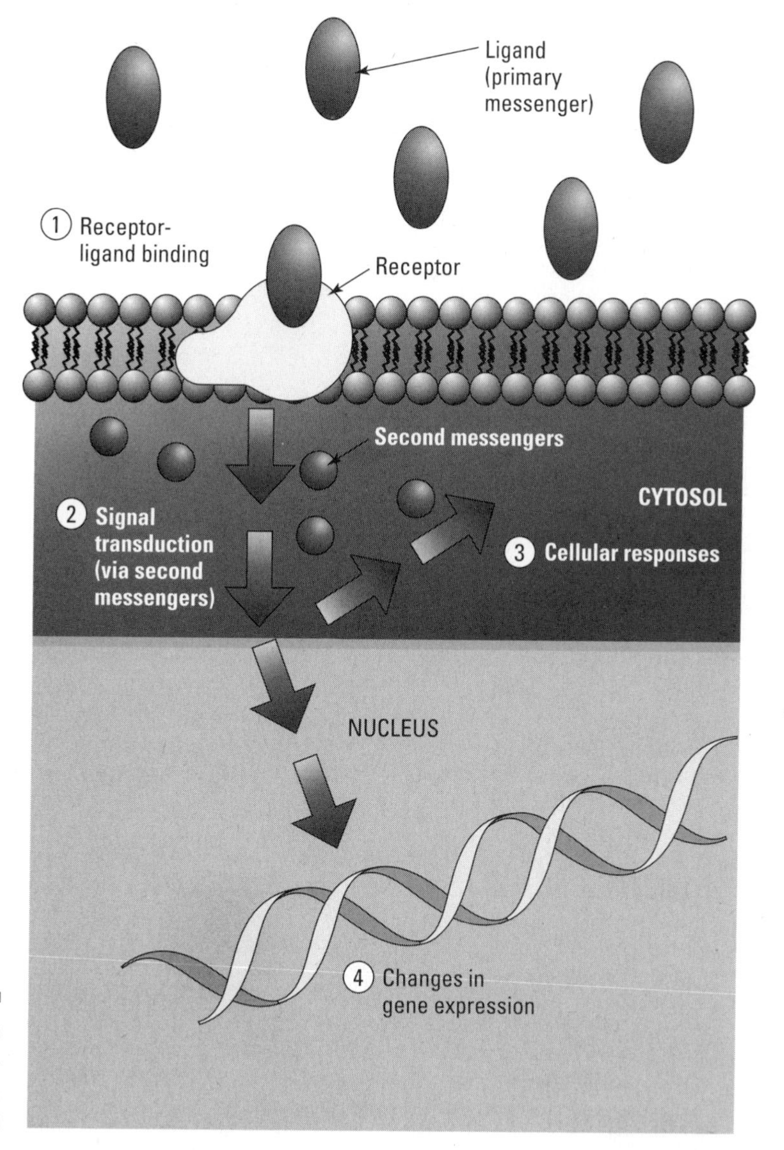

Figure 9-5: Signal transduction.

3. **The signal is amplified when the receptor causes a change inside the cell that activates molecules called *second messengers*.**

   Second messengers are molecules that increase in concentration inside the cell and spread the original signal around the cell. One receptor can trigger the production of many second messenger molecules, which increases the strength of the signal within the cell.

4. **The cell responds when second messengers cause changes in cell behavior.**

   Second messengers may change cellular responses by activating or inhibiting proteins, such as enzymes, or by causing new proteins to be made through changes in gene expression. (For more on gene expression, see Chapter 19.)

Amplifying the signal

Primary messengers, such as hormones, are made in very low levels in the body. When these messengers arrive at cells and bind to receptors, their signal is amplified by a chain of events in cells called a *signal transduction pathway.*

Signal transduction pathways are a little bit like a cellular relay race — one molecule is activated, so it activates the next molecule, which activates the next molecule, and so on, until the signal crosses the cellular finish line and causes a change in cellular behavior. In signal transduction, the molecules in the middle of the pathway are the second messengers. One difference between a traditional relay race and signal transduction is that the primary messenger causes the activation of many secondary messengers at once, rather than just one at a time.

Because one primary messenger causes the activation of many secondary messengers, the original signal is amplified within the cell.

Signal transduction pathways involve the interactions of many different molecules and can get pretty complicated to follow. A couple of basic principles can help you understand these pathways:

- **Proteins can be activated and inactivated when molecules bind to them.** A protein's shape is essential to its function. When something binds to a protein, the proteins shape will change. Thus, the protein's function may change from active to inactive.
- **Adenosine triphosphate (ATP) and guanosine triphosphate (GTP) are often used as a source of phosphate groups.** ATP and GTP are energy molecules that have three phosphate groups as part of their structure. Cells can remove a phosphate group from ATP or GTP by hydrolysis and transfer the phosphate to another molecule.

- **The transfer of a phosphate group to a molecule is called *phosphorylation.*** Enzymes called *kinases* transfer phosphates from ATP or GTP to other molecules.

Enzyme-linked receptors

One type of signal transduction pathway involves *enzyme-linked receptors,* which are receptor proteins that have the ability to catalyze reactions inside the cell. One well-studied group of enzyme-linked receptors is the *receptor tyrosine kinases (RTKs).* The name of this group indicates their function: They are receptors that can phosphorylate a molecule.

The steps of signal transduction for receptor tyrosine kinases are as follows:

1. **The primary messenger binds to the receptor (RTK).**

   The binding of the primary messenger causes the receptor to change shape.

2. **The receptor (RTK) is phosphorylated and becomes an active enzyme.**

   After the receptor's shape changes, it has a binding site for a phosphate group. Once the phosphate group is bound, the enzymatic ability of the receptor is activated.

3. **The receptor (RTK) activates another membrane-associated protein called Ras.**

   When Ras is activated, it binds a molecule of GTP.

4. **Ras kicks off a phosphorylation cascade.**

   Ras activates the first protein in the cascade by transferring a phosphate from its GTP to the protein. The activated protein catalyzes the phosphorylation of the second protein in the cascade, causing it to be activated. The second protein activates the third and so on. Again, the proteins are like runners in a relay race activating each other by handing off a phosphate group. Each activated protein in the pathway activates many copies of the next protein in the pathway which amplifies the signal from the primary messenger. Ultimately, one of the proteins in the cascade causes a change in cell behavior, and the signal transduction pathway is complete.

G proteins

Another important type of signal transduction pathway involves *G proteins.* G proteins get their name because they bind GTP. G proteins wait near the membrane right next to the receptor proteins they work with. The steps of signal transduction involving G proteins are as follows:

1. **The primary messenger binds the receptor.**

   The binding of the primary messenger causes the protein to change shape.

2. **The receptor activates its G protein.**

   The G protein binds to a GTP molecule and splits in half.

3. **Part of the G protein travels along the membrane and binds to an enzyme.**

   The enzyme becomes activated by the binding of the G protein.

4. **The enzyme catalyzes the production of second messenger molecules.**

   The second messengers travel through the cell and trigger a change in cell behavior.

Calming down: Deactivating the signal

Signal transduction pathways are shut down once the primary messenger stops sending the signal. In order to stop signal transduction, cells need to deactivate the molecules that were involved in the pathway. Several types of molecules in cells shut down signaling pathways:

- **Phosphatases** are enzymes that remove phosphates from molecules. When phosphatases remove the phosphate groups from molecules in signal transduction pathways, the molecules are inactivated.
- **Activated G proteins** and **Ras proteins** convert GTP to GDP and deactivate themselves. Once they're inactivated, they no longer activate the enzyme that produces second messengers.
- **Second messenger molecules** are short-lived in the cell. These molecules are inactivated by a number of different cellular mechanisms.

Ras is the gas

Activated Ras proteins kick off a phosphorylation cascade that relays the signal from the primary messenger throughout the cell. Once the primary messenger stops sending the signal, Ras proteins are supposed to convert GTP to GDP and inactivate themselves. However, human cells can develop defects in Ras proteins such that the Ras proteins don't turn themselves off. In this situation, Ras proteins send signals to the cell whether or not the primary messenger is bound to the receptor. This situation is like a gas pedal in a car that stays on whether or not a foot is there to push it down.

The result of defective Ras depends on the signal that is being sent. In many cases, the primary messenger is a growth hormone telling the cell to divide. If Ras is defective, the result may be cells dividing when they aren't supposed to and creating tumors. Approximately 25 percent human cancer cells have defective Ras proteins, which just goes to show how important the control of signal transduction is to normal cell function!

Chapter 10

Metabolism: Transferring Energy and Matter

In This Chapter

- Following the laws of thermodynamics
- Using enzymes to get the job done
- Transferring energy using ATP
- Moving electrons during redox reactions

Cells are constantly converting matter and energy from one form to another in order to get what they need to stay alive. The input of energy into cells allows them to maintain their organization and build new molecules. Cells store energy in food molecules and then break food molecules down again to transfer the energy back out. In this chapter, I present the fundamental components of the vast connected system of reactions that makes up the metabolism of the cell.

Revving Up Your Metabolism

Metabolism is all of a cell's chemical reactions, such as breaking down food molecules or building up muscle proteins, added together. You're probably most familiar with your own metabolism in terms of diet and exercise: the faster your *metabolic rate,* the faster you burn, or break down, your food. Food provides you with the *energy* you need to move and grow, as well as the raw materials, or *matter*, you need to build the cells that make up your body. In order to access the energy and matter in food, your cells take food apart, breaking it down into its components.

Metabolism can be divided into two main categories:

- *Catabolism* includes all the reactions that break larger molecules into smaller ones. Catabolic reactions allow your cells to transfer the energy from food to a form that the cells can use to do their work. For example, your muscle cells do lots of work whenever you exercise, so you burn more food to provide them with the energy they need. Cells also take the components of food and rearrange them into the molecules that build your cells.
- *Anabolism* includes all the reactions that build larger molecules from smaller ones. For example, anabolic steroids increase anabolism, causing the body to build more muscle proteins from amino acids, which is why steroid use increases muscle mass.

Hundreds of different catabolic and anabolic reactions may all happen at one time in the cell, working to provide the cell with what it needs. This incredible complexity is shown in Figure 10-1, a metabolic map showing just some of the many interlocking reactions that make up the metabolism of the cell. In the metabolic map, each dot represents a different molecule and each line between the dots represents a different chemical reaction. As you look at this snapshot of metabolism, just imagine that each and every one of your 10,000,000,000,000 cells is that busy all of the time!

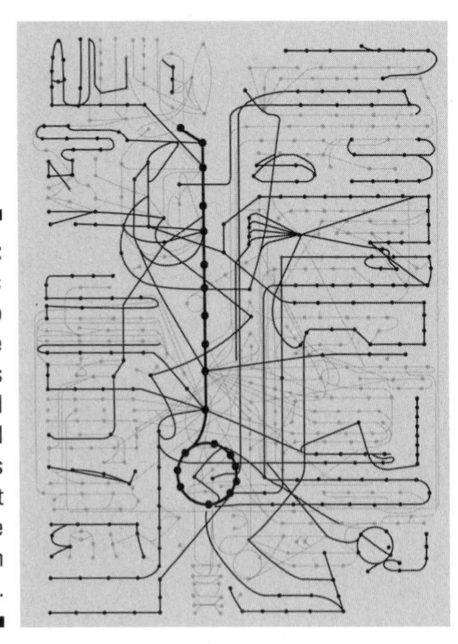

Figure 10-1: A metabolic map showing the molecules (dots) and chemical reactions (lines) that make up the metabolism of a cell.

Stayin' Alive: Cellular Work and the Laws of Thermodynamics

I'm sure you use the words "work" and "energy" all the time in your daily life. In everyday terms, *work* is something that you do that requires an effort, whether it's doing the laundry, mowing the lawn, or typing on a keyboard. *Energy* is something you need in order to get your work done. These ideas about work and energy are actually quite similar to how scientists define these words. To a scientist:

- *Work* is moving something over a distance.
- *Energy* is something you need to do work.

The relationship between energy and work applies to your cells just as much as it applies to your whole body. Your cells do work when they move or build new molecules, and they need energy in order to do that work. Your body transfers energy from the food you eat to a form that your cells can use to fuel their work.

The relationship between work and energy is studied in a branch of physics called *thermodynamics*. Thermodynamics has several laws, but the first two are especially important for understanding how and why cells use energy:

- The first law of thermodynamics states that energy can't be created nor destroyed.
- The second law of thermodynamics states that chemical reactions that happen spontaneously increase the *entropy,* or disorder, of an isolated system.

The first law of thermodynamics

The first law of thermodynamics basically means that you can't make new energy, and you can't just make energy disappear. Any energy you use must come from somewhere and then go somewhere else when you are done.

Moving energy from one place to another is called *energy transfer*. Energy transfers happen every day all around you: If you're using electricity to power a light as you read this text, then energy is being transferred from an electrical system to the light. As the light bulb shines, the light and heat energy from the bulb are transferred into the air all around you.

As energy is transferred from one place to another, it often changes from one type of energy to another. This process is called an *energy transformation*. For example, several forms of energy are involved in producing light from an electrical lamp: light energy, electrical energy, heat energy (from the warm bulb), plus whatever type of energy is used to produce the electrical energy in the first place, such as the flow of water in a hydroelectric plant. If you start with the flow of water in a hydroelectric plant, you can follow the energy through several transformations before it finally becomes light shining from the bulb.

Because energy never disappears, but is just converted from one form to another, the first law of thermodynamics is also sometimes called the *law of conservation of energy*.

In everyday language, *conservation* means to save. However, when scientists refer to conservation of energy, they do not mean saving energy — at least, not in the sense of driving a car that gets better gas mileage or something along those lines. What scientists mean is that the total amount of energy in the universe always remains the same, or is saved. The total amount of energy in the universe remains the same because energy can't be created or destroyed; instead, it just changes form and is transferred from one place to another.

You can organize all the different types of energy into two main categories:

- *Potential energy* is energy that an object or molecule has because of its position or arrangement.
- *Kinetic energy* is energy that objects and molecules have because of their motion.

Potential energy

Potential energy, also called *stored energy,* includes several types of energy:

- **Chemical potential energy,** such as that found in oil, natural gas, or food.
- **Elastic potential energy,** such as that in compressed springs and stretched rubber bands.
- **Gravitational potential energy,** such as that in objects held above ground level.

All these items have energy because of how they're arranged. In the case of chemical energy, the arrangement of electrons within the chemicals determines the amount of stored energy. A compressed spring, on the other hand, has potential energy because of the way it's compressed. Once the spring is released, it will transfer its potential energy to another object — for example, launching a ball in a pinball game. One of the most important kinds of energy in cells is the chemical potential energy of food.

You can easily confuse the scientific definition of potential, which means "stored," with the more common, everyday definition, which means "a possibility." In everyday language, you might say that an object has the potential to store energy, such as a spring with the potential to store energy when compressed. In scientific thinking, however, an object only has potential energy if it's actually storing energy at that particular time. So, in scientific terms, a spring that isn't compressed doesn't have potential energy because it can't do work in its current form. Only after the spring is compressed and ready to do work does it have potential energy. Even though it's possible for an uncompressed spring to store energy if you compress it, to a scientist, a spring only has potential energy when it's actually compressed.

Kinetic energy

Kinetic energy is *energy of motion*. Anything that is moving has kinetic energy, including wind, moving water, the flow of electrons through a wire (electricity), and the movement of electromagnetic waves of radiation, such as light and heat from the sun. The most important kinds of kinetic energy to cells are light, heat, and the movement of charged particles, such as ions and electrons. For example, the movement of charged particles across the membrane of your nerve cells is what allows your nervous system to send signals around your body.

The first law of thermodynamics says that energy can't be created or destroyed; it's just transferred from one form to another. Cells transfer energy from food in order to do cellular work.

The second law of thermodynamics

The *second law of thermodynamics* is complicated but powerful. It explains why things fall down and not up, why things break and don't just fix themselves, and why the Sun will burn out some day. The second law tells you that

- With every energy transfer, the potential energy of the system decreases (if no energy enters or leaves the system).
- Energy spontaneously tends to flow only from being concentrated in one place to becoming diffused or dispersed and spread out.
- In an isolated system, a process can occur only if it increases the total disorder (*entropy*) of the system.
- Heat never moves from a cooler object to a warmer object.
- It's impossible to convert heat completely into work.

Spontaneous reactions

In cellular biology, the most useful expressions of the second law of thermodynamics are those that explain which reactions in cells are *spontaneous* and which are not. Spontaneous reactions happen on their own without requiring input of energy from the cell.

The second law of thermodynamics states that reactions can only be spontaneous if they increase the total amount of disorder, called entropy, in the universe.

The second law of thermodynamics is a very abstract idea, but it's an idea you have seen in action in your everyday life. For example, if you take a spoonful of salt or some colored drink powder and put it into a glass of water, the salt or colored particles will eventually spread out throughout the water even if you don't stir the water. A random distribution of particles is more disordered than having particles concentrated in one part of the water, so the particles spontaneously spread out without any input of energy from you. This same pattern in seen in cells as well — molecules in solution in cells will spontaneously move from areas where they're more concentrated until they're randomly distributed throughout the cell. Chemical reactions that break down larger, more organized molecules into smaller fragments also increase the overall amount of entropy and thus will also happen spontaneously in the cell.

Nonspontaneous reactions

The second law of thermodynamics says that the universe is constantly increasing in disorder. At first glance, living things seem to contradict the second law because living things, and their cells, are very organized systems. Cells maintain their order by constantly using their stored energy to do *nonspontaneous reactions,* which decrease entropy and require the input of energy.

A good analogy for comparing nonspontaneous and spontaneous reactions in cells is to compare a brick wall to a pile of bricks. A brick wall is much more organized than a pile of bricks. If you want to build a brick wall from a pile of bricks, you'll be making things more organized and thus decreasing entropy. This is nonspontaneous — this transformation isn't going to "just happen." You're going to have to put in energy and do work to get the job done.

On the other hand, if you build a brick wall and let it just sit there for a thousand years or so, the wall will gradually fall apart. When the wall breaks down, entropy is increased. So this process is spontaneous and will occur without the input of energy. The difference between building a brick wall and letting one fall apart is similar to the situation in cells (see Figure 10-2) when they're building and breaking down molecules:

- Reactions that build large complex molecules from smaller parts are like building the brick wall. The reactions are nonspontaneous and require the input of energy.

- Reactions that break large molecules down into smaller parts are like letting the brick wall fall apart. These reactions are spontaneous and actually make energy available to the cell.

Although cells must maintain their organization in order to function, they don't really violate the second law of thermodynamics. (It wouldn't be a very good scientific law if evidence revealed that it could be broken!) Cells use stored energy to do nonspontaneous reactions that increase order, or decrease entropy, in the cell. However, in order to store energy in the first place, cells must do spontaneous reactions, such as breaking down food molecules. These reactions increase the entropy in the universe. Overall, the increase in entropy by the spontaneous reactions is greater than the decrease in entropy by the nonspontaneous reactions in the cell. Entropy in the universe continues to increase, and the second law is satisfied.

Cells increase entropy in the universe by doing spontaneous reactions, such as breaking down food molecules into smaller pieces. However, they use the energy made available by these spontaneous reactions to decrease entropy within the cell by building, maintaining, and organizing cellular components.

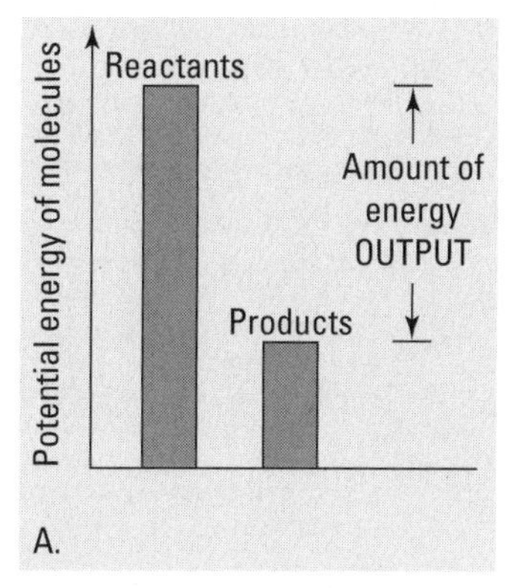

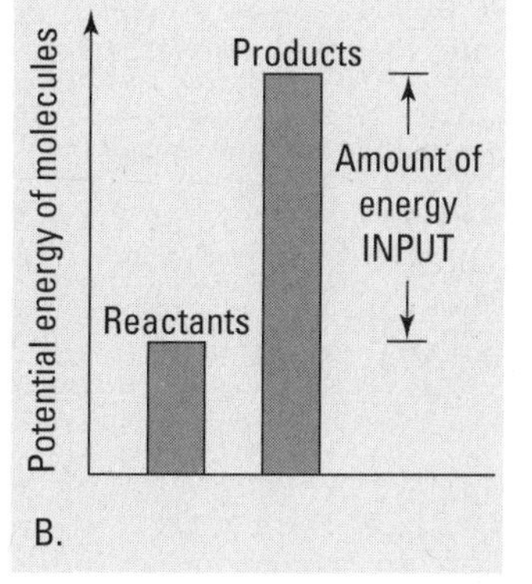

Figure 10-2: Spontaneous versus nonspontaneous reactions.

Free energy

You can figure out whether a chemical reaction will be spontaneous or nonspontaneous by looking at the changes in *free energy* that occur as the reactants are turned into products. *Gibbs free energy* (G) is a measurement that relates entropy and heat and thus gives you a way to evaluate a reaction in terms of the second law of thermodynamics. To determine the changes in free energy (ΔG, called "delta G"), you subtract the free energy of the reactants from the free energy of the products.

$$\Delta G = G_{\text{products}} - G_{\text{reactants}}$$

Entropy = Death

When I was an undergraduate studying biology, I remember seeing a bumper sticker that said entropy=death. I don't think I really got the sticker back then, but now I think it's a very profound way of summing up the relationship between the second law of thermodynamics and living things.

Life is a state of order: for cells to function and thus remain alive, molecules must be built, structures must be maintained, and chemicals and ions must be distributed in certain ways. To maintain this order, living things constantly break down food molecules, transferring energy to the cell that is used to fight the battle against entropy. If this energy transfer stops for any reason, such as when a human heart stops beating, oxygen can no longer travel to your cells, and you can no longer extract energy from your food. Almost the moment your body ceases to provide new energy to your cells, entropy wins. Molecules break down, ions randomize, and your cells cease to function. Life battles entropy within the cell. When that battle is lost, the result is death.

The $\triangle G$ tells you whether the reaction will be spontaneous or nonspontaneous:

- If the reactants have more free energy than the products, the $\triangle G$ will be negative, or less than zero ($\triangle G<0$), and the reaction will be spontaneous.
- If the products have more free energy than the reactants, the $\triangle G$ will be positive, or greater than zero ($\triangle G>0$), and the reaction will be nonspontaneous.

The $\triangle G$ also tells you how much energy is made available to the cell or is needed by the cell as the reaction occurs:

- Spontaneous chemical reactions release energy to the cell. Energy-releasing reactions are called *exergonic reactions*.
- Nonspontaneous chemical reactions require energy input from the cell. Energy-requiring reactions are called *endergonic reactions*.

The graphs of the $\triangle G$ of a reaction in Figure 10-3 show an easy way to remember spontaneous versus nonspontaneous reactions. Spontaneous reactions look like they're going downhill in terms of energy. It's easy to go downhill, so these reactions are spontaneous. Nonspontaneous reactions look like they're going uphill. Going up the energy hill is hard, so these reactions are nonspontaneous, and you're going to have to put in some energy to do it!

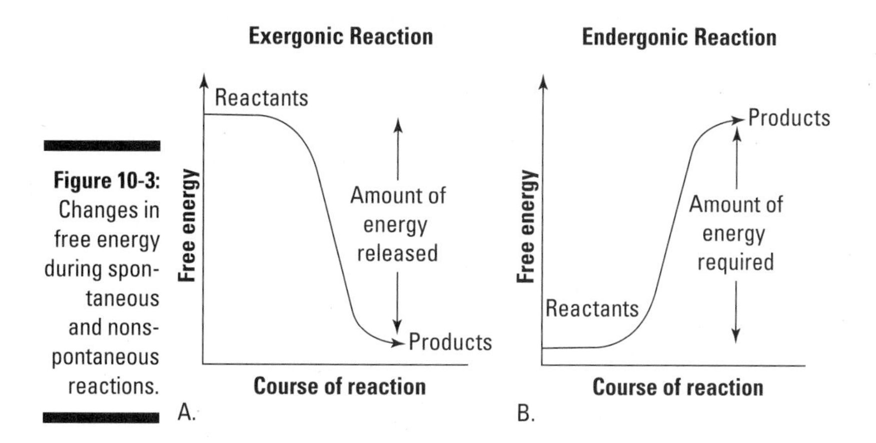

Figure 10-3: Changes in free energy during spontaneous and nonspontaneous reactions.

Going to work in the cellular factory

Cells transfer energy from food molecules in order to do cellular work. You know about the work you do in your life, but you may not be aware of all the work that is going on in your cells every day as you go about your business:

- **Synthesis:** Cells must build large, complex molecules, such as proteins and DNA. They also build organized structures, such as cellular membranes and ribosomes.
- **Organization:** In eukaryotic cells (see Chapter 2), internal cellular membranes create different compartments in the cell that allow organization of materials for specific tasks.
- **Creation of electrochemical gradients:** Cells organize ions on either side of membranes, allowing differences in electrical charge and chemical concentration to be created. These electrochemical gradients help with cellular work, such as signaling, movement, and transport of materials across membranes.
- **Transport and movement:** Cells also transport molecules, such as proteins and lipids, from one part of the cell to another, and some cells swim or crawl using special motility proteins.

Cellular work includes building (synthesis), organizing, transporting, and moving. All these cellular processes are essential to cellular function, and all require the input of energy from the cell.

ATP

Cells break down food molecules for energy to do cellular work. However, they don't use the energy from food directly. Instead, cells use a sort of energy middleman, called *adenosine triphosphate,* as shown in Figure 10-4.

Here's how cells use ATP:

1. **Energy from food is transferred to ATP.**
2. **ATP provides energy for cellular work.**

Adenosine triphosphate (ATP) is a nucleotide, the same type of building block that is used to build DNA and RNA. It has three types of parts:

- The sugar ribose
- A nitrogenous base
- Three phosphate groups

Energy coupling

Cells are constantly building and breaking down ATP, creating an *ATP/ADP cycle,* as shown in Figure 10-5. Cells store energy by building ATP molecules out of adenosine diphosphate (ADP) and phosphate groups (P). Cells obtain energy for cellular work by breaking ATP back down into these components. The breakdown of ATP to ADP and P requires a water molecule and is called *ATP hydrolysis.*

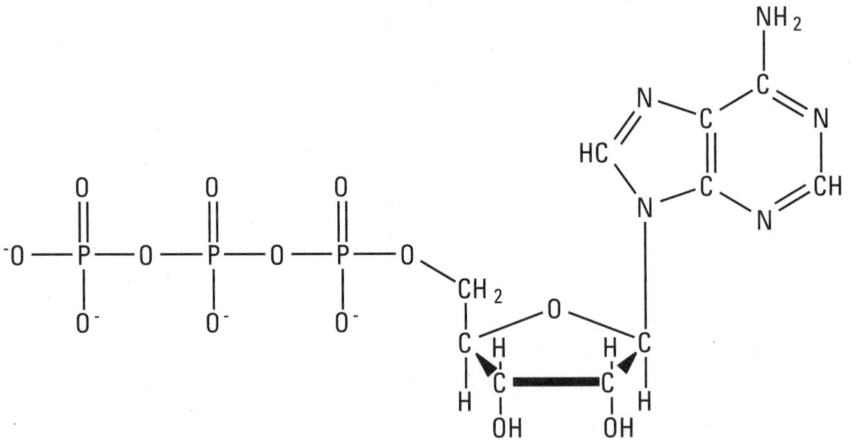

Adenosine triphosphate (ATP)

Figure 10-4: The energy carrier adenosine triphosphate (ATP).

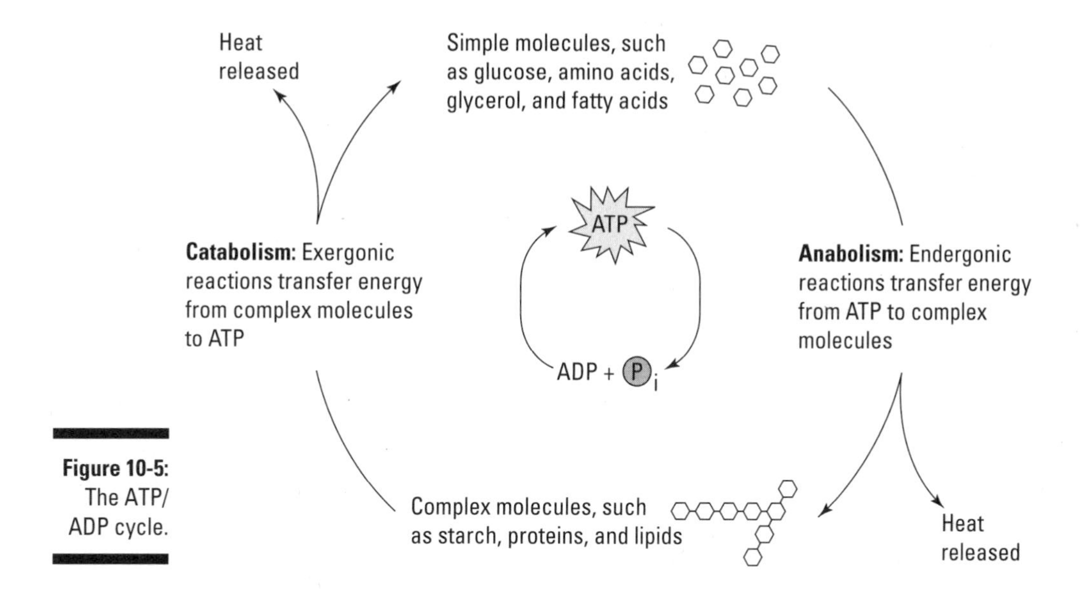

Figure 10-5: The ATP/ ADP cycle.

Cells make and break ATP all the time:

- When cells do exergonic reactions that make energy available to the cell, they capture that energy by forming ATP molecules.
- When cells do endergonic reactions that require energy from the cell, cells provide the energy by breaking ATP molecules.

The ATP/ADP cycle is used to link exergonic and endergonic reactions in a system called *energy coupling.* ATP hydrolysis, which is exergonic, is *coupled,* or done at the same time, as an endergonic process. Some of the energy released from the ATP hydrolysis is used during the endergonic reaction.

When cells do an endergonic reaction by coupling it to ATP hydrolysis, the two reactions do not occur separately. Instead, the two reactions are actually combined and then broken into steps in such a way that all the steps become spontaneous and exergonic. In other words, instead of trying to do a small "uphill" reaction separately from a larger "downhill" reaction, cells change things around to create two smaller "downhill" reactions. For example, if a cell needs to build the amino acid glutamine by adding an amino group to the amino acid glutamate, the reaction would be endergonic and nonspontaneous ($\triangle G=+3.8$ kcal/mol). (The abbreviation kcal/mol stands for kilocalories per mole, which is a measurement of how much energy is produced or needed by the reaction.) However, the hydrolysis of ATP is exergonic and spontaneous and provides more than enough energy for this reaction ($\triangle G=-7.3$ kcal/mol). So, if you consider these reactions together, the overall process would be exergonic and spontaneous. (If you add the two ΔGs, you get –3.5 kcal/mol.) In order to build the glutamine, the cell would combine both reactions and break the process into two exergonic steps:

1. **Glutamate is combined with ATP, and a phosphate group is transferred from ATP to the glutamate.**

   ADP and water are produced.

2. **The glutamate with the attached phosphate is combined with an amino group.**

   The amino group is transferred to the glutamate as the phosphate group is released, forming the amino acid glutamine.

The products are the same as if the two reactions, glutamine synthesis and ATP hydrolysis, were done separately, but because they were combined, they became energetically favorable.

One Step at a Time: Metabolic Pathways

If you refer to Figure 10-1, earlier in this chapter, you can see that metabolism is a vast interlocking web of chemical reactions. If you put your finger on one of the dots in Figure 10-1 that represents a particular chemical, you can trace a path from that dot, along lines, to other dots, and so on. The path your finger travels represents a subset of the many chemical reactions that are occurring in the cell. A single path like the one you traced is called a *metabolic pathway* (see Figure 10-6). Metabolic pathways have several key characteristics:

- Metabolic changes are broken down into small steps, each of which is a single chemical reaction. Several reactions in a series make up a metabolic pathway.
- Enzymes are very important to a functioning metabolism. They speed up chemical reactions by lowering the *energy of activation* so that metabolism occurs quickly enough to support life.
- Electrons are transferred from one molecule to another during many metabolic reactions. Molecules that lose electrons are oxidized; those that gain them are reduced. Electron carriers, such as nicotinamide adenine dinucleotide (NADH), shuttle electrons between reactions.
- Energy is transferred during metabolic reactions. The energy carrier ATP transfers energy to or from reactions.

Taking baby steps during chemical reactions

In cells, chemical reactions usually happen in many small steps rather than one quick change. By doing many small reactions, cells control the energy changes and prevent cellular damage. For example, a cell that needs to make molecule F out of molecule A might do so in five small steps:

$$A \rightarrow B \rightarrow C \rightarrow D \rightarrow E \rightarrow F$$

A represents the starting molecule, or *substrate.* F represents the ending molecule, or *product.* B, C, D, and E all represent molecules that were made during the conversion of A to F; they're called *intermediates.* Every arrow represents one step, or *reaction,* as a chemical change occurred.

Metabolic pathways connect with each other forming a complex interlocking web. The connections and complexities in metabolism occur when:

- The product of one pathway is the substrate of another.
- A pathway may have one or more branches as intermediates connect with other pathways.
- Some metabolic pathways are circular, re-creating the initial substrate during the pathway so that it can repeat, as shown in Figure 10-6. These types of pathways are called *metabolic cycles.*

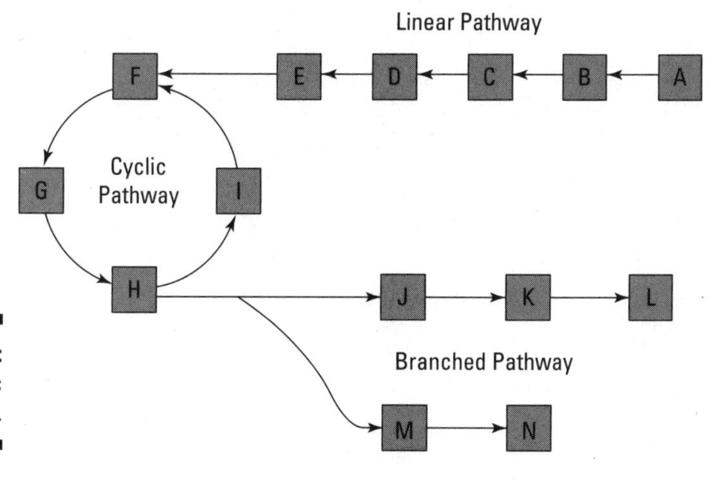

Figure 10-6: Metabolic pathways.

Helping hands from enzymes

Life is very fast paced — too fast to just wait around for necessary chemical reactions to occur, even spontaneous reactions. Although spontaneous reactions are energetically favorable and can occur without energy input from the cell, there is no guarantee on exactly *when* they will occur. In order for a spontaneous reaction to happen, the reactants must find each other and then collide in just the right way and with enough kinetic energy to trigger the necessary changes. In other words, a spontaneous reaction is just waiting to happen, but it needs a little push from the collision of the reactants to get it to go, as shown in Figure 10-7.

The amount of energy that is necessary to trigger the reaction is called the *energy of activation* (E_a). Without the energy of activation, spontaneous reactions may not happen quickly enough to support life. For example, you probably know that if you eat sugar, your body will break it down in order to get some usable energy from the sugar. The breakdown of sugar is a spontaneous, exergonic reaction that releases usable energy to cells. However, if you were to place a bowl of sugar on your kitchen table and stare at it, it's highly unlikely that the sugar would begin to disappear before your very eyes. Outside of your cells, the sugar molecules don't have enough kinetic energy to overcome the barrier (see Figure 10-7).

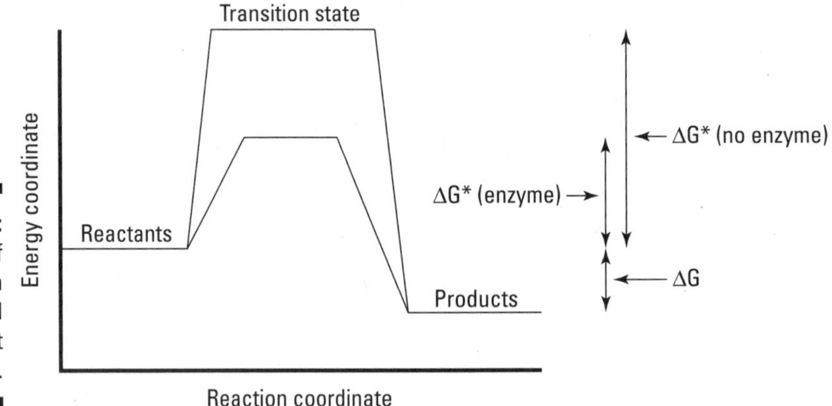

Figure 10-7: Energy of activation with and without enzymes.

Cellular *enzymes* (see Chapter 6) make the difference between the breakdown of sugar inside your cells and outside of your cells. Enzymes are important to cells because

- Enzymes lower the energy of activation for reactions, making it easier for collisions between reactants to overcome the activation energy.
- Every reaction that occurs in cells is catalyzed by an enzyme. Because enzymes are very specific and can bind to only a certain substrate (see Chapter 6), each reaction requires a unique enzyme.

For example, look at this simple metabolic pathway in which molecule A is converted to molecule F by five reactions:

$$A \rightarrow B \rightarrow C \rightarrow D \rightarrow E \rightarrow F$$

Each reaction in this pathway would be catalyzed by a different enzyme, requiring a total of five enzymes to complete the pathway. If you refer to Figure 10-1 and look at all the reactions drawn on the metabolic map, you can get an idea of the hundreds of enzymes that may be needed at any one time in a cell.

Enzymes lower the activation energy of reactions, making it easier for reactants to have productive collisions. *Productive collisions* are those that have enough energy to overcome the energy of activation. In order for productive collisions to occur:

- The particular reactants for a reaction must collide with each other.
- The reactants must be oriented toward each other in the right way so that the correct chemical groups come into contact.
- The reactants must have enough kinetic energy to overcome the energy of activation.

Enzymes do several things to make it more likely that a collision will be productive:

- Enzymes bind reactants (substrates) in their active sites, bringing the necessary reactants for a reaction together in the proper orientation for the reaction to occur.
- The functional groups on the amino acids (see Chapter 6) that make up the enzyme interact with the reactants, altering chemical bonds so that the changes necessary for the chemical reaction are more likely to happen.

The combined effects of these interactions between enzymes and reactants (substrates) result in the lowering of the activation energy for the reaction. Thus, it's more likely that the kinetic energy of the reactants will be enough to cause the reaction to occur.

Giving and taking electrons in redox reactions

During metabolic reactions, electrons are often transferred from one molecule to another.

Molecules change when they give or take electrons:

- When a molecule gives up an electron, it's *oxidized.*
- When a molecule accepts an electron, it's *reduced.*

A good example of oxidation and reduction is the reaction between sodium (Na) and chlorine (Cl^-). When sodium and chlorine come into contact with each other, chlorine steals an electron from sodium. (For more on the interaction between sodium and chlorine, see Chapter 4).

Sodium is oxidized and becomes the sodium ion (Na^+). Chlorine is reduced and becomes the chloride ion (Cl). Because oxidation and reduction reactions occur together — one molecule is oxidized, and one is reduced — these reactions are called *redox reactions.*

It seems backwards to think that when a molecule *gains* an electron, it's reduced. Reduced means less, right, so how can gaining an electron make you reduced? The reason for this contradiction is because elements are compared to their most oxidized state. So, if you think of most oxidized as the max, then every electron a molecule accepts moves it away from the max, or reduces it. Too confusing? Then just remember this friendly little sentence, "LEO the lion goes GER." This stands for Loss of Electrons is Oxidation (LEO); Gain of Electrons is Reduction (GER). This shortcut never fails.

Shuttling electrons with electron carriers

Sometimes electrons are moved directly from one molecule to another during metabolism. Frequently, however, cells use an electron shuttle bus, called an *electron carrier.* Electron carriers accept electrons from one reaction and then transfer those electrons to another reaction. During metabolism, these electrons are often moved as part of hydrogen atoms (H) that are stripped from one molecule and then given to another.

Electron carriers cycle between two forms, an oxidized form and a reduced form, as they shuttle electrons (or hydrogen atoms) around the cell (see Figure 10-8).

- The oxidized form of a carrier *accepts* electrons from reactions. When it accepts electrons, it becomes reduced.
- The reduced form of a carrier *donates* electrons to reactions. When it donates electrons, it becomes oxidized.

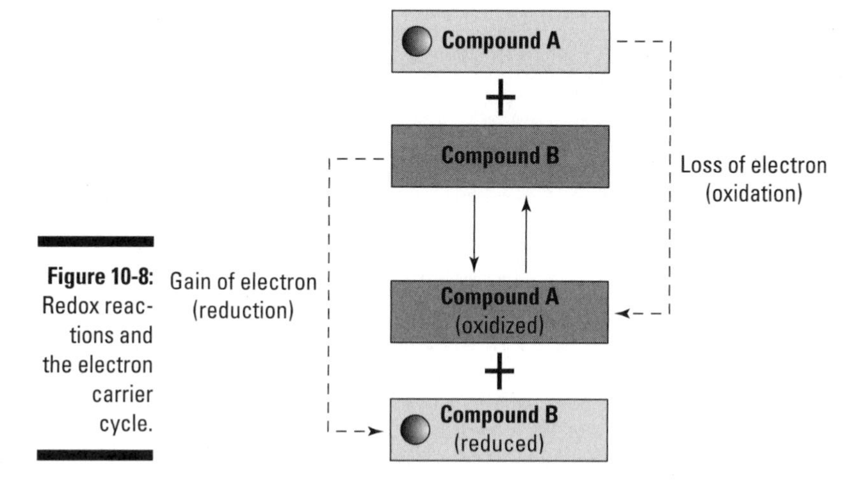

Figure 10-8: Redox reactions and the electron carrier cycle.

Nicotinamide adenine dinucleotide (NAD^+/NADH), shown in Figure 10-9, is a good example of an electron carrier. It's made of two nucleotides hooked together. The nitrogenous base of the upper nucleotide is the part of the molecule that accepts and donates electrons.

As NAD^+/NADH shuttles electrons, it alternates between two forms:

- When the carrier is in its oxidized form, the nitrogenous ring carries a positive charge. This form of the carrier is NAD^+.
- The carrier is reduced when it accepts one electron, which neutralizes the charge, plus one entire hydrogen atom. The reduced form of the carrier is NADH + H^+.

Whenever NAD^+ accepts electrons, it's converted to NADH + H^+. The +H^+ seems like a little brother always following the NADH around. During metabolism, electrons are often stripped off molecules as part of hydrogen atoms. Each hydrogen atom consists of one proton and one electron. When two hydrogen atoms are removed from a molecule during redox reactions, NAD^+ can carry one electron, plus one whole hydrogen atom, in its nitrogenous ring. However, it can't accept one proton. This proton is released into the cell, which is shown by writing $+H^+$, which represents a proton, after the NADH.

You can easily remember which form of an electron carrier is the oxidized form and which one is the reduced just by looking at its name. When electron carriers are reduced, they're carrying their electron passengers in the form of hydrogen atoms. They show that they are carrying passengers by putting the letter H for hydrogen in their name, as in NADH. The oxidized form of the electron carrier doesn't have the H, as in NAD^+, which reveals that it's not carrying any passengers.

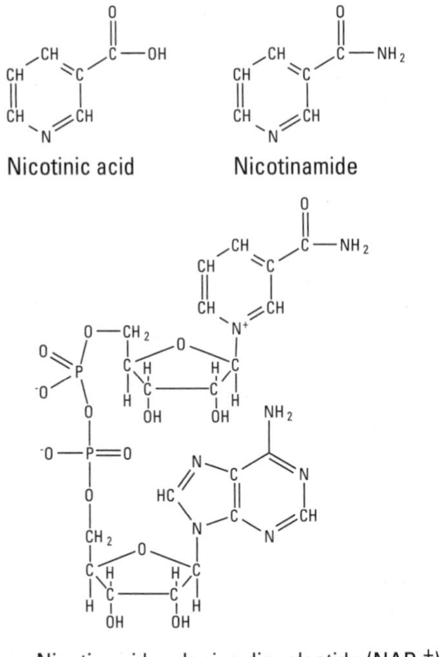

Nicotinic acid

Nicotinamide

Nicotinamide adenine dinucleotide (NAD^+)

Figure 10-9: The electron carrier NAD^+/ $NADH + H^+$.

Getting what you need at the cellular level

Metabolism is all about getting what you need to stay alive. You know what you need on the big scale — good old food, clothing, and shelter. But on the small scale, on the cellular level, what you need to keep your cells functioning comes down to a different three things:

- **Building blocks for growth and repair.** Your cells need a constant supply of materials to obtain the building blocks for the molecules that make up the cell. Food molecules are broken down by catabolism, and their building blocks are rearranged to form the molecules of the cell by anabolism.

- **Energy for cellular work.** Your cells need a constant supply of energy to allow cellular work, such as building molecules, moving things, and organizing cellular components. Food molecules are broken down by catabolism, allowing the transfer of energy from food to the cell.
- **Reducing power.** Your cells need a constant supply of electrons during anabolism in order to build the molecules that make up the cell. During catabolic reactions, electrons from food are transferred to electron carriers. These electron carriers then provide the needed electrons during anabolism.

Chapter 11

Cellular Respiration: Every Breath You Take

In This Chapter

- Burning your food
- Transferring energy to ATP
- Breaking down glucose via cellular respiration

When you eat food, your digestive system breaks down the food into smaller and smaller pieces until it becomes a soup of molecules in your small intestines. Your cells take in food molecules and break them down into even smaller pieces in order to make their stored energy available to the cell. The metabolic pathway that your cells use to rearrange food molecules in order to make them useful as building blocks or to capture their stored energy is called *cellular respiration*. Without cellular respiration, your cells would not be able to obtain the energy and matter they need to function, and you would die. In this chapter, I examine the importance of this metabolic pathway and the details of how it works.

Cellular Respiration: An Overview

Cells break down food molecules — carbohydrates, proteins, and fats — through *cellular respiration*. Cellular respiration is a metabolic pathway that, through a series of small steps, rearranges the atoms in the food molecules, making their stored energy available to the cell.

Cellular respiration, shown in Figure 11-1, is an example of the principles of metabolism discussed in Chapter 10: The pathway is made of many small chemical reactions, redox reactions move electrons from one molecule to another, and energy is transferred. Specifically,

- During cellular respiration, electrons are removed from food molecules, transferred to the electron carrier NADH (see Chapter 10) and then to oxygen (O_2).
- By the end of cellular respiration, the energy from food has been transferred to the energy carrier, ATP.
- Through many chemical reactions, your cells completely rearrange food molecules into the waste products, carbon dioxide (CO_2), and water (H_2O).

Although cells break down all kinds of food molecules via cellular respiration, the simple sugar *glucose* is used as a starting point for studying the pathway. Overall, the pathway is represented by this reaction:

$$C_6H_{12}O_6 + 6\ O_2 \rightarrow 6\ CO_2 + 6\ H_2O + \text{energy}$$

This summary reaction simplifies the process by ignoring all the intermediate steps and showing only the reactants that enter the process (glucose and oxygen) and the waste products that leave the process (carbon dioxide and water). I include energy as a product as a reminder that this pathway is exergonic and thus makes energy available to the cell. The energy released during cellular respiration is transferred to ATP for use in cellular work (see Chapter 10) and also lost to the environment of the cell as heat.

Food has energy already, so why do your cells need to transfer that energy to ATP using cellular respiration? You can think of cellular respiration as similar to the process of refining oil. Crude oil has energy in it, but the energy isn't in the right form to run your car. So, crude oil is refined to gasoline that can fuel your car. Similarly, food is broken down, and energy is transferred to ATP — the right form of energy to fuel your cells.

Sweaty business

Have you ever wondered why you get hot and sweaty when you're working out? The answer is cellular respiration. Your muscles need energy in the form of ATP to do work, so they increase the rate of cellular respiration. Cellular respiration breaks down food molecules and transfers energy to ATP. However, not all the energy from food is captured by the cell. In fact, cellular respiration can transfer only about 40 percent of the energy in food to ATP; the other 60 percent is lost to the environment as heat. As the heat transfers through your body, you heat up, causing your automatic sprinkler system, sweating, to kick in and help cool you back down.

Controlling the burn

If you've every burned a marshmallow, or even a piece of paper, you've seen and felt the energy that was originally stored in the chemical structure of the thing you burned. When you touched the marshmallow to a flame in the presence of oxygen, you provided the energy needed to start a chemical reaction that rapidly converted the sugars to carbon dioxide (CO_2) and water (H_2O) and transformed the chemical potential energy to heat and light (see Figure 11-1).

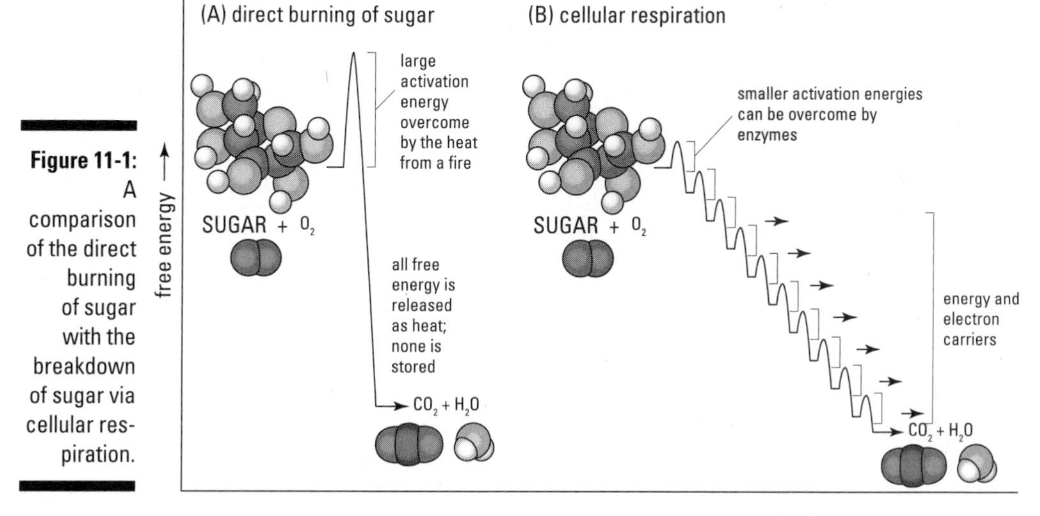

Figure 11-1: A comparison of the direct burning of sugar with the breakdown of sugar via cellular respiration.

The reaction of the direct burning of sugar is identical to the summary reaction for cellular respiration. I'm sure you can appreciate, however, that it would be very painful if your cells broke down sugars as rapidly as this process! Also, I doubt you'd be very eager to touch a flame to your cells to begin the process. Instead, your cells break down sugars by a number of much smaller steps so that the process is compatible with living cells:

- The energy release is done little by little, which allows cells to capture that energy by transferring it to ATP without being harmed.
- Enzymes can lower the activation energy for each of the smaller reactions so that the reactions will occur. (See Chapter 10 for more about enzymes and activation energy.)

Transferring energy to ATP

Cellular respiration is an exergonic process, which means it makes energy available to the cell. If you look at Figure 11-1, you can see that the reactants (glucose and oxygen) have more free energy than the products (carbon dioxide and water). As your cells rearrange the atoms in glucose and oxygen during cellular respiration, the molecules are going downhill in energy: The chemical potential energy in the molecules is slowly being siphoned off as it's transferred into the energy carrier ATP.

Moving electrons to oxygen

During cellular respiration, glucose is oxidized, and oxygen is reduced, which means that electrons are removed from glucose and transferred to oxygen. (You can see these processes in Figure 11-2.) When the electrons are moved, they're moved as part of hydrogen atoms. In fact, if you look again at the summary reaction for cellular respiration, you can see that to turn glucose ($C_6H_{12}O_6$) into carbon dioxide (CO_2), you need to remove the hydrogen atoms. Likewise, to turn oxygen (O_2) into water (H_2O), you have to add hydrogen atoms.

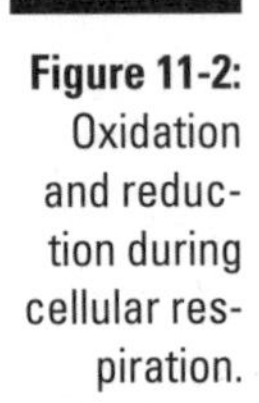

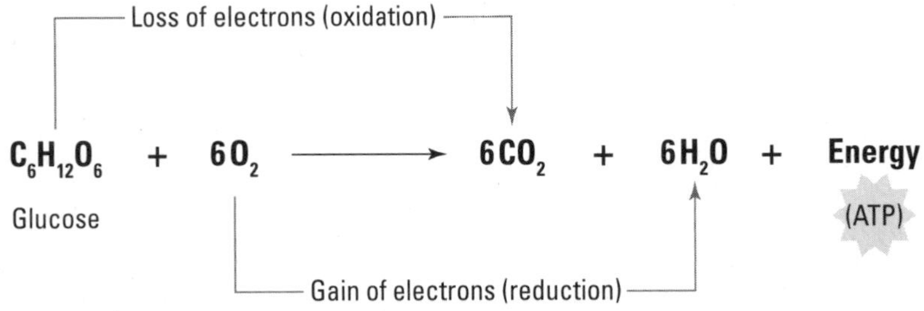

Figure 11-2: Oxidation and reduction during cellular respiration.

The oxidation of glucose is exergonic and makes lots of energy available to the cell. To control the energy release, the oxidation happens as part of the many small steps of cellular respiration, rather than as a single transfer from glucose to oxygen. During several steps of cellular respiration, electrons are transferred from the intermediates to NAD^+, changing it to its reduced form, NADH + H^+. NADH then transfers the electrons to an electron transport chain, where they're passed along through a series of redox reactions before being transferred to oxygen. The electron transport chain is a series of proteins in a membrane that use redox reactions to transfer energy to ATP. Figure 11-3 shows the transfer of electrons and energy via the electron transport chain.

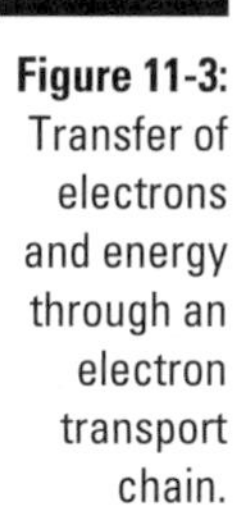

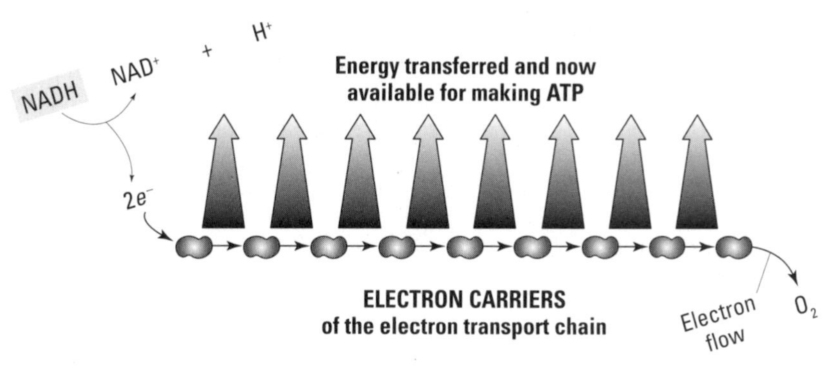

Figure 11-3: Transfer of electrons and energy through an electron transport chain.

Taking things one step at a time

The oxidation of glucose via cellular respiration takes place slowly over a series of small steps and results in the transfer of energy from glucose to ATP. The entire process of cellular respiration occurs in three smaller pathways (see Figure 11-4):

- *Glycolysis* is the first part of cellular respiration. During glycolysis, the oxidation of glucose begins, and some electrons are transferred to NAD^+. Additionally, a small amount of energy is transferred to ATP. Glycolysis converts a single glucose molecule, which contains six carbon atoms, into two molecules of pyruvate, each of which has three carbon atoms. Glycolysis occurs in the cytoplasm of the cell.
- The *Krebs Cycle* (also called the tricarboxylic acid cycle, TCA cycle, or citric acid cycle) continues the oxidation of the intermediates made from glucose, resulting in the production of CO_2. The Krebs cycle transfers electrons to NAD^+ and to another electron carrier, flavin adenine dinucleotide (FAD). The Krebs cycle also transfers a small amount of energy directly to ATP. The Krebs cycle occurs in the mitochondrion (see Chapter 2) of the cell.
- *Oxidative phosphorylation* transfers electrons from electron carriers to an electron transport chain. The electron transport chain is a system of proteins in a membrane that use redox reactions to transfer energy to ATP. The final redox reaction is the transfer of electrons to oxygen, which is reduced to water. Oxidative phosphorylation occurs in the mitochondrion of the cell.

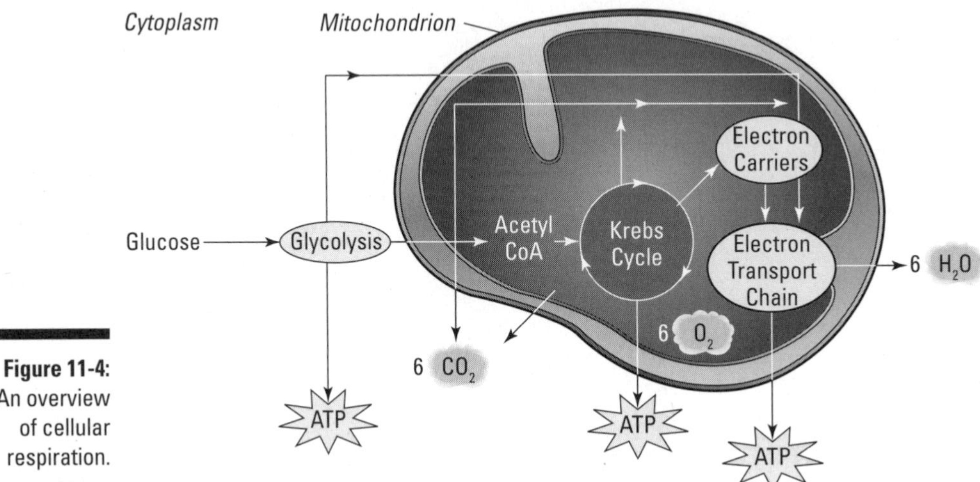

Figure 11-4: An overview of cellular respiration.

Gimme a Break: Glycolysis

Glycolysis is a series of chemical reactions that result in the oxidation of glucose (Figure 11-5), the transfer of energy to ATP, and the transfer of electrons to the electron carrier NAD^+. Overall, the major events of glycolysis are as follows:

- **The bonds between the atoms in glucose break and reform between new combinations of atoms, creating two molecules of pyruvate.** A glucose molecule has six carbon atoms, whereas a molecule of pyruvate has three carbon atoms, so in a way, it's like glycolysis breaks glucose molecules in half.
- **Enzymes transfer energy from intermediates to ATP.** During the first part of glycolysis, enzymes actually transfer energy from ATP to the intermediates, representing the *energy investment phase.* However, enzymes then transfer even more energy from the intermediates to ATP during the second part of glycolysis, representing the *energy payoff phase*. Overall, for every molecule of glucose broken down in glycolysis, enzymes use two ATP molecules, but make four ATP molecules, which results in a net gain of two molecules of ATP for the cell.
- **Enzymes oxidize glucose and reduce NAD+.** Enzymes transfer electrons that were originally in glucose from an intermediate to NAD^+, reducing the electron carrier to $NADH + H^+$. (See Chapter 10 for a detailed discussion of NAD^+.)

Everybody's doing it

Every cell on Earth does some form of glycolysis. Some variations in glycolysis exist among prokaryotes (see Chapter 2 for a discussion of the prokaryotic cell), but it seems that glycolysis is an integral part of life on Earth. When you think about what glycolysis does — transfers energy from food to ATP — I think this makes sense. In order to function, cells must remain organized. In order to remain organized, cells need a constant supply of energy. Glycolysis lets cells capture that energy from food. So, glycolysis is an essential process. And the fact that glycolysis is so widespread among many different types of cells suggests that it developed very early in the history of life on Earth.

Fine print: The steps of glycolysis

Glycolysis has ten chemical reactions, shown in Figure 11-5:

1. **An enzyme transfers a phosphate from ATP to glucose, creating the intermediate glucose-6-phosphate and releasing ADP as waste.**

   This step transfers energy to the intermediate and is part of the energy investment phase of glycolysis.

2. **An enzyme rearranges the bonds in glucose-6-phosphate, creating the intermediate fructose-6-phosphate.**

3. **An enzyme transfers a phosphate from ATP to fructose-6-phosphate, creating the intermediate fructose-1,6-bisphosphate and releasing ADP as waste.**

   This step transfers energy to the intermediate and is part of the energy investment phase of glycolysis.

4. **An enzyme catalyzes the splitting of fructose-1,6-bisphosphate into two 3-carbon molecules, dihydroxyacetone phosphate and glyceraldehydes-3-phosphate.**

   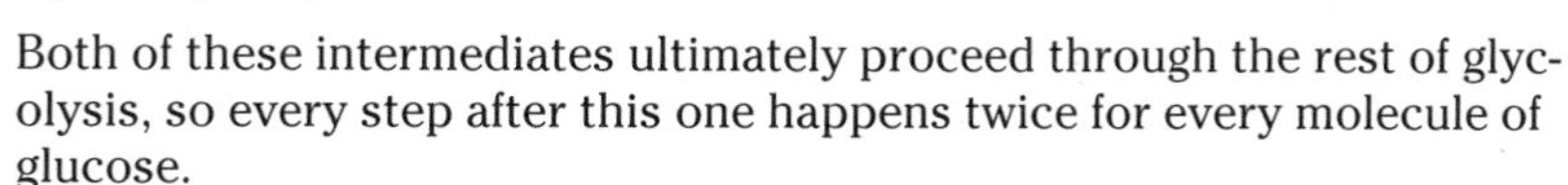
   Both of these intermediates ultimately proceed through the rest of glycolysis, so every step after this one happens twice for every molecule of glucose.

5. **An enzyme rearranges the bonds in dihydroxyacetone phosphate, converting it to glyceraldehyde-3-phosphate, so that it can proceed through the rest of glycolysis.**

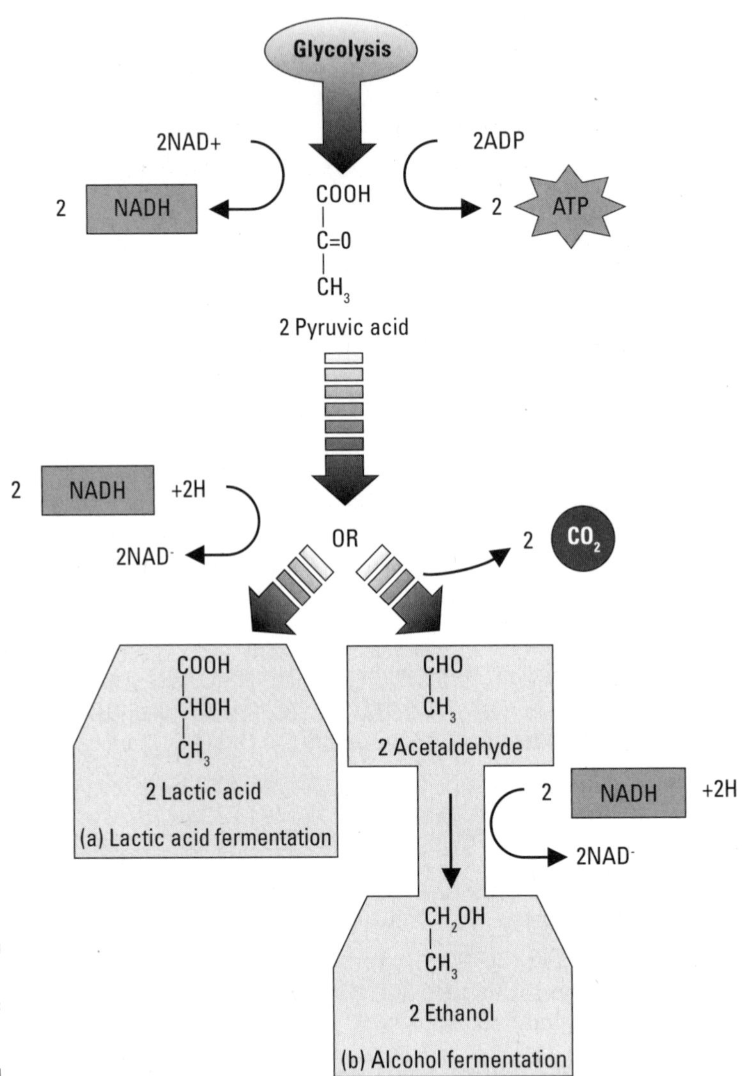

Figure 11-5: The steps of glycolysis.

6. **Electrons are transferred from glyceraldehyde-3-phosphate to NAD+, resulting in the formation of NADH + H+; at the same time, an inorganic phosphate group that was available in the cell is transferred to glyceraldehydes-3-phosphate, resulting in the formation of the intermediate 1,3-biphosphoglycerate.**

7. **An enzyme transfers a phosphate group from 1,3-biphosphoglycerate to ADP, producing ATP and a new intermediate, 3-phosphoglycerate.**

   This step represents a transfer of energy from 1,3-biphosphoglycerate to the energy carrier, ATP.

8. **An enzyme rearranges the bonds in 3-phosphoglycerate, creating 2-phosphoglycerate.**

9. **An enzyme catalyzes the removal of water from 2-phosphoglycerate, forming the intermediate phosphoenolpyruvate.**

10. **Another energy transfer occurs as a phosphate group is transferred from phosphoenolpyruvate to ADP, producing ATP.**

    During the process, phosphoenolpyruvate is converted to pyruvate.

The appearance and names of the molecules in glycolysis may seem strange and hard to remember. However, they do make some sense, and figuring them out may make the whole process easier to understand. For example, glucose with a phosphate added to it is called glucose-6-phosphate. The number 6 indicates that the phosphate group is attached to the number 6 carbon. (See Chapter 5 for a discussion of the numbering of carbons in a sugar.) A molecule with two phosphate groups attached is called a bisphosphate, because the prefix "bi-" means two. So, 1,3-bisphosphoglycerate has two phosphate groups, one attached to the number 1 carbon and one attached to the number 3 carbon. Whenever you see ATP being used or created, the number of phosphate groups on the intermediates will also change. Changes in ATP occur in Steps 1, 3, 7 and 10 of glycolysis.

Making ATP by substrate-level phosphorylation

During Steps 7 and 10 of glycolysis (see the preceding section), enzymes transfer a phosphate group from an intermediate to ADP, forming the energy carrier ATP. During both of these reactions, an enzyme binds the intermediate and ADP in its active site and then transfers the phosphate from the intermediate to ADP (see Figure 11-6). Molecules that bind into the active site of enzymes are called *substrates*, and putting a phosphate on a molecule is called *phosphorylation*, so this method of ATP production is called *substrate-level phosphorylation* to indicate that the phosphate came from a substrate. Substrate-level phosphorylation is possible because the removal of the phosphate from the substrate is exergonic. The cell captures some of that energy for later use by forming ATP.

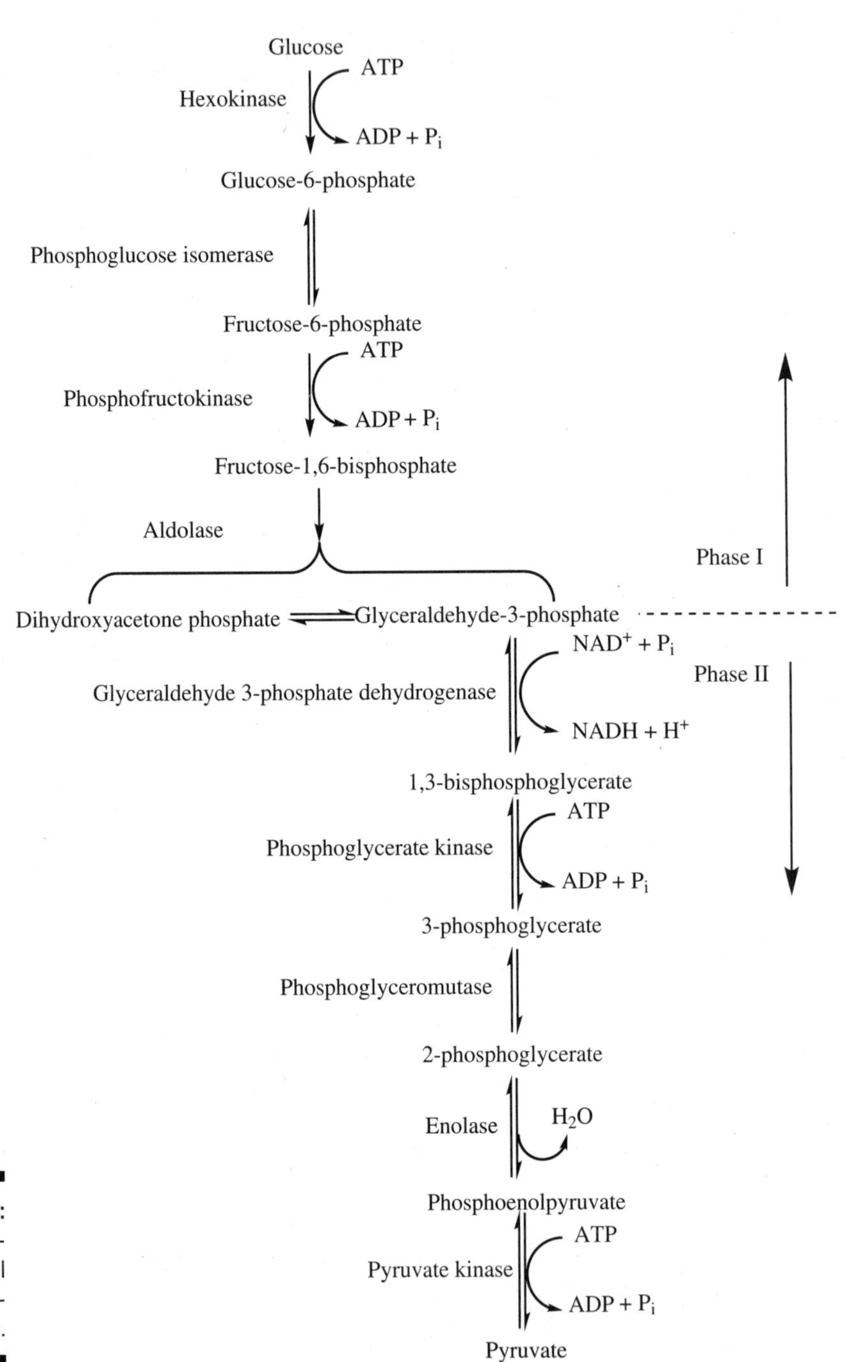

Figure 11-6: Substrate-level phosphorylation.

Living by glycolysis alone: Fermentation

Many cells on planet Earth have the ability to survive using glycolysis alone, without the rest of the cellular respiration. In fact, some cells can't do cellular respiration at all. Cells that rely upon glycolysis combine it with a fermentation step in order to recycle the NADH produced in glycolysis.

Fermentation occurs as either lactic acid fermentation or alcohol fermentation, as shown in Figure 11-7.

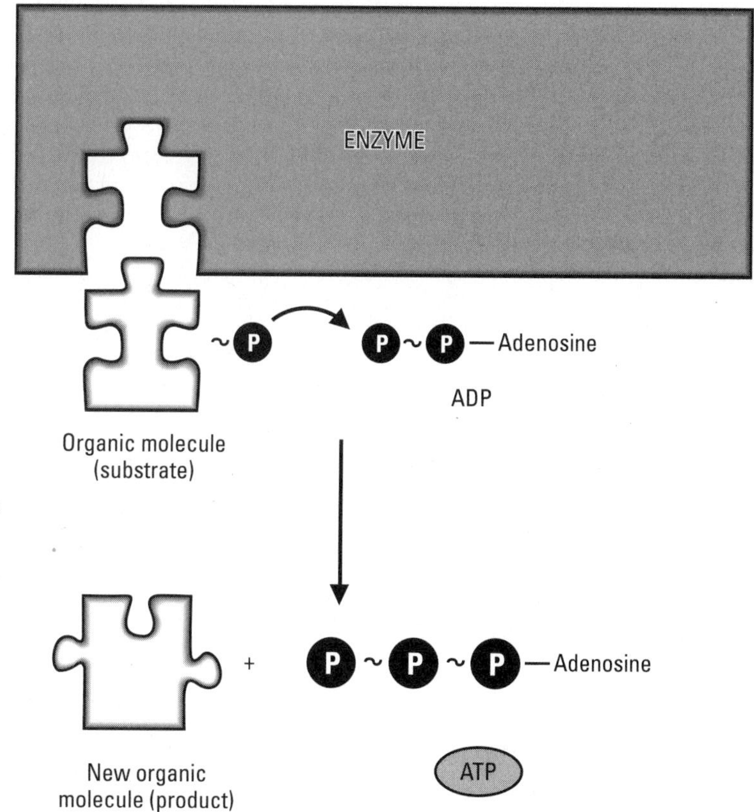

Figure 11-7: A comparison between lactic acid fermentation and alcohol fermentation.

- **Lactic acid fermentation:** Cells that use this type as their primary catabolic pathway do glycolysis, producing 2 pyruvate, 2 NADH + H^+ and 2 ATP. Then enzymes transfer electrons from NADH back to pyruvate, causing it to become lactic acid, and recycling the NADH back to NAD^+ so that glycolysis can proceed again.

Cells that do lactic acid fermentation benefit because they get the ATP they need for cellular work. Lactic acid is a waste product of the cell — one that humans find quite tasty! People use lactic acid bacteria, such as *Lactobacillus*, in the preparation of food such as yogurt. And although most human cells must do cellular respiration to survive, your muscle cells have the ability to do lactic acid fermentation when oxygen is scarce. So, next time you're working out, breathing hard, and starting to feel your muscles tingle, your muscle cells have probably switched into lactic acid fermentation in order to keep making ATP!

- **Alcohol (ethanol) fermentation:** Yeast use this type of fermentation reaction during the production of bread, beer, and wine. It's a little more complicated than lactic acid fermentation, but the basic principle is the same: The cells do glycolysis, which produces 2 pyruvate, 2 NADH + H^+, and 2 ATP. Then, decarboxylation removes a carbon atom and two oxygen atoms from pyruvate, releasing the atoms as carbon dioxide (CO_2). The decarboxylation of pyruvate converts it into a new molecule, called acetaldehyde. (The release of CO_2 is what causes bread to rise and makes bubbles in beer.) Then, enzymes transfer electrons from NADH to acetaldehyde, changing it to ethanol and recycling the NADH back to NAD^+ so that glycolysis can proceed again.

  Ethanol is a waste product to the yeast cells. Although ethanol burns off during the baking of bread, it remains in beer and wine which is what makes them alcoholic beverages.

Lactic acid fermentation and alcohol fermentation are just two examples of the many microbial fermentations that are important in the food industry and other industries as well. Although plants and animals rely upon the entire process of cellular respiration in order to survive, many microbes get the job done with fermentation!

The Wheel of Fire: Krebs Cycle

The Krebs cycle, which occurs in the mitochondrion, picks up where glycolysis left off and has basically the same purpose: The oxidation of intermediates that came from food molecules and the transfer of energy from these intermediates to a usable form for the cell. Overall, the major events of the Krebs cycle are as follows:

- **Enzymes oxidize intermediates and reduce electron carriers.** Oxidation of intermediates is the real work of the Krebs cycle. Multiple redox reactions transfer electrons from intermediates in the cycle to the electron carriers NAD^+ and FAD. These oxidations are exergonic, transferring both energy and electrons to the electron carriers. As cellular respiration continues, the reduced carriers, NADH and $FADH_2$, transfer the energy and electrons to the electron transport chain, helping to make ATP.
- **Enzymes break bonds and rearrange atoms in the intermediates, removing carbon and oxygen atoms and releasing them as CO2.** Decarboxylation happens three times for every molecule of pyruvate that leaves glycolysis and enters the Krebs cycle. Because glycolysis produces two 3-carbon molecules from glucose, the Krebs cycle occurs twice for every molecule of glucose that enters cellular respiration. Thus, two rounds of the Krebs cycle produce six molecules of carbon dioxide. Decarboxylation during the Krebs cycle releases all six of the carbon atoms that were originally part of the glucose molecule. The last step of the Krebs cycle rearranges the atoms in the intermediate malate to recreate the intermediate, oxaloacetate, which is needed for the cycle to repeat again.
- **Enzymes transfer energy to ATP.** Most of that energy is transferred to the electron carriers when they're reduced. However, during one step of the Krebs cycle, energy is transferred directly to ATP by substrate-level phosphorylation. Because the Krebs cycle occurs twice for every glucose molecule that goes through cellular respiration, a total of 2 ATP per glucose are made during Krebs.

When you look at a biochemical pathway, such as those in Figures 11-7 or 11-8, clues tell you what is going on during each reaction. ATP is an energy carrier for cells, so any time you see ATP participating in a reaction, you know that an energy transfer is occurring. If ATP is being used, then energy is being transferred into the reaction. If ATP is being made, then energy is being transferred out of the reaction. Likewise, NAD^+ is an electron carrier for cells, so any time you see NAD^+/NADH participating in a reaction, you know that electrons are being transferred.

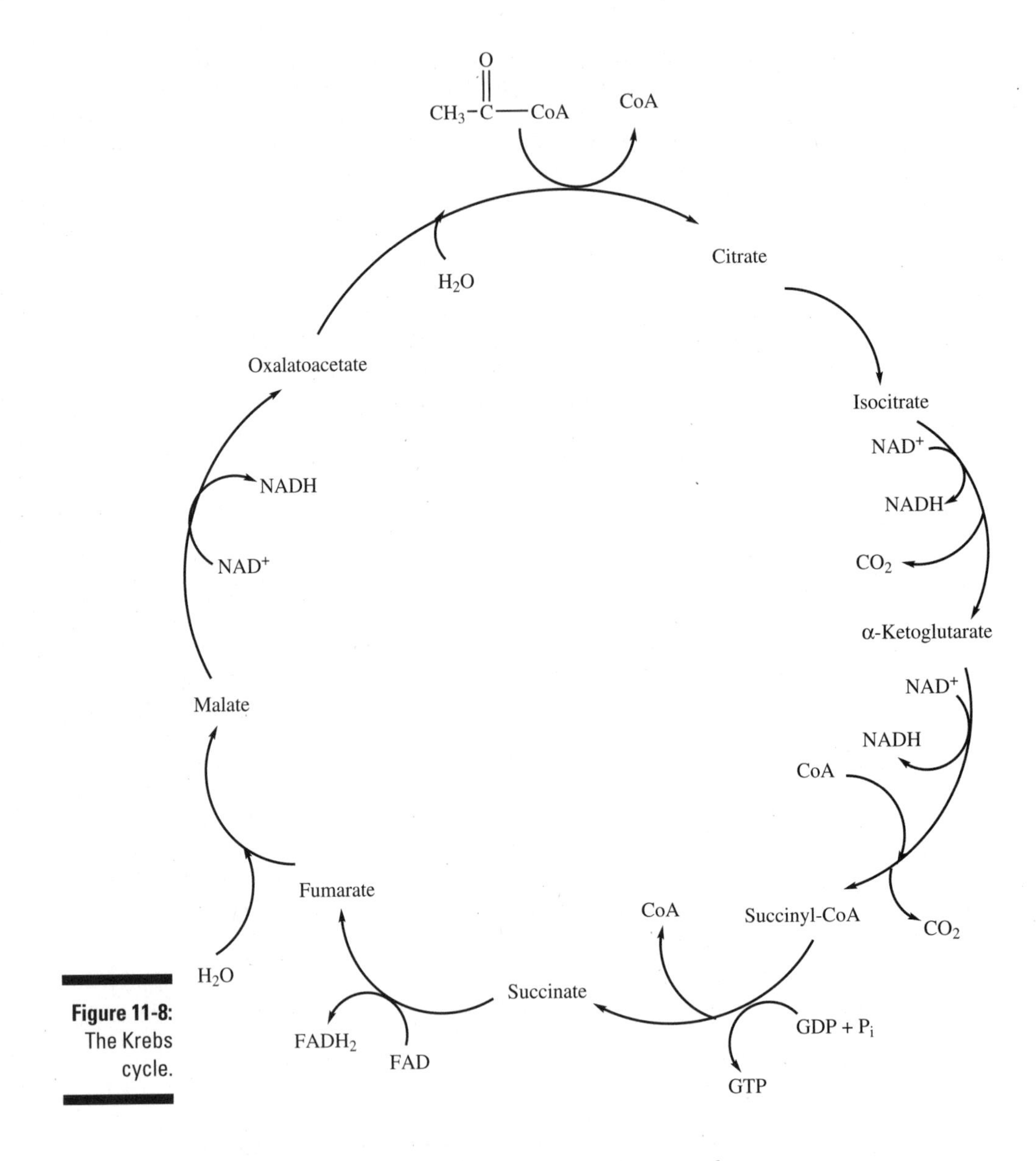

Figure 11-8: The Krebs cycle.

Linking glycolysis and Krebs

The Krebs cycle officially begins when a 2-carbon molecule, acetyl-CoA, is joined with a 4-carbon molecule, oxaloacetate, to form a 6-carbon molecule, citrate. However, glycolysis produces two 3-carbon molecules, pyruvate, from every molecule of glucose that passes through the pathway. So, in order to begin the Krebs cycle, pyruvate must first be converted into acetyl-CoA.

The series of reactions that convert pyruvate into acetyl-CoA have many names, including the linking step, the grooming step, the bridging step, and pyruvate oxidation. I refer to it as *pyruvate oxidation* because I think that term is the most clear about what is actually happening.

Here are the events that occur during pyruvate oxidation:

- Enzymes rearrange the atoms in pyruvate, removing carbon and oxygen, then releasing them asCO_2. This decarboxylation reduces the number of carbon atoms in the intermediates from 3 to 2.
- Enzymes oxidize pyruvate and then transfer the electrons to the electron carrier NAD^+. This reduction of NAD^+ converts it to NADH + H^+.
- Enzymes add coenzyme A to an intermediate, forming acetyl-CoA. Coenzyme A will be released from acetyl-CoA as it enters the Krebs cycle.

Glycolysis occurs in the cytoplasm of the cell, but the Krebs cycle occurs in the interior matrix of the mitochondrion. (See Chapter 2 for the structure of the mitochondrion.) Thus, in order for pyruvate to be oxidized and continue into the Krebs cycle, it has to cross the two membranes that separate the interior of the mitochondrion from the cytoplasm of the cell. The outer membrane of the mitochondrion is relatively easy to cross because it contains proteins called *porins* that allow small molecules to get through. The inner membrane, however, is highly impermeable, which means very few things can cross it unless they have help from a specific transport protein. There is such a transport protein to help pyruvate get across so that pyruvate can be converted into acetyl Co-A and enter the Krebs cycle.

Fine print: The steps of the Krebs cycle

Beginning with the joining of acetyl-CoA and oxaloacetate, the Krebs cycle has eight sets of reactions, shown in Figure 11-8:

1. **Enzymes join the 2-carbon molecule acetyl-CoA with the 4-carbon molecule oxaloacetate, creating the 6-carbon molecule citrate.**

   During this process, a water molecule is incorporated into the intermediate and coenzyme-A is released back to the cell.

2. **An enzyme rearranges the bonds in citrate, creating the molecule isocitrate.**

The number of carbon atoms in the intermediate doesn't change.

3. **Enzymes oxidize and decarboxylate isocitrate, converting it to the 5-carbon molecule α-ketoglutarate. Enzymes transfer the electrons removed from isocitrate to the electron carrier NAD+, reducing it to NADH + H+. Enzymes release the carbon removed from isocitrate as CO2.**

4. **Enzymes oxidize and decarboxylate α-ketoglutarate, converting it to the 4-carbon molecule succinyl-CoA. Enzymes transfer the electrons from α-ketoglutarate to the electron carrier NAD+, reducing it to NADH + H+. Enzymes release the carbon removed from α-ketoglutarate as carbon dioxide CO2.**

   Enzymes require the help of coenzyme A during these reactions and add it to the intermediate.

5. **Enzymes remove coenzyme A from succinyl-CoA and release it back to the cell, changing succinyl-CoA into succinate.**

   This reaction is exergonic and allows the phosphorylation of ADP by substrate-level phosphorylation, producing a molecule of ATP.

   Substrate-level phosphorylation during the Krebs cycle may produce ATP or it may produce a very similar energy carrier called guanosine triphosphate (GTP). Two different enzymes can catalyze substrate-level phosphorylation during the Krebs cycle, and each one prefers either ADP or GDP in its active site.

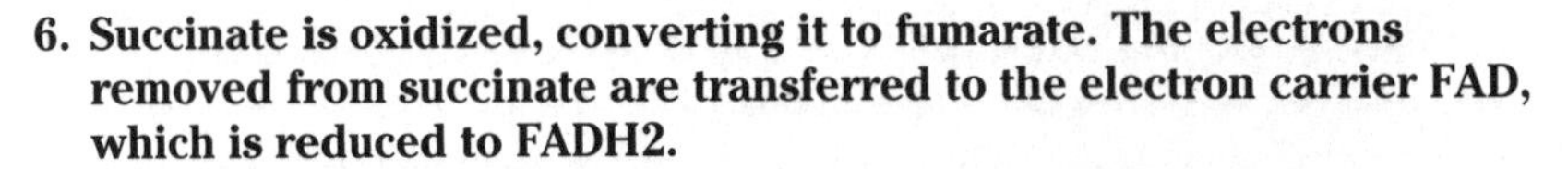

6. **Succinate is oxidized, converting it to fumarate. The electrons removed from succinate are transferred to the electron carrier FAD, which is reduced to FADH2.**

   FAD/FADH$_2$ have a similar structure to NAD$^+$/NADH and perform the same function as electron carriers for the cell. In this case, the carrier FAD works better with the enzyme that catalyzes this reaction.

7. **An enzyme catalyzes the addition of a water molecule to fumarate, converting it to malate.**

8. **Malate is oxidized, converting it to oxaloacetate. The electrons removed from malate are transferred to the electron carrier NAD+, which is reduced to NADH + H+.**

   The re-creation of oxaloacetate brings you back to the beginning of the Krebs cycle, which can occur again with the addition of another molecule of acetyl-CoA.

More is better: Taking advantage of the Krebs cycle

One of the most important functions of cellular respiration is to transfer energy from food molecules to a form that cells can use to do work. To get the energy out of food, cells oxidize food molecules. The oxidation of food molecules is highly exergonic, so, if you're a cell that is trying to get energy out of food, the more oxidation you do, the better, right?

Now, take a look at glycolysis and the Krebs cycle and compare the two pathways (refer to Figures 11-6 and 11-8). Glycolysis has one oxidation step, which you can recognize because NADH + H^+ is formed. Pyruvate oxidation and Krebs have a total of five oxidation steps, which you can recognize because NADH + H^+ and $FADH_2$ are formed. When you consider that Krebs happens twice per glucose molecule, the number of oxidation steps doubles to 10! So, if oxidation of food molecules makes energy available to the cell, it's pretty clear which pathway gives more bang for each glucose molecule — cells that have the enzymes to do the Krebs cycle are much more efficient at extracting energy from their food than are cells that can do only glycolysis!

Taking It to the Bank: Chemiosmosis and Oxidative Phosphorylation

ATP is the most useful energy carrier for cells: Whenever a cell needs to do a job, it's usually ATP that provides the necessary energy. This need for ATP explains why cellular respiration is so important to cells — it's the way to use the energy from food molecules to make ATP. However, if you look at glycolysis and the Krebs cycle, there isn't a whole lot of ATP being made. Even more puzzling, by the end of the Kreb's cycle, an incoming food molecule has been completely broken down and released back to the environment as carbon dioxide (CO_2). So, where's all the ATP?

The answer to that question is found in the *chemiosmotic theory of oxidative phosphorylation* — a really big name for an explanation of how energy and electrons from food help make ATP. You see, by the end of glycolysis and Krebs, a cell hasn't made much ATP per glucose molecule, but it has made a whole bunch of reduced electron carriers, mostly NADH + H^+ but a couple of $FADH_2$, also. These carriers received electrons and energy when intermediates in glycolysis and the Krebs cycle were oxidized, and now they can cash them in to make ATP by using an electron transport chain.

Electron transport chains are made of large protein complexes embedded in a membrane (see Figure 11-9). They accept electrons and pass them from complex to complex, much like the runners in a relay race pass a baton. As the electrons are passed along the chain, some of the protein complexes transfer energy from the redox reactions and use it to do work like moving hydrogen ions (H^+) across the membrane. The electron transport chain that is part of cellular respiration is located in the inner membrane of the mitochondrion.

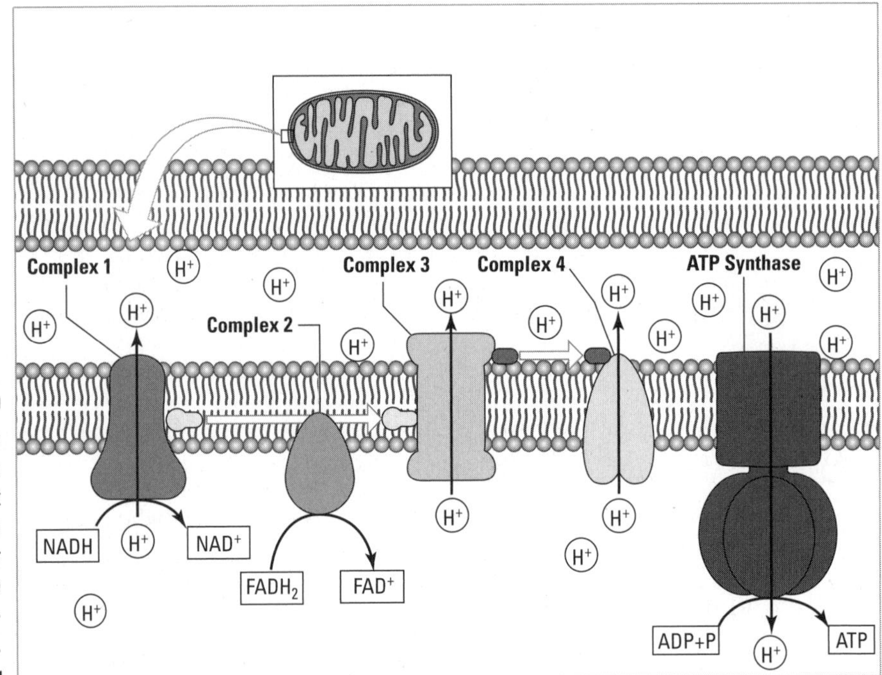

Figure 11-9: The electron transport chain in the inner membrane of the mitochondrion.

The electron transport chains that are embedded in the inner membrane of the mitochondrion have an important partner — an enzyme called ATP synthase. This enzyme gets its name because it makes, or synthesizes, ATP. Most of the ATP that is made by cellular respiration is made by this protein in a process called the chemiosmotic theory of oxidative phosphorylation, or just oxidative phosphorylation for short.

ATP synthase is embedded in the inner mitochondrial membrane right along with the electron transport chain (see Figure 11-9). On one side of the membrane is the intermembrane space; on the other side is the matrix of the mitochondrion. During oxidative phosphorylation, hydrogen ions (H+), which are also called protons, enter a small channel in ATP synthase from the intermembrane space side of the membrane. The protons bind to the ATP synthase and then exit the protein on the matrix side. As they bind to the ATP synthase, the head of the protein rotates. The base of the protein has binding sites for ADP and inorganic phosphate (P). The passage of every three protons through the ATP synthase provides enough energy for the synthesis of one ATP molecule from ADP and P.

Transferring electrons along an electron transport chain

Electron transport chains accept electrons from an electron donor, pass them along through the protein complexes in the chain, and then pass them to a final electron acceptor. When a complex accepts electrons, it's reduced; when it gives them up, it's oxidized: so, as the electrons move along the chain, a series of redox reactions occurs. The electrons move from complex to complex along the chain because each member of the chain is more electronegative, or has more pull for electrons, than the complex before it.

You can think of the proteins in the electron transport chain as a bunch of kids who all want the same piece of candy. A little kid is holding the candy. Only a stronger kid can pull the candy away from the little kid. Then, an even stronger kid can pull the candy away from the second kid, and so on. In an electron transport chain, the candy is the electron and the protein complexes are the bigger and bigger kids. In this analogy, the kid who ends up with the candy will be the strongest kid of all.

In an electron transport chain, the final electron acceptor has to have the most pull of all. During cellular respiration in eukaryotes, that job falls to oxygen (O_2). Oxygen is highly electronegative and pulls the electrons from the chain. When oxygen accepts the electrons, it also picks up some protons (H^+) and is reduced to water (H_2O), which is why water is one of the waste products of cellular respiration.

The donors of electrons to the electron transport chain during cellular respiration are the NADH and $FADH_2$ molecules that were made during glycolysis and the Krebs cycle. These carriers are oxidized as they donate electrons to the beginning of the electron transport chain and are converted back to NAD^+ and FAD. These "empty electron shuttle busses" can now go back to glycolysis and Krebs and pick up more electron passengers.

Transferring energy from food to ATP

You eat food because it contains chemical potential energy. Your cells use cellular respiration to transfer that energy to ATP, which is a more useful form of chemical potential energy for your cells. However, with the exception of the few ATP made by substrate-level phosphorylation during glycolysis and the Krebs cycle, the energy transfer from food to ATP is not direct. In fact, the energy path takes a few twists and turns that can seem downright strange:

1. **Enzymes in glycolysis and Krebs transfer electrons from intermediates to electron carriers, increasing the potential energy of the carriers as they are reduced.**

   NADH and $FADH_2$ are now carrying the energy from food.

2. **Reduced electron carriers transfer energy along with the electrons they donate to the electron transport chain.**

   The energy is now part of the first protein complex in the chain.

3. **As the complexes in the electron transport chain pass electrons along, they're also passing some energy.**

   However, at the same time, some of the complexes use some energy to actively transport hydrogen ions (H^+) across the membrane. The concentrated hydrogen ions, which are also called protons, represent a source of potential energy that can do work. The potential energy of the concentrated protons, called *proton motive force,* now stores the energy from food.

4. **The protein ATP synthase allows hydrogen ions to move back across the membrane.**

   As the hydrogen ions move, they cause part of the ATP synthase to rotate, which increases energy in the ATP synthase. The enzyme uses this energy to join ADP and phosphate groups, making ATP. ATP finally stores the energy from food!

The steps of the chemiosmotic theory of oxidative phosphorylation

The chemiosmotic theory of oxidative phosphorylation is a widely accepted explanation for how scientists think the energy and electrons from food are used to make ATP during cellular respiration. In other words, it is the whole story of the electron transport chain, from NADH to oxygen, and from NADH to ATP.

As you read through the following steps, shown in Figure 11-9, notice that the path for the electrons is different than the path for energy. Use your finger to trace the pathway for each and practice "telling the story" out loud.

1. **NADH donates electrons and energy to the electron transport chain. The first protein complex is reduced, while NADH is oxidized to NAD+.**
2. **Electrons move through the electron transport chain in a series of redox reactions, until they're picked up by oxygen (O_2), the final electron acceptor. Oxygen is reduced to water (H_2O).**
3. **Protein complexes I, II, and III use energy from the redox reactions to pump hydrogen ions (protons) across the membrane.**

   This step creates the potential energy of the proton motive force.
4. **ATP synthase allows protons to recross the inner mitochondrial membrane. As the protons pass through the protein, it uses the energy to join ADP and P into ATP.**

To understand chemiosmosis, think about how people use dams to generate hydroelectric power. The trapping of water behind a dam represents a source of potential energy that can do work, just like the accumulation of hydrogen ions on one side of the membrane. When the water flows from behind the dam, it does work, such as turning a turbine to create electricity. This release of water to do work is similar to what happens when the protons flow through a channel in the protein ATP synthase. The turning of the turbine transfers energy to electricity for your home, just like the turning of ATP synthase transfers energy to ATP for your cells.

Chemiosmosis is a complicated process, but it's hugely important to life on Earth. It's the most efficient way that cells have to make ATP, which is an absolutely essential energy source for cells. A cell that can do chemiosmosis is like a fuel-efficient car — the cell makes more ATP from its energy source, just like a hybrid car can go farther on a gallon of gasoline.

Chemiosmosis uses electron flow through an electron transport chain coupled with the transport of hydrogen ions (H^+) across a membrane to transfer energy to ATP. Chemiosmosis occurs as part of cellular respiration and also photosynthesis (see Chapter 12).

Heavy breathing

Have you ever wondered why you start to breathe harder when you're working out? The answer, once again, is cellular respiration. When your muscles are doing lots of work, they need lots of ATP. Your cells make ATP by doing cellular respiration. In order to make ATP, you need oxygen to accept electrons at your electron transport chain. So, as you use up your ATP in your muscles, you breathe faster to bring in more oxygen, so you can have more oxygen in your mitochondria to accept more electrons, to make more ATP. This is why you breathe.

Everything you already knew about breathing, such as bringing oxygen to your lungs and having your red blood cells carry it around your body, is all true, but that's really more about how you get oxygen to your cells, not why your cells need it. The why is all about electron transport chains. Really. And if you're denied oxygen for some reason, you die because no oxygen = no final electron acceptor = no ATP = no cellular work = cells cease to function = death.

Doing the math: How many ATP can you make from the energy in a glucose molecule?

The exact number of ATP molecules you'd get for breaking down a glucose molecule through cellular respiration would be pretty hard to actually count, but you can use the following rules to estimate what is theoretically possible:

- Every time a pair of electrons from NADH passes through the electron transport chain, ATP synthase gets enough energy to produce approximately three ATP molecules.
- Every time a pair of electrons from $FADH_2$ passes through the electron transport chain, ATP synthase gets enough energy to produce approximately two ATP molecules.
- In eukaryotes, the NADH produced in the cytoplasm has to cross into the mitochondrion in order to donate electrons to the electron transport chain. Different cell types have different ways of moving this NADH into the mitochondrion, which can require energy comparable to the amount in one ATP molecule.

With these rules, you can count up the products of cellular respiration and calculate the theoretical number of ATP molecules you could produce in cellular respiration from the energy in one glucose molecule:

- Two ATP per glucose are produced during glycolysis.
- Two ATP per glucose are produced by the Krebs cycle.
- Two NADH per glucose are produced during glycolysis. If it costs one ATP to move these into the mitochondrion, then they're worth only two ATP each ($2 \times 2 = 4$ ATP for these NADH).
- Two $FADH_2$ per glucose are produced by the Krebs cycle. The $FADH_2$ molecules are worth two ATP each, so $2 \times 2 = 4$ ATP for these $FADH_2$.
- Eight NADH per glucose are produced by pyruvate oxidation and the Krebs cycle. The NADH molecules are worth three ATP each, so $8 \times 3 = 24$ ATP for these NADH.

When you add up all the ATP molecules, that is $2 + 2 + 4 + 4 + 24 = 36$ ATP per glucose. Because some cells, such as bacterial cells, don't have to "pay" to transport NADH from the cytoplasm to the mitochondrion, these cells would get 2 more ATP. So, the theoretical max for the number of ATP per glucose molecule that is complete oxidized during cellular respiration is 36 – 38 ATP.

Breaking Down Complex Carbohydrates, Proteins, and Fats

In order for food molecules besides glucose to be broken down in cellular respiration, enzymes need to change the food molecules into intermediates in glycolysis or the Krebs cycle. Once the food molecules become intermediates in these pathways, they will be oxidized and their energy will be transferred to ATP.

Finding an on-ramp to the superhighway

Cellular metabolism consists of hundreds of chemical reactions that interconnect to make a vast network. The product of one pathway may be a reactant or an intermediate in another. The molecules that move through this metabolic network are like little cars driving along the freeway. They can go down one freeway for a long time, or they can exit a freeway and move onto another one just like a molecule can exit one pathway and get picked up by another. Enzymes control the traffic flow — in order for a molecule to exit or enter a metabolic pathway at a particular point, there must be an enzyme that can turn it into the correct metabolic intermediate.

If you can think metabolism as a vast network of interconnected highways, then cellular respiration is the biggest interstate of them all. It's like I-5 on the west coast of North America, running North to South from Canada to Mexico. Many other pathways connect to cellular respiration, just like many highways connect to I-5. Figure 11-10 shows the connections between some important metabolic pathways and cellular respiration. In order for food molecules other than glucose to enter cellular respiration, enzymes in these connecting metabolic pathways break bonds and rearrange the atoms in the food molecules, turning them into intermediates in glycolysis or the Krebs cycle. At that point, an enzyme from glycolysis or Krebs can pick them up and feed them into the system.

Feeding complex carbohydrates into the system

Of all the food molecules, carbohydrates enter cellular respiration most easily because many of the intermediates in glycolysis are carbohydrates:

- Simple sugars, such as glucose, either fit right in or require few reactions to convert them into an intermediate in glycolysis.
- Cellular respiration easily breaks down the complex carbohydrate starch, which is a common human food. Starch is a polymer of glucose, so once the enzyme amylase breaks the chain down into individual glucose molecules, the glucose molecules fit right into glycolysis.
- Enzymes break down complex carbohydrates other than starch into their individual sugar components. Then, enzymes convert the simple sugars into intermediates in glycolysis.

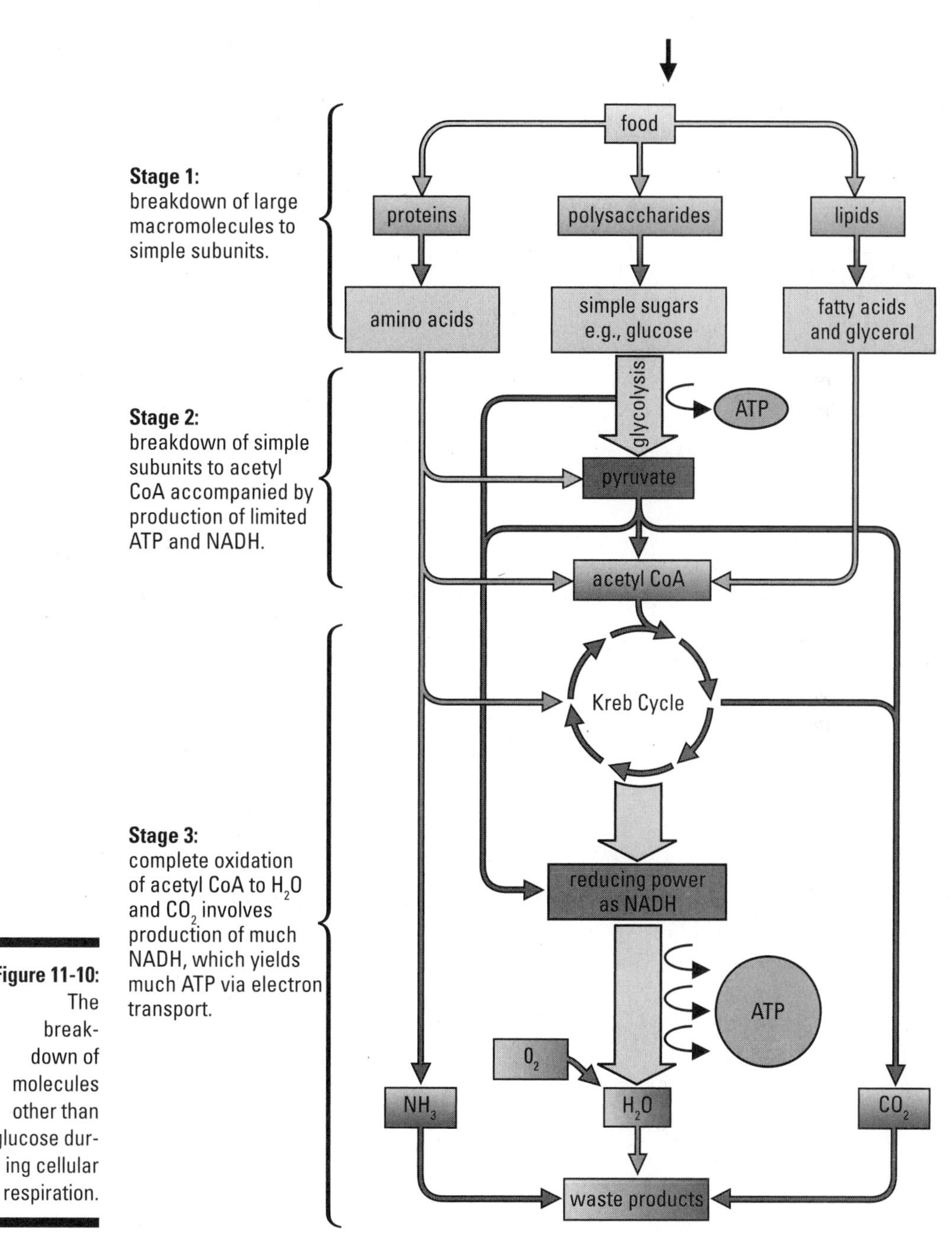

Figure 11-10: The breakdown of molecules other than glucose during cellular respiration.

Burning fat

Enzymes called lipases break the fats and oils into their two components: glycerol and fatty acids. Then, glycerol and fatty acids feed into cellular respiration at two different locations:

- Glycerol is a 3-carbon compound. An enzyme converts it into glyceraldehyde-3-phosphate, an intermediate in glycolysis.
- Enzymes break fatty acids up into fragments containing 2 carbons each. Then, enzymes break these 2-carbon fragments into acetyl-CoA, an intermediate in the Krebs cycle.

Breaking down proteins

Because of the unique features of proteins, it's a little more work to get them ready for cellular respiration. Like other polymers, they're broken down into their building blocks, so enzymes called proteases break the protein chains into their component amino acids. Then, enzymes make two major alterations to amino acids so that they become intermediates in glycolysis or the Krebs cycle:

- Amino acids have nitrogen-containing amino groups. The intermediates in cellular respiration don't have nitrogen-containing groups, so the amino group on amino acids must be removed by *deamination* before amino acids can enter glycolysis or Krebs.
- After deamination occurs, enzymes convert the remaining part of the amino acid into an intermediate in glycolysis or Krebs. The 20 different amino acids found in cells enter cellular respiration at various points. Where the converted amino acids enter the pathway depends largely upon their number of carbon atoms, but most amino acids enter as intermediates in the Krebs cycle.

It's a Two-Way Street: Connections Between Metabolic Pathways

A major purpose of cellular respiration is to enable the transfer of energy from food to ATP. However, cellular respiration has another purpose: Through all the little reactions that convert one molecule into another into yet another, cellular respiration generates a wide diversity of metabolic intermediates. These intermediates come in handy when a cell needs to build a carbohydrate, protein, lipid, or nucleic acid for growth or repair. In other words, metabolism — cellular respiration included — is a two-way street: Catabolism breaks down food molecules, but anabolism builds new molecules from the metabolic intermediates in cellular respiration. The pathways for drawing off intermediates for anabolism are the reverse of the ones used to feed macromolecules into cellular respiration (see Figure 11-11).

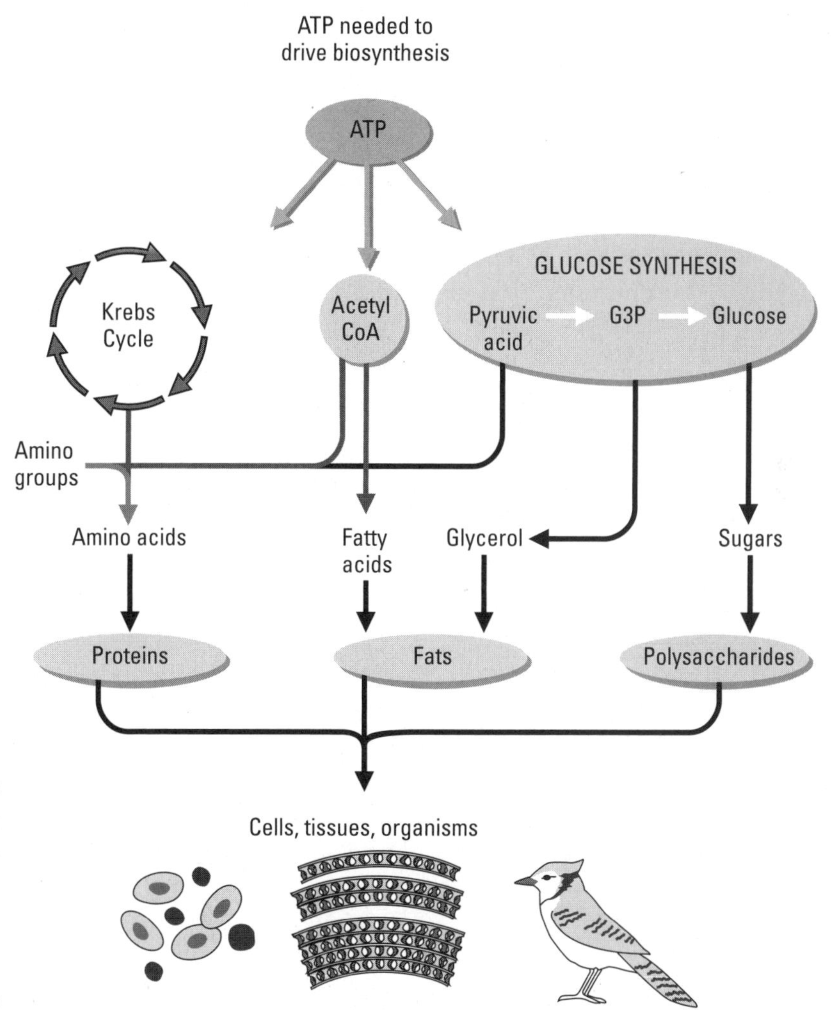

Figure 11-11: Connections between catabolism and anabolism.

Reversing the flow of matter and energy

Enzymes break down macromolecules through cellular respiration, providing a source of energy for the cell. Or, if a cell needs to build a macromolecule, then enzymes use intermediates from within cellular respiration as building blocks during anabolism. Because all metabolic reactions rely upon enzymes, regulation of enzymes controls the direction of traffic through metabolic pathways. Enzymes are most commonly regulated through through feedback inhibition. (See Chapter 6 for a description of feedback inhibition.) This type of regulation is a simple feedback loop — if you have plenty of something, you don't need to make more.

Because the most important product of cellular respiration is ATP, it makes sense that ATP would act as an inhibitor of the pathway. And, in fact, ATP inhibits several enzymes of cellular respiration:

- Phosphofructokinase, which catalyzes the third step in glycolysis
- Pyruvate dehydrogenase, which catalyzes the oxidation of pyruvate into acetyl-CoA
- Citrate synthase, which catalyzes the synthesis of citrate from oxaloacetate and acetyl-CoA

When a cell has plenty of ATP, ATP binds to enzymes in glycolysis and the Krebs cycle, inhibiting their function and slowing the process of cellular respiration until ATP supplies decrease.

Similarly, if a cell has plenty of NADH, cellular respiration is no longer needed. NADH inhibits the function of several enzymes in cellular respiration, including pyruvate dehydrogenase and citrate synthase.

Packing on the fat

Imagine you've been hanging out on the couch, eating snacks and watching TV. You've supplied your body with plenty of food, so your cells have made lots of ATP via cellular respiration. However, you haven't been doing a lot of cellular work to use up that ATP. So, your cells contain a lot of ATP. That ATP is going to bind to enzymes in cellular respiration, inhibiting the enzymes and slowing down the process. In addition, ATP is going to bind to enzymes in fat synthesis pathways and cause their activity to speed up. Intermediates from glycolysis and the Krebs cycle are going to be rerouted into fat synthesis pathways:

- Enzymes use glyceraldehyde-3-phosphate from glycolysis to produce glycerol.
- Enzymes use many molecules of acetyl-CoA to synthesize fatty acids.

The result? The food you're eating is going to be converted to fat molecules.

Building muscle

In order to build proteins, whether for muscles or other cells, amino acids must be strung together in long chains. Adult humans can make 11 of the 20 amino acids necessary for protein synthesis; the other amino acids must be obtained from the diet and thus are called the *essential amino acids.* Enzymes synthesize most amino from intermediates in the Krebs cycle.

Cellular respiration in the real world

Anyone who has ever tried to lose weight knows the story: "Eat less, exercise more." This advice helps a person lose weight for several reason, but one main reason is the effect that it has on cellular respiration. By exercising more, cells use up their supplies of ATP and increase the amount of ADP. ADP stimulates enzymes in cellular respiration, triggering an increase in the breakdown of food molecules. If you combine this stimulation with a decrease in the amount of food you're eating, then your cells have to turn to your stored food molecules to fuel cellular respiration. In other words, they have to turn to your stored fat. Lipases break fats down into glycerol and fatty acids, then enzymes in cellular respiration break these building blocks down even further. The amount of fat stored in your fat cells, or *adipose tissue,* decreases, and you lose weight.

Where does the weight go?

At some point in your life, you've probably lost at least a pound or two, or maybe even more. When you lost weight, where did it go? You probably know that when you ate less and exercised more, your body turned to your stored fat for energy. However, energy doesn't weigh anything, so that doesn't explain where the weight went. Fat is made of glycerol and fatty acids — in other words, lots of carbon, hydrogen, and oxygen atoms. It's the atoms that actually weigh something, so where did they go?

When your body broke down the fat, it used cellular respiration. The waste products of cellular respiration are carbon dioxide (CO_2) and water (H_2O) — also made of carbon, hydrogen, and oxygen atoms. So, when your body breaks down fat in cellular respiration, you lose the weight through the carbon dioxide you breathe out and the water you lose in your sweat and urine. Hard to believe you can actually breathe away fat molecules, but it's true. Of course, you still need to follow the old "eat less, exercise more" advice that is needed to ramp up fat breakdown — too bad deep breathing alone couldn't get the job done!

Chapter 12

Photosynthesis: Makin' Food in the Kitchen of Life

In This Chapter

- Bringing matter and energy together to make food
- Transforming light energy to chemical energy
- Capturing carbon

Plants, along with algae (seaweeds) and some bacteria, make the food on which all living things depend. They do photosynthesis, using energy from the sun and matter from carbon dioxide (CO_2) and water (H_2O) to form carbohydrates (sugars). In this chapter, I journey into a plant cell to show how photosynthesis makes the carbohydrates that serve as the building blocks of life.

Photosynthesis: An Overview

Photosynthesis transfers light energy (from the Sun) into chemical energy (stored in carbohydrates) that is usable by living things. With these carbohydrates and some minerals, living things can build all the molecules needed for life: proteins, carbohydrates, lipids, and nucleic acids.

Photosynthesis is usually represented by the following chemical reaction:

$$6\ CO_2 + 6\ H_2O + \text{light energy} \rightarrow C_6H_{12}O_6 + 6\ O_2$$

This reaction summarizes all the events of photosynthesis into one chemical sentence with reactants on the left of the arrow and products on the right side of the arrow. Because it ignores the gory details, this summary reaction is useful for thinking about the big picture of photosynthesis:

- Carbon, hydrogen, and oxygen atoms from carbon dioxide and water are rearranged to form sugar (glucose) and oxygen. This reaction represents a transfer of matter from the environment (carbon dioxide from the air, water from the soil) into matter inside the plant body (sugars) and waste back to the atmosphere (oxygen).
- Light energy from the sun is transformed into chemical energy and stored in the bonds of sugar for later use by the plant (or, if the plant is unlucky, by something that eats the plant).

The overall purpose of photosynthesis is to combine matter and energy into one tidy package: food in the form of carbohydrates.

As an analogy, compare a baker making a cake to a plant doing photosynthesis. A baker would put the cake's building blocks, such as eggs, flour, and milk, into a bowl. A plant would bring matter in the form of carbon dioxide (CO_2) and water (H_2O) into its cells. The eggs, flour, and milk contain all the necessary ingredients to make the cake, but at this point, they don't look much like a cake. Similarly, the carbon dioxide and water contain the carbon, hydrogen, and oxygen atoms that are needed to build carbohydrates, but they're not yet in the right arrangement. In order to create the changes they want in the cake, the baker places it into the oven. The heat energy from the oven helps the cake batter change. Likewise, plants absorb light energy from the sun and use it to power a series of reactions that build carbohydrates from carbon dioxide and water. The light energy from the sun acts like the heat from the oven, changing those raw materials into something different — carbohydrates!

Getting what plants need

You may have noticed that plants don't get around much. They're pretty much stuck where they put down their roots. Sometimes, I think people overlook plants because they don't seem very exciting. I mean, plants are just sitting there, not doing anything, right? Most people don't realize just how amazing plants really are. For example, imagine if I buried you in dirt up to your knees and told you to hang out that way for a few weeks. You could have any water that fell from the sky, and whatever else you bring into your body without actually moving from that spot. I'm guessing you wouldn't be happy for very long.

So, how do plants manage to stay in one spot yet do so well? The main reason is that they're *autotrophs,* meaning they can make their own food. (Autotroph literally means "self-feeding.") While you would slowly starve if not allowed to move from one spot, plants make all the food they need through photosynthesis. All they need in order to do photosynthesis is water, carbon dioxide, and sunlight, and these things essentially come to them.

Plant structures (see Figure 12-1) are specialized for taking in the materials that they need:

- Roots take water from the soil.
- *Xylem,* tubes of cells specialized for water transport, moves water from the roots throughout the plant.
- *Phloem,* tubes of cells that are specialized for sugar transport, move sugars from the green leaves (where the sugars are made) to the rest of the plant.
- *Stomates* — little pores in the leaves — let carbon dioxide into the leaves.
- *Chlorophyll*, the green pigment in the leaves, absorbs light.

Plant leaves are full of the green pigment *chlorophyll,* which is very good at absorbing blue and red light, but not very good at absorbing green or yellow light. So, when you look at a leaf on a plant, you see green because the chlorophyll in the plant leaf is absorbing blue and red light and letting the green bounce off, or reflect, from the surface of the leaf.

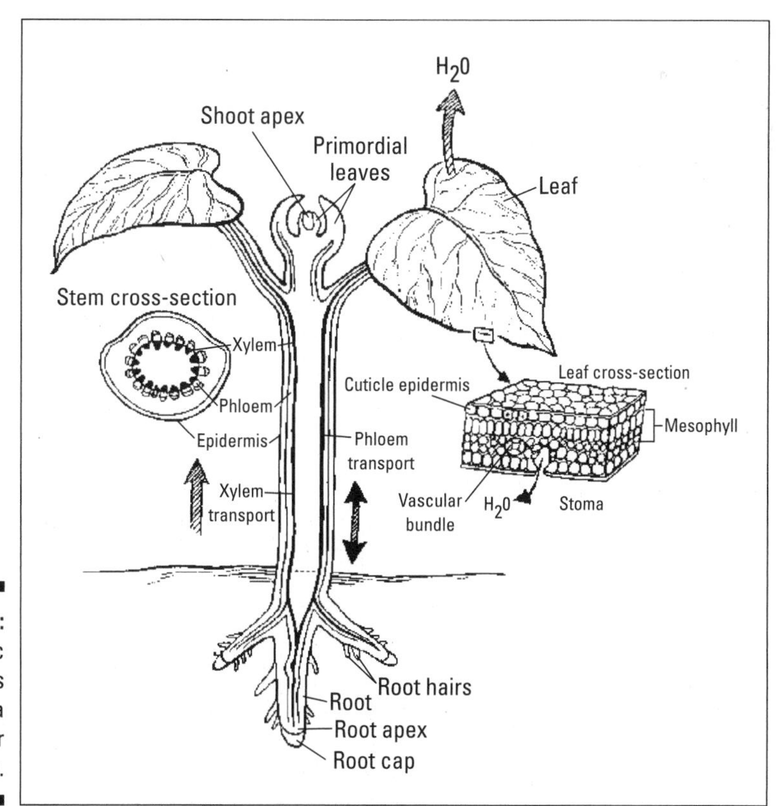

Figure 12-1: Basic structures of a vascular plant.

Examining the role of soil

You may be thinking, "What about soil? Since kindergarten, I've thought that plants need soil!" to survive. Well, my answer is that plants don't, really. Soil is useful to plants because it gives them something sturdy to stand in, it may hold water for them, and it may be full of lots of lovely minerals that plants need to keep their metabolism running smoothly. However, if you give them other means of support and fertilizer, plants will do just fine without soil. Hydroponic gardeners, who grow plants without soil, rely upon this fact.

Soil and commercial plant fertilizers typically contain nitrogen, phosphorous, potassium, and small amounts of other minerals, such as calcium, magnesium, and iron. Nitrogen and phosphorous are needed for the construction of macromolecules other than carbohydrates. For example, proteins contain large amounts of nitrogen. If a plant cell needs to build proteins, it takes some of its stored carbohydrates (which contain C, H, and O), adds nitrogen, and rearranges the atoms to form a protein. Minerals such as calcium, magnesium, and iron are necessary for the proper function of proteins. So, soil basically does for plants what a vitamin and mineral supplement does for your body.

Some fertilizers are called plant food, but they're not really food! They contain the same minerals as other fertilizers, but someone thought "plant food" sounded better than "plant fertilizer." Plants make their own food and then use that food to grow. When a plant grows bigger, most of the matter it uses to build its tissues comes from carbon dioxide from the air, not from the small amount of minerals it takes up through its roots!

Basking in the sun

Everyone knows that the Sun sends light to the Earth, but did you know that the Sun's radiation contains more than the light you can see? In fact, the light you can see, called *visible light,* is only a tiny part of the wide band of *electromagnetic radiation* that the sun sends to the Earth. The different types of radiation from the sun are shown in Figure 12-2 and probably sound familiar to you — things like x-rays, ultraviolet radiation (the stuff that causes skin cancer), and microwaves. Plants capture electromagnetic radiation within the visible light band for photosynthesis.

The different types of electromagnetic radiation travel from the Sun to the Earth in waves. Each type of electromagnetic energy is identified by its *wavelength,* the distance from one energy peak to another. (A nanometer is 1/100,000,000 of a meter, so wavelengths of light are pretty small!) The radiation captured by plants for photosynthesis has wavelengths between 400 and

700 nanometers. Radiation with wavelengths of 400 nanometers is visible as blue light, while radiation with wavelengths around 700 nanometers are visible as red light.

To understand wavelength, imagine looking at the wake of a boat and seeing the spread of ripples over the water. If you water-skied across the boat wake, you'd go up and down as you hit each crest. If you measured the distance from one crest to another, you'd know the wavelength between the ripples.

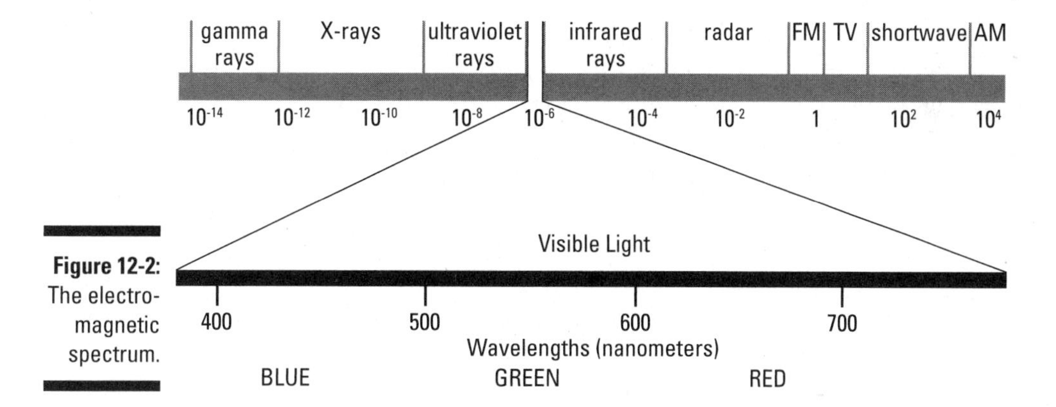

Figure 12-2: The electromagnetic spectrum.

Capturing the Sun's energy with pigments

All living things that do photosynthesis have some form of chlorophyll, as well as other *photosynthetic pigments.*

Figure 12-3 shows the different kinds of light absorbed by each of these pigments:

By having chlorophyll a, plus these other accessory pigments, photosynthetic organisms increase the wavelengths of light energy they can absorb for use in photosynthesis.

Yin and yang: The light reactions and the Calvin cycle

Here's another look at the summary reaction for photosynthesis:

$$6\ CO_2 + 6\ H_2O + \text{light energy} \rightarrow C_6H_{12}O_6 + 6\ O_2$$

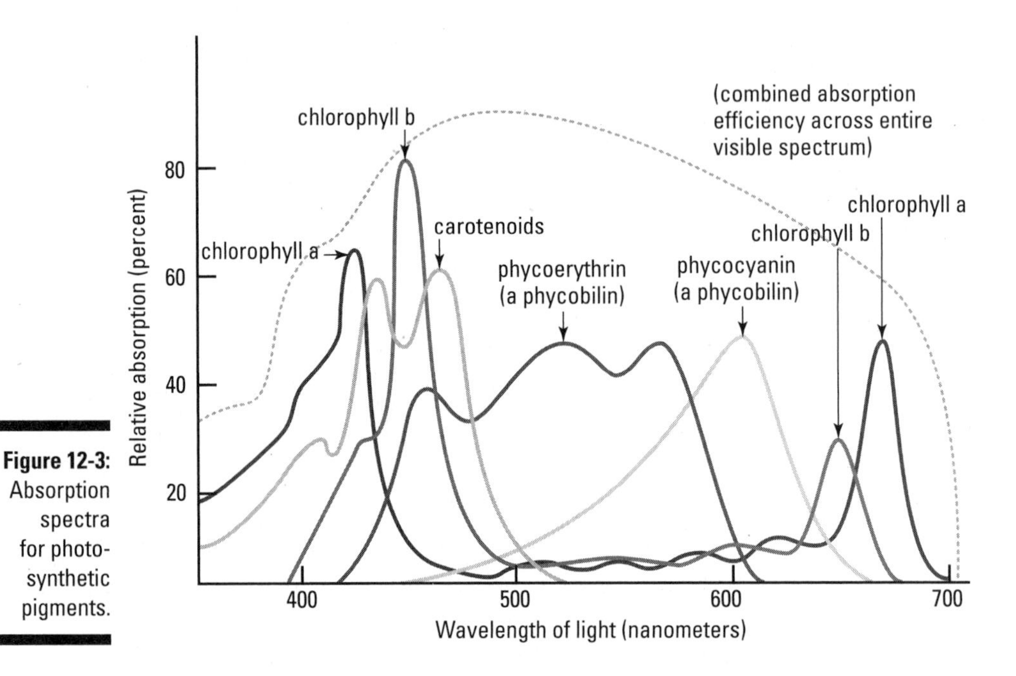

Figure 12-3: Absorption spectra for photosynthetic pigments.

Written in a kind of chemical shorthand, this reaction is a good reminder of the overall purpose of photosynthesis. It shows the transfer of matter (carbon dioxide and water) and energy (light) from the environment into the plant where they're combined to produce food in the form of carbohydrates (glucose), while oxygen is released back to the environment as waste. This combination of matter and energy into food represents the "why" of photosynthesis.

Now, turn your attention to the details of the process of photosynthesis — in other words, the "how." Like the metabolic processes described in Chapter 10, photosynthesis is complicated, involving the transfer of electrons, atoms, and energy from one form to another. I will try to keep this explanation manageable by starting with the bigger picture and moving toward the details.

First of all, photosynthesis doesn't occur in one step as indicated by the summary reaction. As is typical in metabolism, lots of little steps make up the whole. During photosynthesis, these little steps are organized into two separate sets of reactions in the plant: the *light reactions* and the *Calvin cycle*. Figure 12-4 gives an overview of these two pathways and shows how they're connected. (The rest of this chapter covers the details of these pathways.)

- In light reactions, plants capture light energy from the Sun and transform it into chemical energy in the form of the energy molecule ATP. Plants also take electrons from H_2O and transfer them to the electron carrier *nicotinamide adenine dinucleotide phosphate* (NADPH). NADPH has a very similar structure and function as the electron carrier NADH. (See Chapters 10 and 11 for more about NADH.)

- In the Calvin cycle, plants capture carbon dioxide (CO_2) and convert it into carbohydrates. To do this conversion, they need the molecules that were made during the light reactions: ATP provides chemical energy, and NADPH provides the necessary electrons.

To understand the role of electrons and NADPH in the process of photosynthesis, compare the chemical formulae for carbon dioxide (CO_2) and the carbohydrate glucose ($C_6H_{12}O_6$). If you want to turn CO_2 into $C_6H_{12}O_6$, what do you need? In other words, what is missing from CO_2? You probably noticed that what is missing is hydrogen (H), which is what NADPH provides to the process of sugar construction during the Calvin cycle. Essentially, this hydrogen is taken from water (H_2O) during the light reactions and carried by NADPH to the Calvin cycle. NADPH donates the H needed to build $C_6H_{12}O_6$ out of CO_2.

When $NADP^+$ accepts electrons, it's converted to its reduced form, NADPH. You can tell that NADPH is the reduced form because you can see that it's carrying a hydrogen atom at the end of its name. (Electrons are usually carried as part of hydrogen atoms.) The H in the name NADPH tells you that the electron carrier is bringing electron passengers to a new location. In the case of photosynthesis, it's carrying electrons from the light reactions to the Calvin cycle.

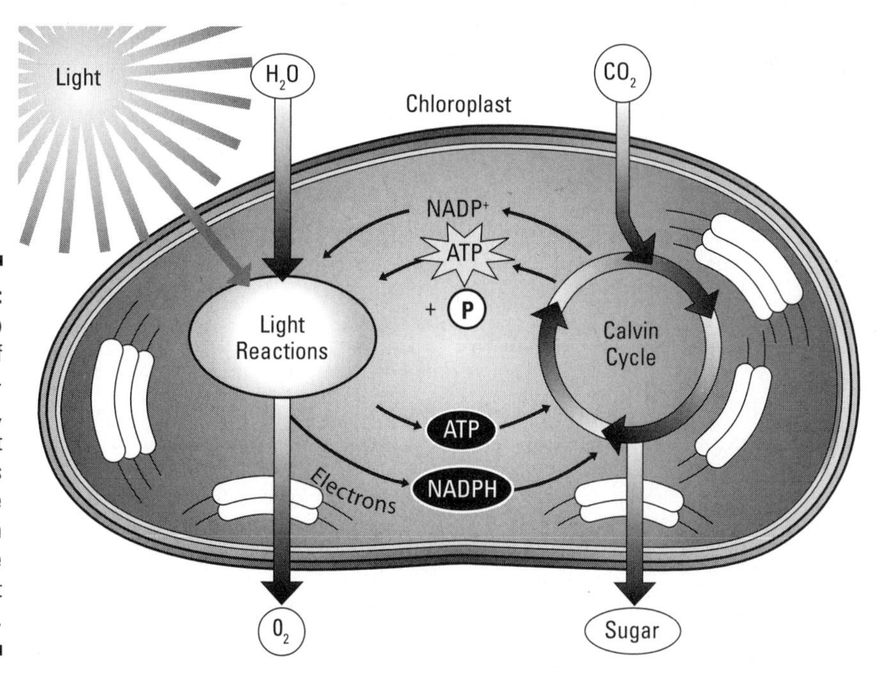

Figure 12-4: The two halves of photosynthesis, the light reactions and the Calvin cycle, are separate but linked.

Shine on Me: The Light Reactions

Every day, the sun beams a steady stream of energy to the planet, free for the taking to anyone who can use it. Plants may be the cleverest of all living things because they've evolved a way to do just that! During the light reactions of photosynthesis, they spread their leaves to the sky and absorb some of that energy, transforming it into chemical energy in the form of the energy molecule ATP. Cells use ATP as a convenient source of chemical energy to power the daily work of the cell.

Transferring light energy to chemical energy

Photosynthesis is possible because the photosynthetic pigments that absorb the light are hooked up to a little cellular machine that transfers the light energy into chemical energy stored in the bonds of the energy molecule ATP. The cellular machine that makes this energy transfer possible is an *electron transport chain.* (For more on electron transport chains, see Chapter 11.) Figure 12-5 shows the electron transport chain used in photosynthesis.

The electron transport chain for photosynthesis essentially works the same way as the one that is used in cellular respiration (see Chapter 11). Proteins in a membrane pass electrons in a series of redox reactions, using energy from this transport to pump protons (H^+) across the membrane. Then, ATP synthase uses the potential energy of the protons to make ATP. If you understand this process, called chemiosmosis, from cellular respiration, you've already got it down for photosynthesis.

When plant cells use light energy and electron transport chains to make ATP, it's called *photophosphorylation.* In this case, plant cells are adding a phosphate to ADP, which is the building block for ATP.

Photophosphorylation is really just another way of saying "the light reactions of photosynthesis."

The electron transport chains for photosynthesis are embedded in the thylakoid membranes of chloroplasts and contain both chlorophyll a molecules and accessory pigments to absorb light energy. The chlorophyll a molecules form *reaction centers,* key locations where the light energy is transferred into the electron transport system. All the other pigments within the membrane form an *antenna complex* that absorbs light energy and transfers it into these reaction centers. When the reaction center chlorophyll receives light energy, its electrons are excited to a higher energy state, causing them to move first to outer electron orbitals and then to jump to a protein in the electron transport chain.

When energy flows from the antenna complex to the reaction center chlorophyll, it's a little bit like people "doing the wave" at a sporting event. Imagine yourself as the reaction center chlorophyll. Light energy is absorbed by the other pigments, causing them to be energized like a line of people standing up to do the wave. The wave travels nearer and nearer to you and then arrives at your location. You're excited and jump up to do the wave! Lifting your arms is like your electrons moving to a higher orbital. If you were the reaction center chlorophyll, you'd then pass one of these electrons to the electron transport chain.

When the excited electron from chlorophyll is transferred to the electron transport chain, the plant cell has successfully transferred light energy to chemical energy!

The steps of photophosphorylation

Photophosphorylation means making ATP using the energy from light. Although this process and the making of ATP during cellular respiration (see Chapter 11) are similar, the processes also have several differences. In particular, the sources of energy and electrons for photosynthesis are different than they are for respiration. In photosynthesis, light provides the energy, and water provides the electrons. In cellular respiration, both the energy and the electrons come from food molecules.

Live like a plant

Solar panels transform light energy to electrical energy that people use to light and heat homes. Solar panels are actually more efficient than plants at transforming light energy — plants transform about 6 percent of the available energy to chemical energy, whereas solar panels can transform 10 to 20 percent of the available energy. As the amount of fossil fuels on Earth continues to decrease, people need to find new ways to fulfill their energy needs. Sunlight is clean and free for the taking, so solar energy could help solve both the human energy crisis and the human pollution problem. Solar power can provide enough energy for many human needs even in cloudy parts of the world — Germany is the world's leader in solar energy even though it doesn't have any more sunny days than cloudy Seattle.

Two different types of photophosphorylation occur in cells:

- In *noncyclic photophosphorylation*, which is also called the *Z-scheme,* electrons from chlorophyll travel through an electron transport chain then reduce $NADP^+$ to NADPH (see Figure 12-5).
- In *cyclic photophosphorylation,* electrons from chlorophyll travel a circuit, leaving chlorophyll when they get excited and returning to chlorophyll after their energy has been transferred to ATP (see Figure 12-6).

Cyclic photophosphorylation gets its name from the cycle that the electrons make from chlorophyll to an electron transport chain and back to chlorophyll again. Noncyclic photophosphorylation gets its name because the electrons take a one-way ride from H_2O to NADPH.

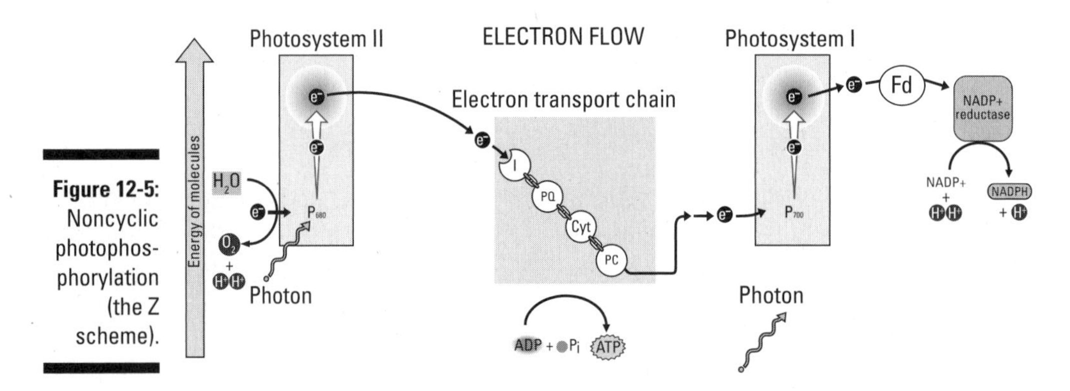

Figure 12-5: Noncyclic photophosphorylation (the Z scheme).

The complete steps of noncyclic photophosphorylation, shown in Figure 12-7, show how the energy and electrons are harvested in this process:

1. **Light energy is absorbed by chlorophyll and the accessory pigments in the plant leaves, and the energy is transferred to reaction center chlorophylls.**
2. **The electrons from reaction center chlorophylls are excited and move to outer orbitals.**
3. **Proteins in the electron transport chain pull the electrons from chlorophyll and pass them along the chain of proteins.**
4. **Chemiosmosis is used to make ATP.**

   During this process, energy from the movement of electrons is used to transport hydrogen ions (H^+) across the thylakoid membrane. The hydrogen ions flow back across the membrane like water through a dam, and the energy of their movement is used to make ATP from ADP and inorganic phosphate. The protein ATP synthase produces the ATP.

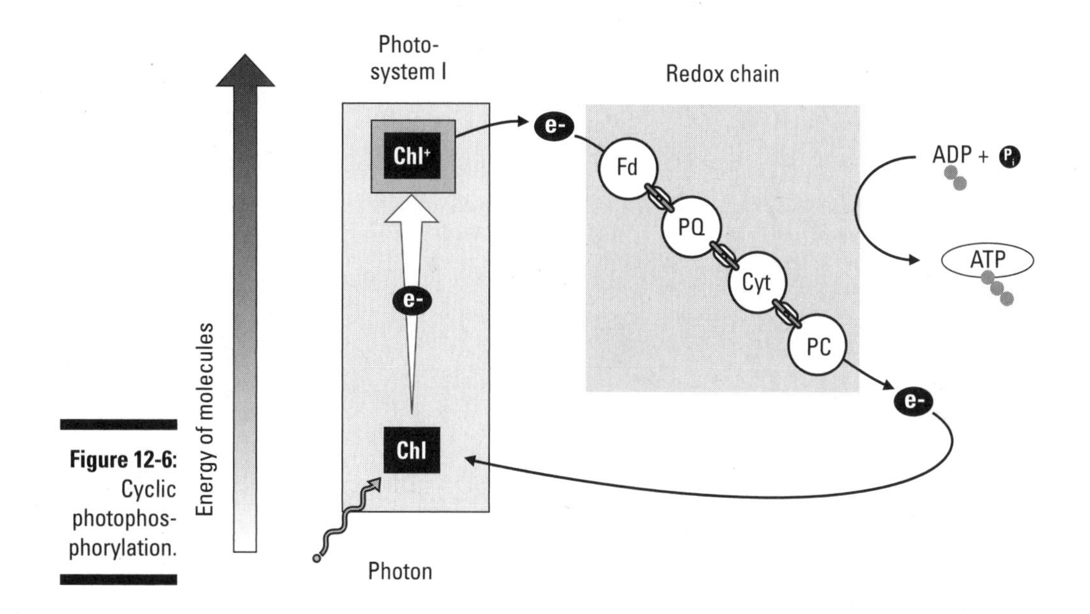

Figure 12-6: Cyclic photophosphorylation.

5. **The electrons are accepted at the end of the chain by the electron carrier NADP+, which** changes NADP+ to its reduced form, NADPH + H^+.
6. **As light energy is absorbed by chlorophyll, some light energy is used to split water (H2O) molecules by photolysis into protons (H+), electrons (e-), and oxygen gas (O2).**

   The electrons are transferred to chlorophyll to replace the electrons lost to the electron transport chain. The protons are released into the plant cell. The oxygen is released to the cell and then the atmosphere as a waste product of photosynthesis.

Oxygen (O_2) produced by photosynthesis doesn't come from carbon dioxide (CO_2). Oxygen is produced when water is split to provide electrons to the light reactions. Thus, the O_2 comes from the splitting of two H_2O molecules.

Electrons from water don't end up in the ATP made during the light reactions. If you were to trace the flow of electrons during the light reactions, it would look like this: Electrons from water replace lost electrons in chlorophyll, then go to the electron transport chain, and then are accepted by $NADP^+$ to form NADPH + H^+. The pathway of electrons during the light reactions is different from the flow of energy during the light reactions: Light energy absorbed by chlorophyll, transfers to the electron transport chain, then transfers to proton motive force, and finally transfers to ATP.

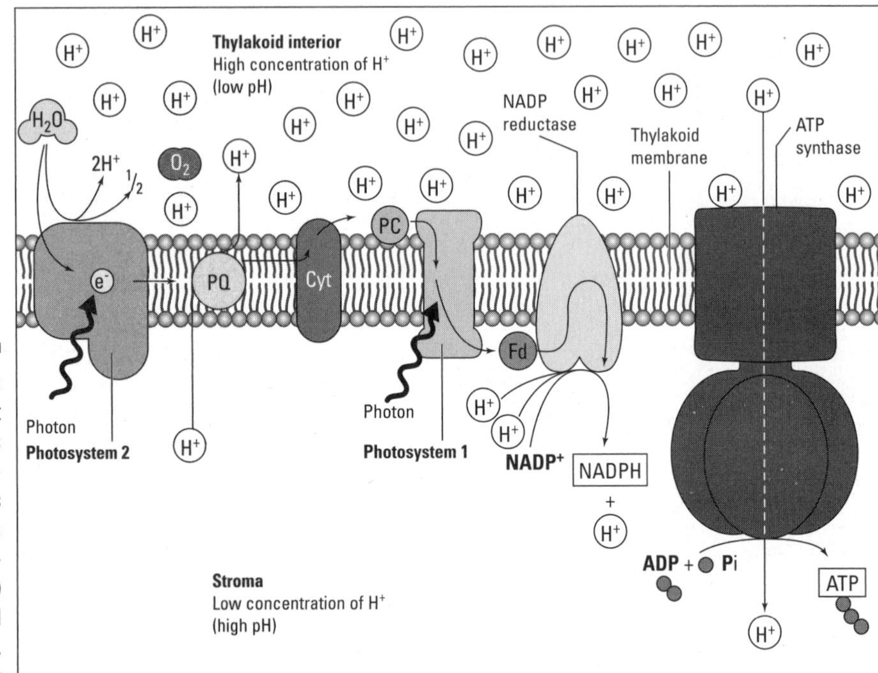

Figure 12-7: The light reactions of photosynthesis (noncyclic photophosphorylation) in a thylakoid membrane.

The Circle of Life: Calvin Cycle

The light reactions of photosynthesis transfer light energy to chemical energy, producing the energy molecule ATP and the electron carrier NADPH + H^+. ATP provides usable energy to cells for the cellular work that is necessary for survival. NADPH is an important source of electrons for reactions that build molecules for the cell. However, these two chemicals alone are not enough to provide the materials necessary for life. For living things to grow, repair themselves, and reproduce, they need a source of building materials. The ultimate source of building materials for most life on Earth traces back to one process: the Calvin cycle. The *Calvin cycle* transforms inorganic carbon in the form of carbon dioxide (CO_2) into organic carbon in the form of carbohydrates. All living things build their cells using organic carbon that ultimately traces back to the Calvin cycle or similar processes.

The Calvin cycle is also called the Calvin-Benson cycle, the Light Independent Reactions, and the Dark Reactions.

The steps of the Calvin cycle

The Calvin cycle takes place in the stroma of the chloroplast inside of plant cells. Like most metabolic pathways, the Calvin cycle is complicated and contains many small steps. The ultimate result of this pathway is to produce carbohydrates from carbon dioxide. The process to make carbohydrates from carbon dioxide includes four important events, shown in Figure 12-8:

1. **The enzyme Rubisco attaches carbon dioxide molecules to a 5-carbon sugar called ribulose bisphosphate (RuBP).**

   Rubisco is a much-needed nickname for ribulose-1,5-bisphosphate carboxylase/oxygenase.

   This critical first step captures the carbon dioxide from the environment and is called *carbon fixation*. The 6-carbon molecules that are formed by this step immediately split into two copies of a 3-carbon molecule. These 3-carbon molecules are called *3-phosphoglycerate*.

2. **3-phosphoglycerate is phosphorylated when it's given a phosphate group from ATP, which transfers energy from ATP to the 3-phosphoglycerate. Next, electrons from NADPH are transferred to the molecule, reducing it to the sugar *glyceraldedehyde-3-phosphate* (G3P).**

   The plant has now officially made carbohydrates. The energy and electrons for this phase, called the *reduction phase,* came from the light reactions.

3. **Some G3P is used to make glucose and other sugars.**

   This process is called *biosynthesis,* or just *synthesis,* because the plant is making sugars.

4. **Some G3P and ATP are used to regenerate RuBP, the 5-carbon sugar the plant started with.**

   The plant needs to move carbons around to reproduce the starting molecule so that it can start over again. This step is called the *regeneration phase*.

Rubisco: Earth's most important enzyme

Rubisco has been called the most important enzyme on Earth because it takes the first step in food production upon which all life depends. Rubisco is the most abundant protein in leaves and may be the most abundant protein on Earth. Scientists are trying to genetically engineer plants to either make more Rubisco or make more efficient forms of Rubisco. Their goal is to increase the production of food and decrease the amount of carbon dioxide in the atmosphere.

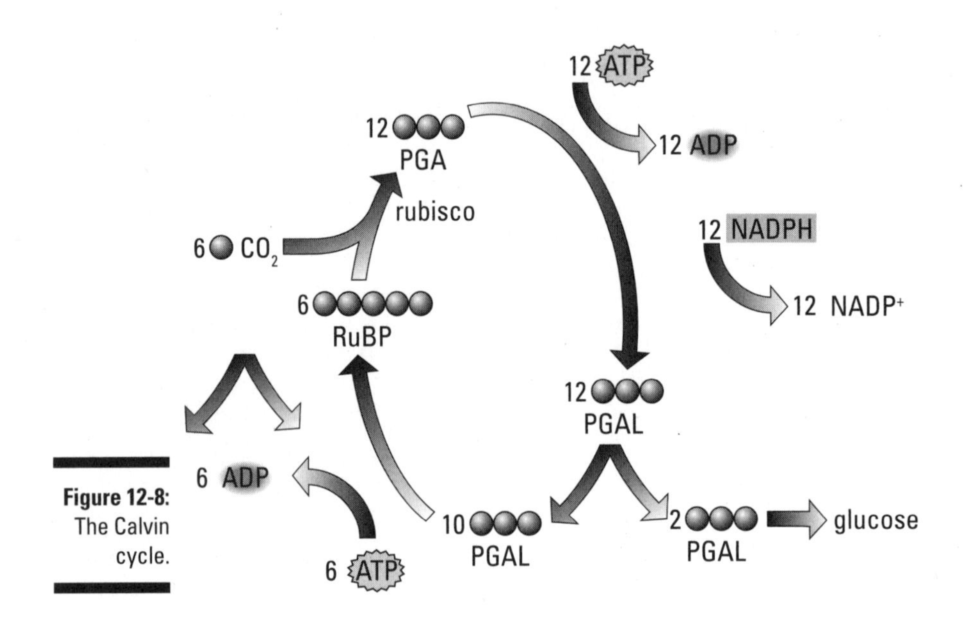

Figure 12-8: The Calvin cycle.

Plants use energy from the Sun, electrons from H_2O, and carbon from CO_2 to make sugars. The energy and electrons in these sugars will ultimately be the electrons that are captured in cellular respiration to make ATPs. Therefore, the energy from the Sun, through the processes of photosynthesis and respiration, will be used to run your cells.

Got Food? Photosynthesis in the Real World

The carbohydrates made by plants, algae, and bacteria serve as the food for all life on Earth, which is why you'll find one of these autotrophs at the bottom of every food chain. Food chains are a visual way to show how energy flows through living things. A simple food chain may include an *autotroph* that makes its own food, followed by a *heterotroph* that eats the autotroph. Heterotrophs, or other-feeders, are living things that get their matter and energy by eating others. For example, a cow eating grass would represent a simple food chain. The grass is the autotroph, and the cow is the heterotroph. If you ate a hamburger made from the cow, you could be added to the food chain as another heterotroph. This simple food chain would look like

Grass → Cow → You (human)

The arrows represent the flow of energy from the grass to the cow to you. The grass stored energy from the Sun in the form of carbohydrates. The cow got energy for its own metabolism and growth by eating the stored carbohydrates in the grass. You got energy for your metabolism and growth by eating the stored protein and fat from the cow. You and the cow made the chemical energy stored in the food available to your body by breaking down food molecules transferring energy to ATP. (For more on cellular respiration, see Chapter 11.)

In addition to getting energy from food, you also get the raw materials, or matter, you need to build your bodies. Your body can rearrange food molecules and use the atoms in those molecules to build new molecules for your cells. (Yes, you are what you eat!) This adage is true for every living thing on Earth. For example, as the cow digests the carbohydrates from the grass and grows, it uses the raw materials from the carbohydrates to build the protein and fat molecules that it needs for its body. Carbohydrates provide the carbon, hydrogen, and oxygen atoms that are also found in proteins and fats. The cow also uses nitrogen and phosphorous atoms that it got from minerals in its diet in order to construct the protein and fat molecules.

This simple food chain is an example of how photosynthesis plays a role in your life every day. Obviously, the interactions between living things get much more complicated than this example. If you look outside, you can see many kinds of plants that are all doing photosynthesis. You can imagine all the different kinds of living things that are eating these plants and then the next layer of living things that eat the first layer! These types of complicated interactions, called *food webs,* are part of the study of ecology and are not my focus here. However, I think you can appreciate the importance of photosynthesis to the web of life that exists all around you. Photosynthesis is the cellular machine that is capturing the matter and energy from the environment and turning it into food that can be used by the entire web. Without the matter and energy from this food, there would be no life on Earth.

Photosynthesis isn't the only food-building process on Earth. Some bacteria do chemosynthesis, transferring energy from the oxidation of inorganic chemicals to ATP. Then, these bacteria convert carbon dioxide to carbohydrates by using the Calvin cycle or other similar processes. All organisms that make food, whether by photosynthesis or chemosynthesis, are autotrophs. Chemosynthetic autotrophs live in many different environments, but they're especially important in environments without light, such as at the bottom of the ocean near hydrothermal vents.

Photosynthesis and global warming

Carbon dioxide is one of the most significant greenhouse gases because it traps heat that is reflecting off the Earth's surface, redirecting it down toward Earth and contributing to global warming. Because it removes carbon dioxide from the atmosphere and converts it to plant tissue, photosynthesis can lead to fewer greenhouse gases in the atmosphere and thus has the potential to lower surface temperatures.

Chapter 13

Splitsville: The Cell Cycle and Cell Division

In This Chapter

- Understanding why cells divide
- Multiplying bacteria
- Discovering the phases of the cell cycle
- Exploring the events of mitosis

Cells divide in order to grow, repair tissues, and reproduce themselves. Cell division in bacteria is a fairly simple process, whereas cell division in eukaryotes is much more complicated. In eukaryotes, cells progress through a cell cycle that includes dividing and nondividing phases. In this chapter, I present the steps of cell division in bacteria and eukaryotes as well as the mechanisms of regulation of the eukaryotic cell cycle.

Reproducing the Cell

Cells make copies of themselves in order to make new cells. In fact, an important theory in cell biology called the Cell Theory says that all cells come from pre-existing cells. Cells make new cells for three reasons:

- **Growth:** Many living things begin as a single cell made from the fusion of a sperm and egg. That single cell becomes thousands of cells when cells copy themselves over and over again. For example, you began life as a single cell that was invisible to the naked eye, but now you're much larger and quite visible. Most of that growth is the result of cell division.
- **Repair:** If a multicellular organism is wounded, cells surrounding the wound will reproduce themselves to repair the tissue. Next time you get a cut, watch the process as it heals. You'll see new pink skin growing around the cut. That new skin is formed from the division of the skin cells that surrounded the cut.

- **Reproduction:** Single-celled organisms, such as bacteria and yeast, increase from single cells to populations by making exact copies of their cells. Some multicellular organisms can also produce new individuals by copying themselves. For example, you can make a cutting of some plants and use it to make an entirely new plant that will be genetically identical to the original. The process of making new individuals by copying the cells of existing organisms is called *asexual reproduction.*

Drifting Apart: Binary Fission

Bacteria have a simple process of copying their cells called *binary fission* (see Figure 13-1). Binary fission involves the following steps:

1. **The bacterial cell makes a copy of its chromosome.**
2. **The bacterial cell gets larger as it makes copies of the ribosomes and molecules in the cytoplasm.**
3. **New plasma membrane and a cell wall are built to divide the cell into two.**

Some bacteria can complete the whole process of binary fission in as little as 10 minutes, reproducing from one cell to more than a thousand cells in just 2 hours. Think about that fact the next time you leave food out on the kitchen counter!

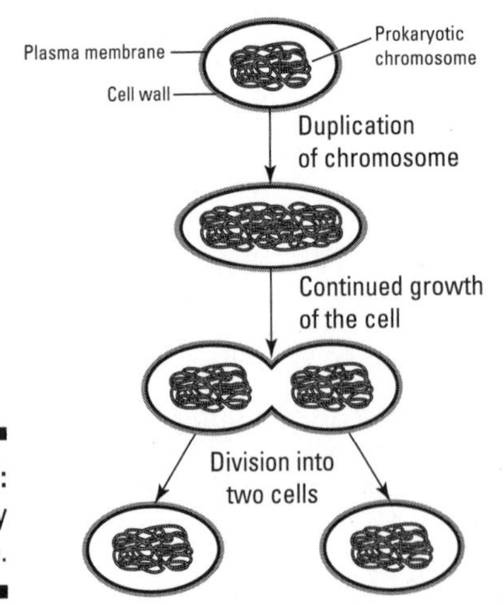

Figure 13-1: Binary fission.

Red Light, Green Light: The Cell Cycle

Eukaryotic cells divide at different rates. Single-celled eukaryotes typically divide rapidly when food is available. In multicellular organisms, some cells divide frequently, while others rarely divide. Different types of human cells, for example, have different division behaviors:

- **Some cells divide all the time.** Cells on surfaces like skin and mucous membranes are constantly being shed from the body and need to be replaced so that cells in these tissues are always dividing to replace the lost cells.
- **Some cells divide when signaled to divide.** Some cells in organs like the liver don't normally divide, but may be triggered to divide if the organ is damaged.
- **Some cells don't usually divide**. Most cells in the nervous tissue of humans don't divide. For example, if you have an injury that involves nerve damage in the spine, the nerves can't be repaired.

The dividing phase of eukaryotic cells is called *mitosis,* and the nondividing phase is called *interphase.* Interphase contains three subphases: G_1, S, and G_2. The alternating cycle of mitosis and interphase is called the *cell cycle* (see Figure 13-2).

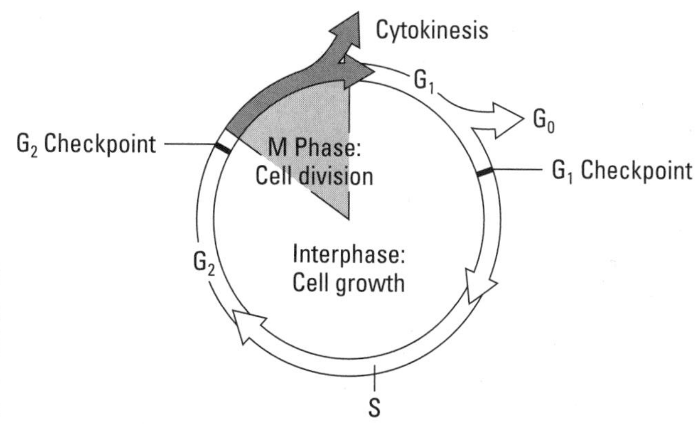

Figure 13-2: The cell cycle.

Pausing during Gap 1

Cells spend most of their time in the first phase of interphase, which is called Gap 1 (G_1). Cells in G_1 are active, functioning cells. They grow and copy all the cell contents except for the DNA. Mature cells, such as nerve cells, that won't divide remain at rest in this phase at a point called G_0 (pronounced "Gee Zero").

Cells that are going to divide must pass a test, called a *checkpoint,* before they can exit G1 and enter the next phase of interphase. Checkpoints are points in the cell cycle where cells check to make sure that everything is proceeding normally. If cells can't pass a checkpoint, repairs will be made, if possible. If not, the cell may be signaled to commit cell suicide, called *apoptosis.* In order for cells to pass the G1 checkpoint, several conditions must be met:

- Signals tell the cells to divide.
- Cells must have plenty of nutrients.
- The DNA must be in good condition.
- Cells must be large enough to divide.

The S phase and Gap 2

During S phase of interphase, cells copy their DNA by DNA replication. (See Chapter 17 for more on DNA.) In fact, S stands for synthesis because synthesis means "to make," and cells are making new DNA. After the DNA is copied, cells proceed to the next part of interphase.

During Gap 2, cells check the work they did during S phase at the G_2 checkpoint. Before cells can proceed out of G_2 and into mitosis, several conditions are checked:

- The DNA isn't damaged.
- The cell copied all the chromosomes.
- Signals tell the cell to proceed into mitosis.

Again, if cells can't meet these conditions, they will be stuck in G_2. If not, programmed cell death, called *apoptosis,* will be triggered, and the cell will commit suicide.

The importance of quality control

The protein p53 is the quality control agent for the G_2 checkpoint. If the DNA is damaged, p53 puts the brakes on the cell cycle and stops the cell in G_2 so proteins can either repair the DNA or tell the cell to commit cell suicide. If p53 isn't working properly, cells with damaged DNA may continue through the cell cycle, leading to the development of cancer. Because proteins like p53 help prevent cancer when they're functioning normally, they're called *tumor suppressors.*

The Dance of the Chromosomes: Mitosis

Cells that enter mitosis, shown in Figure 13-3, have successfully copied their DNA and the rest of the contents of their cell. They're ready to divide into two new cells, each of which has a complete copy of everything. Cells need to be particularly careful when they divide up the two copies of DNA so that each cell gets a complete set with one of every chromosome. The main purpose of mitosis is to make sure that the chromosomes are divided up correctly.

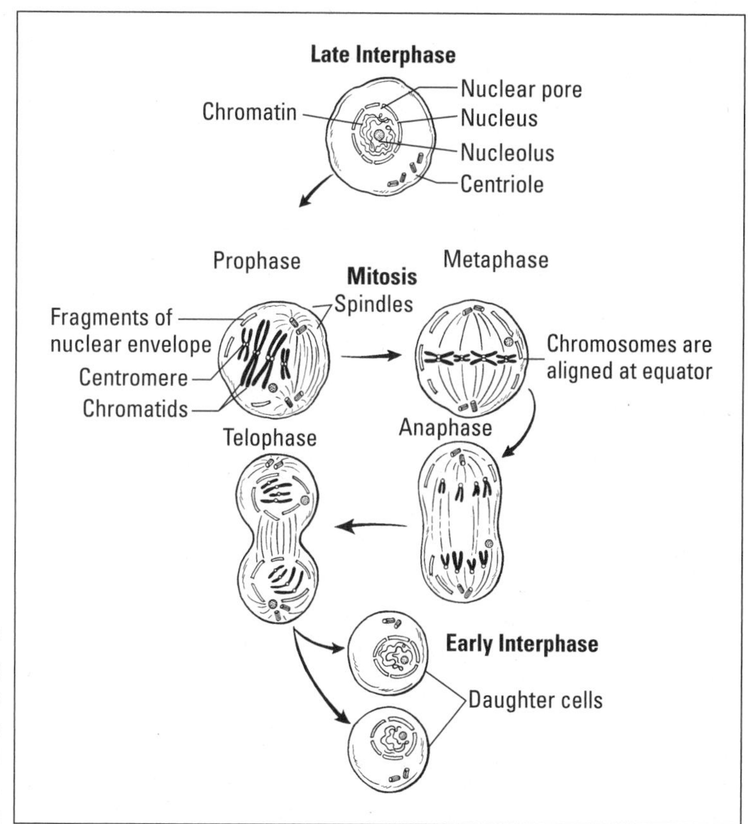

Figure 13-3: The process of mitosis and interphase.

As an analogy for thinking about how mitosis works, imagine you have 23 pairs of socks that are really special to you because they were hand-knitted for you by your mom. Your mom made the socks out of multicolored yarn so that each sock is unique. Even the two socks in a pair aren't perfectly identical to each other. These socks are so special to you that you decide that you want to make exact copies of the socks so that you can give an identical 46 socks to your child. To accomplish this task, you take each sock and use it as a pattern to knit an exact replica. Then you line up the 46 socks, with each sock having its identical twin right next to it. You carefully separate the identical socks so that the one complete set of 46 socks goes into one pile for you to keep, and one complete set of 46 socks goes into another pile for your child.

You have 46 chromosomes, or 23 pairs, in each cell of your body. When your chromosomes are copied during S phase, each chromosome is used to make an exact copy of itself. The two identical copies — *sister chromatids* — remain attached to each other, as shown in Figure 13-3, at the *centromere*, forming *replicated chromosomes* that often look a little bit like the letter X. During mitosis, the cell organizes the pairs of chromosomes and carefully separates the sister chromatids from each other so that each new cell gets a complete set of 46 chromosomes.

The cell structure that has the responsibility for separating the pairs of chromosomes is the *mitotic spindle*. The mitotic spindle is made of microtubules, which are cytoskeletal proteins. (For more on the cytoskeleton, see Chapter 2.) Microtubules work with the partner protein dynein to move structures in cells. Dynein and other proteins form protein clusters called *kinetochores* that attach to the replicated chromosomes at their centromeres. The proteins in the kinetochores walk along the microtubules of the mitotic spindle in order to move the chromosomes within the cell.

Mitosis is a carefully controlled process designed to organize and separate the chromosomes correctly. The four phases of mitosis have unique events that contribute to the careful sorting of the chromosomes:

- **Prophase:** During interphase, the DNA is uncoiled, and the chromosomes aren't visible, even under a microscope. To get ready to be moved around, the chromosomes coil up tightly. Coiling of the chromosomes, called *condensation*, happens during prophase. The important events of prophase are
    - The chromosomes coil up (condense) and become visible.
    - The nuclear membrane breaks down so that the mitotic spindle can reach into the center of the cell.
    - The mitotic spindle attaches to the chromosomes.
    - The nucleoli break down and become invisible.
- **Metaphase:** During metaphase, cells organize the chromosomes by lining them up in the middle of the mitotic spindle. At the *metaphase checkpoint*, cells make sure that all the chromosomes are attached to the mitotic spindle. Cells can't continue through mitosis unless they pass the metaphase checkpoint.

- **Anaphase:** During anaphase, the replicated chromosomes are separated so that each sister chromatid goes to opposite sides of the cell. The identical sister chromatids are separated carefully to ensure that each new cell will get one. After the sister chromatids are separated from each other, they're again called chromosomes.
- **Telophase:** Telophase wraps up mitosis by reversing the events of prophase. During telophase, the following events occur:
  - The chromosomes uncoil (decondense) and become invisible.
  - The nuclear membrane reforms.
  - The mitotic spindle breaks down.
  - The nucleoli reform and become visible.

Breaking Up Is Hard to Do: Cytokinesis

After the chromosomes separate and the nuclear membrane reforms, the cytoplasm separates to form two distinct cells. The separation of the cytoplasm is called *cytokinesis* (see Figure 13-4).

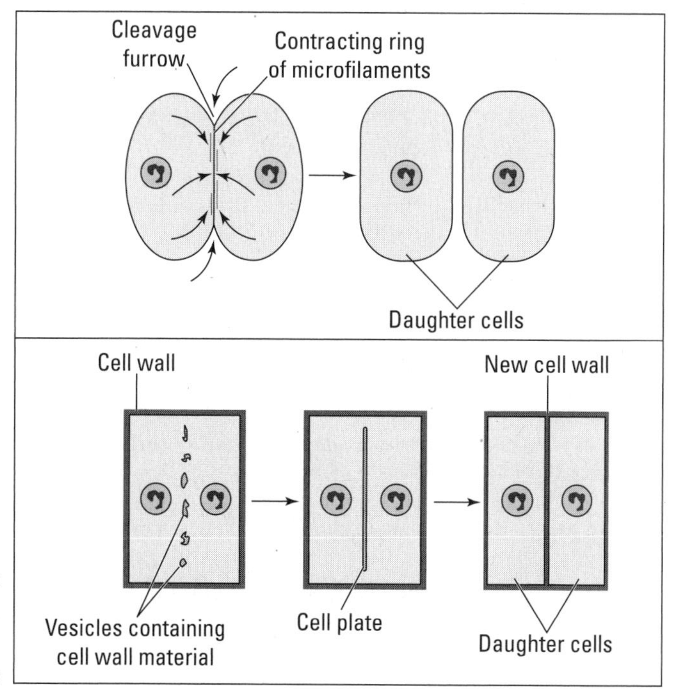

Figure 13-4: Cytokinesis.

Because plants have a cell wall and animals don't, cytokinesis occurs differently in the two types of cells:

- **Cytokinesis in plant cells:** Once the chromosomes move to opposite sides in a plant cell, new plasma membrane and cell wall are laid down in the middle of the cell. Small vesicles containing wall material line up in the middle of the cell. The vesicles then fuse together, bringing together the vesicle membranes and the wall material that was inside the vesicles. Finally, the membranes of the vesicles form the new plasma membranes for each cell. The wall material joins together to form the *cell plate*, which becomes the new cell wall between the two cells.
- **Cytokinesis in animal cells:** After the chromosomes move to opposites of an animal cell, the cell pinches in two to form two new animal cells. A band of actin microfilaments forms in the middle of the cell, and the actin microfilaments contract, forming a *cleavage furrow* down the middle of the cell. The actin microfilaments then act like a belt that is being tightened around the cell, pinching it into two new cells.

Keeping It Under Control

The progress of eukaryotic cells through the cell cycle is regulated by proteins called *cyclins*. The amounts of cyclin proteins increase and decrease during the cell cycle. Cyclins bind to partner proteins called *cyclin-dependent kinases* (Cdks) to form activated complexes that trigger events in the cell cycle. During each phase of the cell cycle, the amount of different cyclins increases and binds to their partner protein:

- In G_1, cyclins bind to a Cdks and form a complex that triggers cells to get the chromosomes ready for replication.
- In S phase, cyclins bind to Cdks and form a complex called S-phase promoting factor that enters the nucleus and helps trigger DNA replication.
- In G2, cyclins bind to Cdks to form M-phase promoting factor, which triggers the formation of the mitotic spindle, breakdown of the nuclear envelope, and condensation of the chromosomes. In metaphase of mitosis, M-phase promoting factor activates the process that triggers sister chromatids to separate.

Each type of cyclin is destroyed soon after it does its job in the cell cycle. When the cyclins from a particular phase of the cell cycle are destroyed, the phase can end, which helps prepare the way for the next phase to begin.

Part IV
Genetics: From One Generation to the Next

In this part . . .

Genetics tracks the inheritance of traits from one generation to the next. Living things that reproduce sexually sort their chromosomes and pass copies to their offspring. The combination of genetic information from Mom and Dad determines the traits of offspring and can be predicted using the tools of genetics.

Mistakes and surprises sometimes happen on the way to making new offspring as well — chromosomes may not sort correctly, or traits once hidden suddenly surface in a new generation. In this part, I explain how cells sort their chromosomes when they make eggs and sperm and present the basics of how traits are passed from parents to offspring.

Chapter 14

Meiosis: Getting Ready for Baby

In This Chapter

- Reproducing sexually
- Producing gametes by meiosis
- Making mistakes during meiosis

Organisms that reproduce sexually produce eggs and sperm that join together to make new individuals. To produce these sex cells, cells must undergo a special form of division, called meiosis, that cuts the genetic information of cells in half. In this chapter, I present the steps of meiosis. I also discuss its importance in generating genetic variability in species that practice sexual reproduction

Let's Talk About Sex, Baby: Reproduction

When living things reproduce sexually, each parent contributes a cell to make a new organism. Sperm and egg join together, combining their genetic information. If sperm and eggs were like any other cells, this combination would create a problem because each new generation would have twice the genetic information as the generation before. So, sperm and eggs need to be made in a special way, a unique type of cell division that cuts the genetic information of the cell in half. That way, when sperm and egg combine, the new organism has the right amount of genetic information. The special division that creates sperm and eggs is called *meiosis.*

Meiosis is a special type of cell division that occurs in gonads of sexually reproducing organisms.

Riding the life cycle

In sexually reproducing organisms, cells alternate between having the full amount of genetic material and having half the amount of genetic material. In humans, for example, body cells, called *somatic cells,* have twice the genetic material of cells that participate in sexual reproduction, called *gametes.* Body cells produce gametes by meiosis, and then gametes combine by *fertilization* to create somatic cells, creating a *life cycle* (see Figure 14-1).

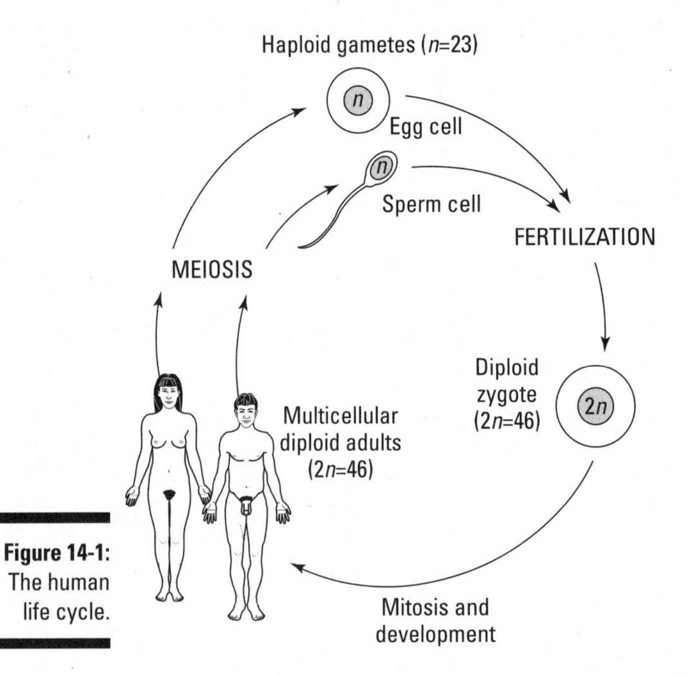

Figure 14-1: The human life cycle.

The life cycle for humans has several steps:

1. **Somatic cells undergo meiosis to produce gametes.**

   In humans, meiosis occurs in glands called *gonads*: *testes* in males and *ovaries* in females. Male gametes are called *sperm,* and female gametes are called *eggs*.

2. **Sperm and eggs join together in fertilization, creating a first cell called a *zygote.***

   The nuclei of the sperm cell joins with the nucleus of the egg cell, combining the chromosomes into the nucleus of the zygote.

3. **The zygote divides by mitosis to create a multicellular organism.**

   Development occurs as cells specialize to create different tissues and organs. (For more on mitosis, see Chapter 13.)

Counting chromosomes

Gametes have half the genetic material, or half the number of chromosomes, as somatic cells. One complete set of chromosomes is called the *haploid number* of chromosomes and is represented by the letter *N*. Gametes have one set of chromosomes, making them *haploid* (1N or N). Somatic cells usually have two sets of chromosomes, making them *diploid* (2N).

Human cells have 23 different kinds of chromosomes. Human gametes have 23 chromosomes because they have one complete set. Human somatic cells have 46 chromosomes because they have two complete sets ($2 \times 23=46$). The number of chromosome sets a cell has is called its *ploidy*.

Many organisms besides humans reproduce sexually. Animals, plants, and many other eukaryotes have a life cycle that is similar to humans. Each organism has a unique set of chromosomes. For example, pea plants have 7 different types of chromosomes. Thus, the haploid number of chromosomes in peas is 7 (N=7). Pea gametes are haploid (1N) and have 7 chromosomes, while pea somatic cells are diploid (2N) and have 14 chromosomes. In fruit flies, the haploid number of chromosomes is 4. Thus, fly gametes (1N) have 4 chromosomes, while fly somatic cells (2N) have 8 chromosomes. The number of chromosomes in living things varies widely and can be as few as 6 chromosomes in a diploid mosquito cell to thousands of chromosomes in some plants that are *polyploid,* which means they have many sets of chromosomes. One fern, *Ophioglossum*, has 1,200 chromosomes per cell!

Homologous Chromosomes

Cells can have matching pairs of chromosomes, called *homologous chromosomes.* For example, when sperm and egg join together to form a human zygote, each gamete contributes one set of 23 chromosomes. Each of the 23 chromosomes is unique and contains genes for certain traits. A zygote receives two of each kind of chromosome, one from each parent.

Homologous chromosomes are a pair of chromosomes that contain the same types of genes as each other. Diploid organisms, such as humans, get one of each pair of homologous chromosomes from mom and the other one from dad.

You can identify the differences among chromosomes by looking at chromosomes through a microscope. When cells are about to divide, the chromosomes condense and become tight little bundles that are visible in a light microscope. Several physical differences help you identify the different types of chromosomes:

- **Chromosome length:** Chromosomes can be many different lengths, from very small to fairly long.
- **Position of the centromere:** Chromosomes are most visible when cells are just about to divide, so the chromosomes have been replicated. Two sister chromatids attached at the centromere are visible. Centromeres can occur in the middle of a chromosome or toward either end.
- **Staining pattern:** Some stains will stick to chromosomes differently, creating striped patterns of light and dark that are unique to each chromosome. Also, antibody-linked fluorescent stains can mark each type of chromosome with a unique color.

Scientists identify chromosomes by the differences in the way they look. When scientists examine the chromosomes in a diploid cell, they see pairs of chromosomes that look alike. Each pair of chromosomes is a homologous pair. Scientists sort the chromosomes into pairs and lay them out in a chromosome map called a *karyotype,* shown in Figure 14-2.

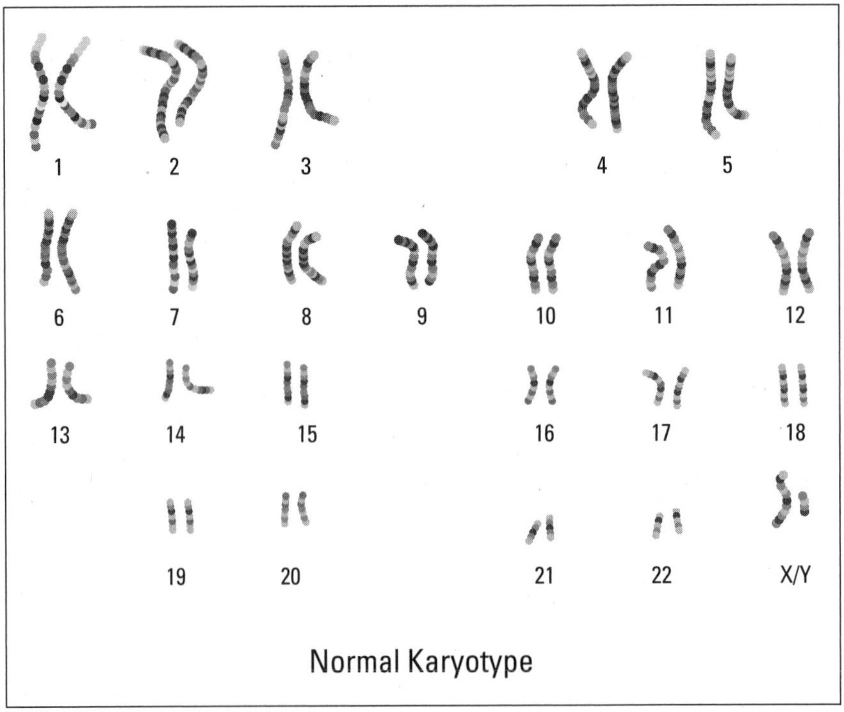

Figure 14-2: A human karyotype.

By examining the karyotype of a cell, you can tell whether the cell has the right number of chromosomes. A normal human cell will have 23 pairs of chromosomes. Twenty-two pairs are exactly the same in men and women and are called *autosomal chromosomes*. One pair, called the *sex chromosomes*, is different between men and women. Men have a larger chromosome, called the X chromosome, and a smaller chromosome, called the Y chromosome. Women have two X chromosomes.

Going Separate Ways: Meiosis

Meiosis is a special type of cell division that carefully separates homologous pairs of chromosomes so that gametes receive one of each type of chromosome. That way, when a man and a woman make a child, each parent donates 23 chromosomes to the new baby, making the normal number of 46 chromosomes in the baby's cells.

Each of the 23 chromosomes has slightly different types of information on it. So, the cells in the gonads that are making sperm and egg can't just pile up all 46 chromosomes and randomly divide them into two piles. If the process of splitting chromosomes was random, gametes could end up having extras of some chromosome types and missing others. A baby made from these gametes may get 46 chromosomes, but may not get the two of each kind needed to be a normal human. So, the ultimate goal of meiosis is to separate the homologous chromosomes carefully so that every gamete gets one complete set. In humans, that means a set that contains 1 of each of the 23 kinds.

Following the plan

Meiosis occurs as part of the cell cycle, just like mitosis. (For more on mitosis and the cell cycle, see Chapter 13.) In fact, meiosis has many similarities to mitosis:

- Cells that are going to undergo meiosis receive a signal that causes them to leave G_1 of interphase and enter S phase.
- During S phase, the cells copy all the chromosomes, creating replicated chromosomes that have identical sister chromatids attached at the centromere.
- The cells then proceed through G_2 and into meiosis.
- During meiosis, chromosomes are moved around by the spindle.
- The chromosomes are moved and sorted in very similar ways as they are in mitosis and the same terms, prophase — metaphase, anaphase, and telophase — describe these movements.

An overview of meiosis

The goal of meiosis is to cut the amount of DNA in half to make haploid gametes. However, before meiosis occurs, the DNA is doubled during S phase. Thus, in order to get the DNA down to the haploid amount, meiosis includes two divisions, shown in Figure 14-3B. The part of meiosis that leads up to the first division is called *meiosis I.* The part of meiosis that leads up to the second division is called *meiosis II.*

The details of meiosis may seem less overwhelming if you remember the purpose of each division:

- The purpose of meiosis I is to separate pairs of homologous chromosomes.
- The purpose of meiosis II is to separate sister chromatids.

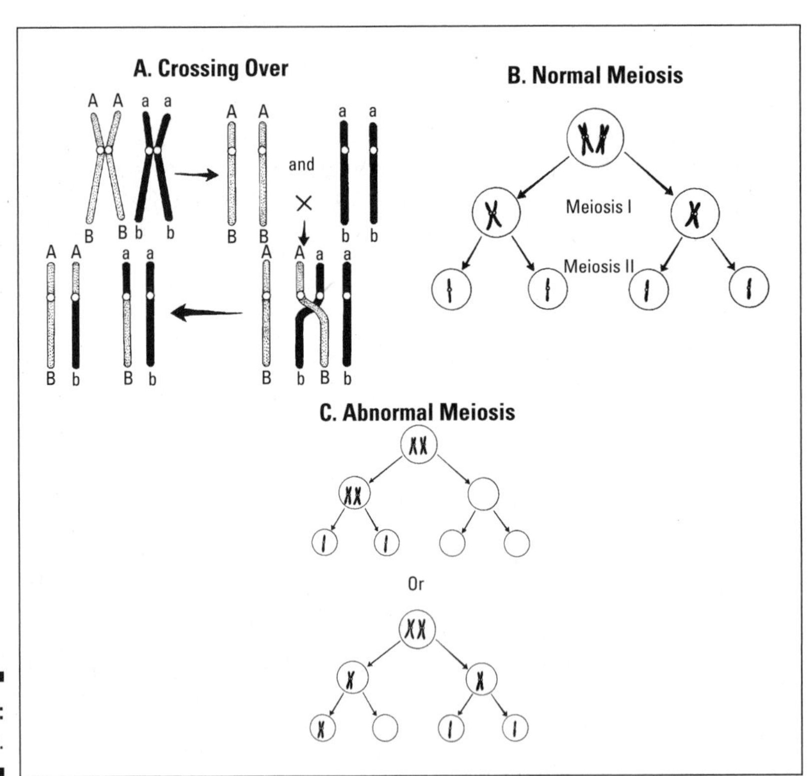

Figure 14-3: Meiosis.

The best way I know of to understand meiosis is to practice the phases using pipe cleaners or string for chromosomes. Create a cell with a diploid number of six or eight chromosomes and take them through the entire process beginning with G_1, going through S phase, and proceeding on through both divisions of meiosis. Try doing this walkthrough for mitosis as well and compare the results side by side. By comparing the same cell going through mitosis and meiosis, you'll soon recognize the phases based on the arrangement of chromosomes. Biology instructors have a nasty habit of showing you random pictures of cells undergoing cell division and asking you what phase you're looking at!

The events of meiosis I

Meiosis I separates each pair of homologous chromosomes. The phases of meiosis I have many similarities to the phases of mitosis and are called prophase I, metaphase I, anaphase I, and telophase I. The events of meiosis I are as follows:

- During **prophase I**, several events occur:
  - The nuclear membrane breaks down.
  - The nucleoli disappear.
  - Homologous chromosomes find each other and pair up. The two replicated chromosomes of each pair actually stick together, forming a structure called a *tetrad.* Tetrads have four arms because each replicated chromosome has two sister chromatids.
  - The chromosomes condense, or coil up, and become visible.
  - The spindle attaches to the chromosomes.
- During **metaphase I,** homologous pairs of chromosomes are lined up in the middle of the cell.
- During **anaphase I**, homologous chromosomes are separated from each other, and one from each pair goes to opposite sides of the cell.
- During **telophase I**, nuclear membranes form in some organisms, creating two nuclei. These nuclei are now haploid because they only have one of each type of chromosome. The spindle breaks down.
- **Cytokinesis** (see Chapter 13) occurs, resulting in the formation of two cells.

After meiosis I is complete, both cells proceed directly to meiosis II without going through the stages of interphase (for more information on interphase, see Chapter 13).

The events of meiosis II

Meiosis II separates the sister chromatids of each replicated chromosome and sends them to opposite sides of the cell. Once again, the cells pass through several phases similar to those of mitosis and meiosis I. During meiosis II, the phases are called prophase II, metaphase II, anaphase II, and telophase II. The events of meiosis II are as follows:

- During **prophase II**, a spindle forms in each cell and attaches to the chromosomes. If nuclear membranes formed during telophase I, they break down again.
- During **metaphase II**, the chromosomes are lined up in the middle of the cell.
- During **anaphase II**, sister chromatids are separated and move to opposite sides of the cell.
- During **telophase II**, several events occur:
    - Chromosomes uncoil.
    - The spindle breaks down.
    - Nuclear membranes reform.
    - Nucleoli reappear.
- **Cytokinesis** occurs in both cells, resulting in the formation of four cells from the original cell. These four cells develop into gametes. In females of some species, only one of the four cells will actually become an egg. The other three may break down or become a tissue that supports the egg.

Shuffling the Genetic Deck: Crossing Over

When homologous chromosomes are paired up during prophase I of meiosis, something interesting happens. Little bits of DNA are switched between the two homologous chromosomes in a process called *crossing-over* (refer to Figure 14-3A). Crossing-over actually exchanges pieces of the chromosomes that came from Mom with the ones that came from Dad, creating chromosomes that have bits of DNA from each parent.

The homologous chromosomes contain the same types of genes, so normal crossing-over doesn't add or take away any genetic information; it just changes the source of that information on a particular chromosome. Because each cell undergoing meiosis can have different patterns of crossing-over, crossing-over helps increase the genetic diversity of gametes.

Crossing-over involves several steps:

1. **Homologous chromosomes are attached all along their length.**

   Proteins called the *synaptonemal complex* act like glue to stick the homologous chromosomes together.

2. **Proteins make small cuts in the DNA backbone of the homologous chromosomes.**

3. **Proteins reseal the breaks in the DNA backbone, attaching pieces of one homologous chromosome to the other.**

As an analogy for crossing-over, imagine that you have one deck of cards that has blue backs and one deck of cards that has red backs. If you shuffle these decks together and then divide them into two complete decks again, ignoring the color of the backs, you'd end up with two complete decks containing cards with some red backs and some blue backs. Because the two decks of cards matched each other exactly card for card, your new decks are still complete with the right number of kings, queens, and so on; the cards just came from different places.

Why Two Divisions Are Better Than One

Organisms that reproduce asexually by mitosis make offspring that are exact copies of themselves, whereas organisms that reproduce sexually make offspring that are different than themselves. In the long run, the genetic variability that results from sexual reproduction makes it more likely that a species will survive. After all, if everyone is the same and life is good, then there is no problem.

But all that changes if disease strikes. If a disease sweeps through that is very good at killing the type of organisms that are present, and everyone is the same, everyone dies. On the other hand, if everyone is a little bit different and a disease sweeps through, some individuals might survive and be able to continue the species. Thus, the genetic variability generated by sexual reproduction provides an advantage to species.

Sexual reproduction increases genetic diversity in three ways:

- Crossing-over between homologous chromosomes in prophase I creates new patterns of genetic information in the chromosomes.
- Every cell that undergoes meiosis sorts its homologous chromosomes during anaphase I independently from all other cells in the gonads undergoing meiosis. Every gamete gets a complete set of chromosomes, but each gamete can have a slightly different assortment of which homologous chromosome it got from each pair.

- Sexual reproduction requires that gametes from one parent combine with gametes from another parent to produce offspring. Fertilization is random and creates many new combinations of genes in offspring.

It Was All a Mistake: Nondisjunction

Meiosis must sort the chromosomes very carefully to ensure that each gamete gets a complete set of chromosomes. Checkpoints in metaphase ensure that all chromosomes are attached to the spindle and lined up correctly.

Despite these quality control measures, meiosis doesn't always proceed correctly. Sometimes chromosomes that are supposed to separate and go to opposite sides of the cell fail to separate and travel together. Chromosomes that fail to separate correctly undergo *nondisjunction*. Nondisjunction leads to the creation of gametes that have the wrong number of chromosomes. (Refer to Figure 14-3C.)

Nondisjunction can occur in meiosis I. During anaphase I, the members of each pair of homologous chromosomes are supposed to separate and go to opposite sides of the cell. If this separation fails to happen, both members of a pair can travel to the same side of the cell. If meiosis II then proceeds normally, the result will be two gametes that have an extra copy of one of the chromosomes, and two gametes that are missing one of the chromosomes.

Living with Down Syndrome

Down Syndrome, or Trisomy 21, is the most common chromosomal defect among humans, occurring in 1 of every 733 births. People born with Down Syndrome experience a range of physiological and mental effects. The most serious effects are increased risk for heart disease, Alzheimer's disease, and childhood leukemia. Other effects include greater risk for respiratory and thyroid conditions and mild to moderate mental impairment. Medical treatment of the physical conditions associated with Down Syndrome has improved greatly in recent years, increasing the expected life span of people with Down Syndrome to almost normal levels.

The risk of having a child with Down Syndrome increases with the age of the mother, beginning to rise quickly at about age 35. However, because younger women have more babies than older women, the total number of babies affected with Down Syndrome is actually greater for babies born to women under 35. Scientists are working to identify which genes cause Down Syndrome and hope to someday be able to block the effect of having extra copies of those genes so that people with Trisomy 21 will be physiologically and mentally normal.

Nondisjunction can also occur in meiosis II. During prophase II of meiosis, sister chromatids are supposed to separate and go to opposite sides of the cell. If this separation fails to happen correctly, both sister chromatids can travel to the same side of the cell. Nondisjunction in meiosis II results in two normal gametes, one gamete that has an extra copy of one chromosome and one gamete that is missing a copy of one chromosome.

When a gamete with an abnormal number of chromosomes undergoes fertilization, the result is an *aneuploid* individual that has the wrong number of chromosomes in all of their cells.

In humans, many types of aneuploidy don't result in living offspring. However, some aneuploid individuals are born and live with some physiological abnormalities. One of the most common aneuploid conditions in humans is *Trisomy 21,* where a person has three copies of chromosome 21. Trisomy 21 results from fertilization between a gamete that has an extra copy of chromosome 21 and a normal gamete. A range of physiological and mental defects result from Trisomy 21. This syndrome of effects is commonly known as *Down Syndrome*.

Chapter 15

Mendelian Genetics: Talkin' 'Bout the Generations

In This Chapter

- Discovering Mendel's Laws of Inheritance
- Tracking alleles with Punnett squares
- Linking inheritance with meiosis

Genetics is the study of how parents pass traits onto their offspring. The modern understanding of inheritance began with the work of an Augustinian monk named Gregor Mendel who studied the inheritance of traits in peas. By carefully tracking the inheritance of traits in peas, Mendel established fundamental laws of inheritance that are still used today. In this chapter, I present Mendel's Laws of Inheritance along with some tools that can help you analyze the inheritance of traits.

Pass the Peas, Please: Mendel and Segregation of Single Gene Traits

People have probably always had some kind of awareness that parents pass traits to their children. After all, family resemblances to grandparents, aunts, and uncles were common even in the ancient past. Farmers have known for a long time that traits are inherited in animals, and even crop plants, and planned their breeding programs to capture the traits they were interested in.

However, for most of history people didn't really understand why traits were passed from parents to offspring, or why some traits seemed to appear more often than others. The first person who actually did a statistical analysis on the inheritance of traits was an Austrian monk named Gregor Mendel.

Mendel's work laid the foundation for the current understanding of inheritance as is represented as three main ideas called *Mendel's Laws:*

- **The Law of Dominance** basically says that one form of a trait can hide another form of a trait. In other words, organisms have two genetic messages for every trait, but only one message may be visible in the organism. For example, if a brown-eyed person has children with a blue-eyed person, their children may have brown eyes even though they inherited messages for blue eyes from one of their parents. The effect of the messages for brown eyes hides the fact that messages for blue eyes are even there. Mendel said that the form of the trait that is visible is *dominant* to the hidden form of the trait, which is *recessive.*
- **The Law of Segregation** basically says that when adults make *gametes* (eggs and sperm), they give only one of their two messages for each trait to each gamete. In other words, the two messages for each trait in adults separate from each other, or *segregate,* when gametes are made.

  Gametes get together with other gametes to make a new organism, so it makes sense for gametes to get only one copy of each genetic message. The other copy of each genetic message will come from the other gamete.
- **The Law of Independent Assortment** basically says that each pair of genetic messages sorts itself independently from other pairs of genetic messages. In other words, when pairs of genetic messages are getting split up and sent to gametes, each pair is doing this process on its own. When you think about the fact that humans have thousands of different genes that are sorting themselves randomly from each other every time a gamete is made, you can imagine that a single person could make lots of gametes with different combinations of genetic messages. You can see this variation in families with more than child — even though the children were made by the same parents, they don't all look alike. The gametes that made each child were a little bit different from the gametes that made the child's sisters and brothers.

Test anxiety

Gregor Mendel is famous as the father of genetics. His careful mathematical analysis of his plant breeding experiments revealed patterns of inheritance that we still study today. It's hard to believe that Mendel may never have done his groundbreaking work if he hadn't flunked out of becoming a university professor! Perhaps because he lacked a formal university education, Mendel didn't pass his first university exam. He took more courses and was preparing to take the exam for a second time when he became ill. He withdrew his application and continued his plant breeding experiments instead!

Living like a monk

Gregor Mendel was an Augustinian monk with interests in science, nature, and math. He became very interested in the heredity of traits and spent seven years carefully observing, counting, and analyzing the results of breeding experiments in pea plants. Pea plants have many different traits that can be studied, including flower color, pea shape, pea color, pod shape, and plant height. Also, pea flowers contain both male and female parts, and they can make offspring when they're mated to themselves; in other words, they're *self-fertile*. Thus, Mendel was able to *outcross* pea plants by mating pea plants to other pea plants and also *self-cross* pea plants by mating them to themselves. Although studies on inheritance were done before Mendel's work, his work was especially important for several reasons:

- **Mendel developed lines of pure-breeding plants**. He mated plants with a trait he wanted to study with themselves until he had plants that always gave the same version of the trait. For example, he self-crossed purple-flowered pea plants and removed any white offspring that were produced. Then he took the purple-flowered offspring and self-crossed them again. He kept self-crossing the purple-flowered offspring until he always got purple-flowered pea plants from a self-cross of a purple plant. Then, Mendel did the same thing with white-flowered plants — self-crossed them until he had pure-breeding white-flowered plants. By creating pure-breeding plants, Mendel knew he was starting any experiment with a plant that had only one variation in the genetic message he was interested in.
- **Mendel kept detailed numerical records and analyzed his experimental crosses using statistics.** By applying math to analyze his crosses, Mendel was able to discover patterns of inheritance that led to his understanding of how inheritance works.

Mendel worked on inheritance in the 1850s, long before anyone knew anything about meiosis (see Chapter 14) or DNA (see Chapter 7). It's really amazing he figured out that traits separate during gamete formation without knowing anything about the physical basis for inheritance! In fact, he was so far ahead of his time that when he presented his results, people just didn't get it. His work was ignored by the scientific community until it was rediscovered almost 50 years later and people were ready to understand.

Speaking the lingo

Mendel's work on inheritance occurred before scientists knew about DNA or meiosis and around the same time that chromosomes were first being discovered. When Mendel described his conclusions, he talked about factors that control the inheritance of traits. Although his ideas are still used today, they're usually presented with the modern language and terms that have developed since his work. In order to understand the science of heredity, called *genetics,* you need to understand the language:

- A message for a particular trait is called a *gene.* The different variations of the message are called *alleles.* For example, pea plants have a gene for flower color. One allele is for purple flowers, while another is for white flowers.
- The total of all the genetic messages an organism has is called its *genotype.* The way the organism appears and functions because of its genes is called its *phenotype.* The genotype is like the total blueprint for the organism; the phenotype is what is built from that genotype.
- When an organism has two identical alleles for a gene, the organism is *homozygous* for that gene. When the alleles for a gene are different, the organism is *heterozygous* for that gene. Homo means same, so homozygous is having the same two messages. Hetero means different, so heterozygous is having two different messages. Another word that is commonly used to describe organisms with two different alleles is the term *hybrid.*

Round pea meets wrinkled pea

It's easiest to understand heredity when you follow just one trait at a time. For example, Mendel did a cross between a pure-breeding plant that produced round peas and a pure-breeding plant that produced wrinkled peas. The offspring of this mating all produced round peas. However, when the round pea offspring were mated among themselves, both round and wrinkled pea plants were produced. Because the wrinkled trait appeared again in the offspring of the first generation, Mendel knew that the wrinkled pea trait must be hidden by the first generation plants even though he couldn't see it.

Mendel's experiment makes sense if you think about it in the following way:

- **The parental plants are pure-breeding, so they have only one kind of allele for the trait.** Round pea plants can give messages only for round peas to their offspring, and wrinkled pea plants can give messages for only wrinkled peas to their offspring.
- **Each pea plant has two alleles for each trait.** Plants can have two copies of the round-shaped allele, two copies of the wrinkled-shape allele, or one copy each of round and wrinkled.
- **Each pea plant gives only one allele to its offspring.** When pea plants make gametes, they separate the two alleles for each trait and give one allele for each trait to each gamete.

- **The first generation plants get an allele for round peas from one parent and an allele for wrinkled peas from the other parent.** Thus, the first generation plants have two different messages for pea shape. Because their two alleles for this trait are different, they're heterozygous for this trait.
- **The allele for round peas is dominant to the allele for wrinkled peas.** Although the first generation plants had two different messages, you see only the effect of the round pea message. Thus, the round pea message is hiding the wrinkled pea message.
- **When the first generation plants reproduce, they can give either their allele for round peas or their allele for wrinkled peas to their offspring.** Thus, some plants in the second generation had wrinkled peas even though all the first generation plants had round peas.

When you're analyzing the results of a genetic cross, the different generations can get a little confusing. To keep track of the generations in a cross like this, geneticists use a specific notation:

- **The parental generation is identified as P1.** P_1 plants are always purebreeding for the trait that is being studied.
- **The first generation that results from a cross between two parentals (P1) is called the F1 generation.** F stands for filial, which refers to children, so the F_1 generation is the *first filial generation* or the first generation of offspring from the parental cross.
- **The second generation that results from a cross between two members of the F1 generation is called the F2 generation.** F_2 stands for *second filial generation.*

The odds are 3:1

When Mendel did his plant breeding experiments and analyzed the results, he looked at the numbers of different types of offspring very carefully. The patterns he saw in the numbers led him to his conclusions about inheritance. For example, when he crossed parental round pea plants with parental wrinkled pea plants, all the F_1 generation had round peas. When he self-crossed the F_1, he got 5,474 round pea plants and 1,850 wrinkled pea plants in the F_2 generation. In other words, there was a ratio of 5,474 round pea plants to 1,850 wrinkled pea plants. If you divide both numbers by 1,850, you get a reduced ration of 2.96:1. In other words, for every wrinkled pea plant in the F_2, there were about 3 round pea plants.

To reduce a ratio, divide both sides by the smaller number. For example, a ratio of 75:25 reduces to a ratio of 3:1.

The 3:1 ratio of round to wrinkled pea plants is exactly what you'd expect if Mendel's laws of inheritance are correct. To understand why this ratio occurs, you can use a grid called a *Punnett square,* shown in Figure 15-1, to track the alleles through this cross:

- On the edges of a Punnett square, you write symbols for the alleles in the gametes that are combining in the cross.
- By filling in the square, you create all the possible combinations of gametes that may occur in the next generation.
- In plant genetics, letters are used to represent the different alleles. Usually, the letter used is the first letter of the trait. The same letter is used for both alleles of the trait. For example, in Figure 15-1, the letter R is used to represent pea shape because one possible shape is round, which starts with the letter R.
- Capital letters are used to represent dominant alleles; lowercase letters represent recessive alleles. For example in Figure 15-1, R is used for round and r is used for wrinkled.

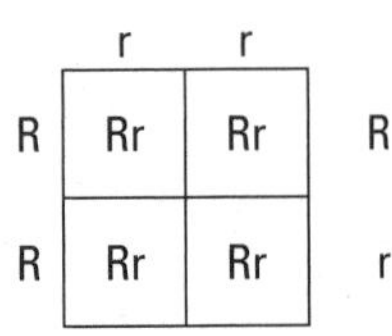

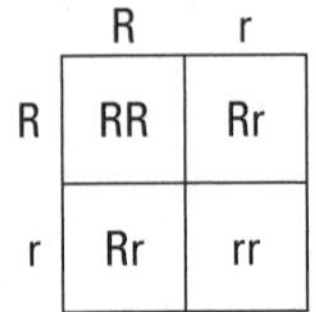

Figure 15-1: Punnett squares showing single gene crosses.

Making a prediction

Punnett squares make predicting the outcome of a particular genetic experiment easier. For example, the two Punnett squares in Figure 15-1 predict the outcome of Mendel's experiment that crosses pure-breeding round pea plants with pure-breeding wrinkled pea plants and then self-crosses the F_1 generation. The Punnett square on the left predicts the cross between the parental generation:

- The round pea parental is pure-breeding for the round phenotype, so it can produce only gametes that contain the allele for round peas. These gametes are represented by the capital R's that are written down the left side of the Punnett square.

- The wrinkled pea parental is pure-breeding for the wrinkled phenotype, so it can produce only gametes that contain the allele for wrinkled peas. These gametes are represented by the lowercase r's that are written across the top of the Punnett square.
- In this case, you have only one possible combination of the two types of gametes — all the F_1 offspring have the genotype Rr. This generation is shown in the smaller squares inside of the Punnett square.

The allele for round peas is dominant to the allele for wrinkled peas, so this Punnett square leads to the prediction that all F_1 plants will be heterozygous and will all have the round pea phenotype. Mendel's actual experiment with round and wrinkled pea plants exactly matches this prediction — all his F_1 plants produced round peas!

A Punnett square also makes predicting the outcome of the self-cross of the F_1 generation easier. Because the F_1 generation are all heterozygous, or hybrid, for one trait, this type of cross is called a *monohybrid cross.* Each member of the F_1 generation is heterozygous, so they can make two different types of gametes — gametes that contain the dominant allele and gametes that contain the recessive allele. When one F_1 plant is crossed with another F_1 plant, the meeting of their gametes is random. The Punnett square on the right in Figure 15-1 models random mating by creating all possible combinations of gametes:

- The possible gametes produced by two individuals from the F_1 generation are shown along the side and top of the Punnett square.
- The predicted F_2 generation is shown in the small squares inside the Punnett square.
- Some F_2 offspring are expected to be homozygous for the dominant allele (RR), some are expected to be heterozygous (Rr), and some are expected to be homozygous for the recessive allele (rr). The predicted *genotypic ratio* in the F_2 generation is 1:2:1 for RR:Rr:rr.
- Because the round pea allele is dominant to the wrinkled pea allele, any individual in the F_2 who is homozygous dominant (RR) or heterozygous (Rr) will have the same phenotype and make round peas. When phenotypes are compared, the predicted *phenotypic ratio* in the F_2 generation is 3:1 for round peas:wrinkled peas.

The Punnett square for the monohybrid cross between individuals in the F1 generation predicts a 3:1 ratio of round pea plants to wrinkled pea plants. In other words, for every one plant with wrinkled peas, you predict that there will be three plants with round peas. This estimate is extremely close to the 2.96:1 ratio of round to wrinkled peas that Mendel saw in his experiments, confirming Mendel's laws, which the Punnett square models.

Testing an idea

Mendel's work with peas led to his ideas about inheritance, known as *Mendel's laws* because they've been supported by many experiments over time. Mendel himself, of course, didn't know that his ideas were correct, and he wanted to test them for himself. He tested them by taking a closer look at the F_2 generation from his monohybrid crosses. He predicted that the F_2 had two types of round pea plants — some that were homozygous for the dominant allele (RR) and twice as many that were heterozygous (Rr). This prediction is modeled in the Punnett square to the right in Figure 15-1.

Mendel tested his prediction about the round F_2 pea plants by crossing them with parental pea plants that had the recessive trait. The parental pea plants were pure-breeding, homozygous recessive (rr).

A cross between an unknown individual and a pure-breeding recessive is called a *test-cross*.

Test-crosses allow you to identify the genotype of the unknown individual because the recessive parental can donate only recessive alleles to the test-cross offspring. Mendel made the following predictions about the outcome of his test-crosses:

- **Some of the test-crosses would yield offspring that were all round.** This result would represent the crosses between homozygous dominant plants (RR) with the test-cross parent (rr). Because the homozygous F_2 plant could donate only dominant alleles to the next generation, all offspring would show the dominant trait.
- **Twice as many of the test-crosses would yield offspring that were 50:50 round to wrinkled.** This result would represent the crosses between the heterozygous F_2 plants (Rr) and the test-cross parent (rr). Because the heterozygous plants could make equal amounts of two types of gametes (half with the dominant allele and half with the recessive allele), half the offspring of the test-cross would be heterozygous (Rr) and half would be homozygous recessive (Rr).

Well, you can probably guess that since Mendel is considered the father of genetics and his ideas are now called laws, his predictions proved correct. The ratios of types of offspring that he got in his test-cross experiments were very close to his predictions, again supporting his ideas.

Remembering meiosis

Mendel's ideas about inheritance were supported by his own work and by the work of many scientists since Mendel's time. Scientists today, of course, know a lot more about the details of how inheritance works. Since Mendel's time, they've discovered that the factors of inheritance that Mendel proposed are genes located on chromosomes made of DNA. Mendel proposed that parents have two copies of each factor (gene) and that these factors segregate from each other when parents make offspring. Scientists now know exactly when that segregation occurs — during anaphase I of meiosis (Chapter 14). During anaphase I, homologous chromosomes separate from each other and go to opposite sides of the cell, becoming destined to end up in different gametes. Those homologous chromosomes each carry the same set of genes.

The two alleles for each gene carried by the parent segregate from each other when the two homologous chromosomes separate from each other during anaphase I.

Only one copy of each gene ends up in the gametes that result from meiosis. If Mendel were alive today, I bet he'd be really excited to see how his factors segregate!

Playing by the rules

Gametes combine randomly during sexual reproduction. In other words, certain sperm don't seek out certain eggs — it's just a sort of reproductive free-for-all to see who gets lucky. Because fertilization is random, inheritance follows the rules of probability.

You can use the rules of probability to predict the outcome of a cross in the same way that you use a Punnett square to make predictions. Here are the rules of probability that apply to problems of inheritance:

- **The rule of multiplication says that if two events are independent of each other, then the probability that they will both occur at the same time is the product of their individual probabilities.** You can use the rule of multiplication to predict the outcome of a cross, such as a monohybrid cross (Rr × Rr). For example, you can answer the question "What is the chance that a homozygous recessive offspring (rr) will be produced from a monohybrid cross?" by completing the following steps:

   1. **Determine the probability of each gamete type that the parents can produce.** The parents are heterozygous (Rr), so half their gametes will receive the recessive allele and half the gametes will receive the dominant allele. The probability that each parent would donate each type of allele is one-half (½).

   2. **Figure out how many ways this cross could produce the offspring in the question.** In order for a homozygous recessive offspring to be produced, each parent would have to donate a recessive allele. So, these parents could produce a homozygous recessive offspring in only one way.

3. **Multiply the independent probabilities together.** The probability that one parent would donate the recessive allele is independent of the probability that the other parent would donate the recessive allele. (After all, they're both making gametes independently in their own bodies.) So, the total probability of creating a homozygous recessive offspring (rr) in this cross is the probability that one parent would donate the recessive allele (½) times the probability that the other parent would donate the recessive allele (½), which equals ¼. The rules of probability predict that one out of four offspring from this cross would be homozygous recessive. If you compare this result to the Punnett square on the right side of Figure 15-1, you can see that the two methods make the same prediction!

- **The rule of addition says that if an event can occur in more than one way, then the probability that it will occur is the sum of the independent probabilities for each way that it can occur.** Again, you can use the rule of addition to predict the outcome of a cross, such as a monohybrid cross (Rr × Rr). For example, you can answer the question "What is the chance that a heterozygous will be produced from a monohybrid cross?" by following these steps:

  1. **Determine the probability of each gamete type the parents can produce.** As in the preceding example, the probability that each parent will donate each type of allele is one-half (½).

  2. **Figure out how many ways this cross could produce the offspring in the question.** In this case, the parents can produce a heterozygous offspring (Rr) in two different ways: If Mom gives a dominant allele and Dad gives a recessive allele, or if Mom gives a recessive allele and Dad gives a dominant allele.

  3. **Calculate the probabilities for the different ways the parents can produce this offspring.** The probability that Mom gives a dominant allele is one-half (½) and that dad gives a recessive allele is one-half (½). These events are independent, so you multiply: ½ × ½ = ¼. The probability the heterozygote is produced by Mom giving the recessive allele and dad giving the dominant allele is the same: ½ × ½ = ¼.

  4. **Add the independent probabilities to calculate the total probability of the event: ¼ + ¼ = 2/4 = ½.** So, the rules of probability predict that half the offspring from a monohybrid cross will be heterozygous. Again, if you compare this result with the Punnett square on the right in Figure 15-1, you can see that both methods give the same prediction.

- **The rule of numbers says that, for random events, the larger the sample size, the more likely that actual events will correspond to predictions.** You can easily see an example of this rule in real life. You can predict that whenever you flip a coin, you have a 50:50 chance that it will come up heads. However, if you flip a coin just two times, you may get heads both times. That's 100 percent heads — not at all close to your prediction.

However, if you flip the coin ten times, you might get six heads and four tails. This 60:40 split is much closer to your prediction. Flip the coin 100 times or 1,000 times and you may get very close to 50:50. The same rule applies in genetics. You have a 50:50 chance that a human baby will be a boy or a girl, but you probably know families that have two or even three boys and no girls. To get samples that come close to their predictions, geneticists need organisms that produce lots of offspring, or they need to look at the same cross many times in order to increase their numbers.

Tracing a trait: Pedigrees

Humans don't make especially good subjects for genetic studies because they don't produce very many offspring (compared to, say, a fruit fly) and they don't generally mate on command. However, humans are very interested in their own traits, especially genetic diseases. So, geneticists study human traits by examining the pedigrees of families that display a certain trait.

For example, hemophilia is a disease that affected the descendants of Queen Victoria of England, and geneticists have used that pedigree (and many others) to study the inheritance of hemophilia. Geneticists use certain symbols when drawing human pedigrees (see Figure 15-2A):

- Males are shown as squares, females as circles.
- A mating is shown by a line drawn between a female and a male.
- The children of a mating are drawn below the parents in birth order, with a line connecting the children to the line of the mating that produced them.
- The symbol for a person who has the trait that is being studied is shaded. For example, if you look at a pedigree for hemophilia in the royal families of Europe, anyone who has hemophilia will be shaded.

Geneticists study the pattern of inheritance in pedigrees in order to figure out how a particular trait is inherited. Certain key patterns are important to look for when trying to interpret a pedigree:

- **Traits that are passed by recessive alleles can skip generations.** For example, look at the pedigree in Figure 15-2 C. The first few generations don't show the trait, but the last generation does. Children get their genes from their parents, so the parents in generation V must have the alleles for the trait, but they don't show it. Thus, the alleles for the trait are being hidden in this generation, indicating that the alleles for the trait are recessive to other alleles.
- **Traits that are passed by dominant alleles show up in every generation.** For example, look at the pedigree in Figure 15-2B. Other alleles can't hide a trait that is carried by a dominant allele. Thus, if a child has the trait, at least one of the parents of the child must show the trait as well.

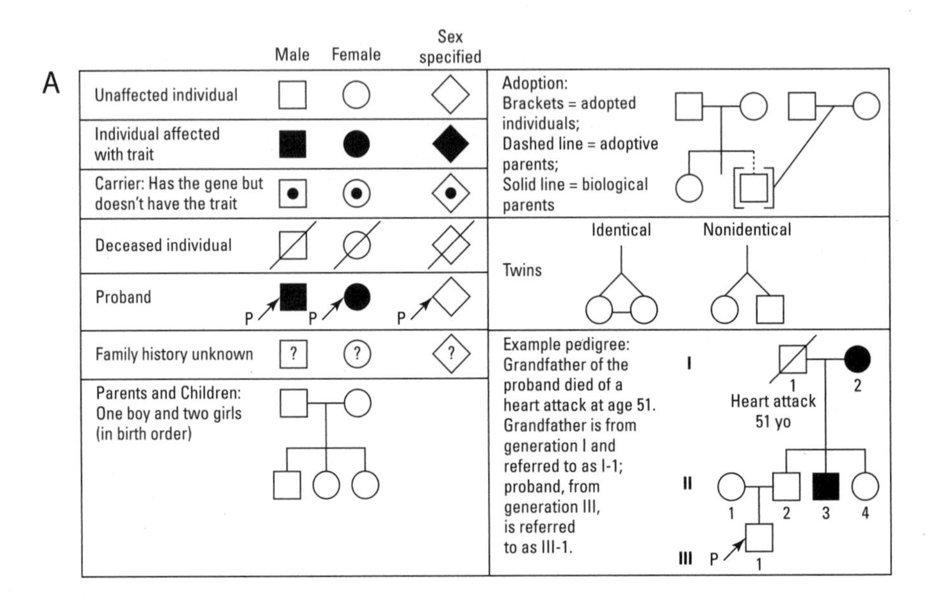

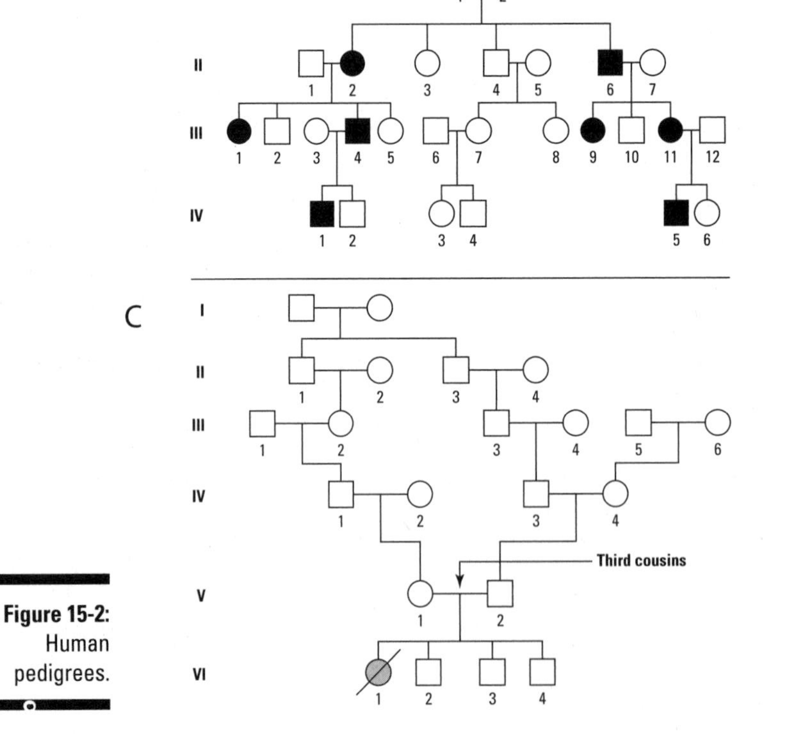

Figure 15-2: Human pedigrees.

Geneticists look at the pedigrees of many families in order to try to determine the inheritance of a particular trait. The information in one pedigree may not be enough to absolutely determine the inheritance pattern, but another one may hold the key. To try to determine which types of inheritance are possible based on a pedigree, you can propose a particular type of inheritance and then try to assign everyone in the family a genotype. For example, if you didn't know what type of inheritance was represented by the pedigree in Figure 15-2C, you could try to figure it out:

1. **Decide which type of inheritance you want to test.**

   For example, you could propose that you want to test whether the pedigree would fit if the trait was recessive.

2. **Assign genotypes for the people that are certain.**

   For example, if the trait is recessive, then anyone who has the trait must be homozygous recessive for the trait (for example, aa). In the pedigree in Figure 15-2C, the first child in generation VI with the shaded circle would have to have the homozygous recessive genotype (aa).

3. **Working from the people whose genotypes you know, try to assign genotypes to the rest of the family.**

   For example, the parents of the affected child in Figure 15-2C must both be heterozygous (Aa). They didn't have the trait, but they must have had the allele to pass it on to their children that do have the trait.

4. **Continue working on the pedigree, trying to assign genotypes.**

   Sometimes, you can't be sure whether a person's genotype is homozygous dominant or heterozygous (AA or Aa). In this case, you can record their genotype as A– to indicate the uncertainty of the second allele. If you ever find a spot where your proposed type of inheritance can't possibly work in this family — for example, homozygous dominant offspring (AA) from homozygous recessive parents (aa) — then you know that type of inheritance is impossible, and you can cross it off your list. Some families won't have enough information for you to figure out the inheritance. In that case, you'll have to leave more than one possibility open unless you have other pedigrees with more information.

If a trait skips a generation, it must be carried on a recessive allele. However, if a trait shows up in every generation, it may not be carried on a dominant allele. (It may be a recessive trait, but everyone in the family has it.)

I Can Go My Own Way: Independent Assortment

Mendel's formed his Law of Independent Assortment based on his observations of the inheritance of two traits at the same time. He saw that each trait followed the law of segregation independently of the other, which led him to conclude that the inheritance of one gene doesn't affect the inheritance of the other.

You can understand the inheritance of two traits at the same time by using the same tools you use to track a single gene, such as Punnett squares and the rules of probability; it just gets a little more complicated to follow. (For more on Punnett squares, see the section "Making a prediction," earlier in this chapter. For more on the rules of probability, see the section "Playing by the rules.")

Round yellow pea meets wrinkled green pea

To follow the inheritance of two traits, Mendel selected pure-breeding plants that always produced round yellow peas and pure-breeding plants that always produced wrinkled green peas. He crossed these parentals and then self-crossed the F_1 generation. Here's what Mendel discovered:

- The cross between the pure-breeding round yellow peas and pure-breeding wrinkled green peas produced all round yellow peas in the F_1 generation.

  a. The parental plants have two copies of each gene but only give one to their gametes. The pure-breeding parentals were homozygous for both traits, so they each could contribute only one type of message for pea color and one type of message for pea shape to the offspring. The round yellow parents gave messages for round yellow peas, and the wrinkled green parents gave messages for wrinkled green peas.

  b. The round allele is dominant to the wrinkled allele and the yellow allele is dominant to the green allele. Because the F_1 were all round and yellow, the allele for round peas must be dominant to the allele for wrinkled peas, and the allele for yellow peas must be dominant to the allele for green peas.

  c. The plants of the F_1 generation are heterozygous for both traits. You can write the genotype for the round yellow parents as RRYY and the genotype for the wrinkled green parents as rryy. When the parent plants make gametes, they send one copy of the gene for pea shape and one copy of the gene for pea color. Round yellow parents make gametes with the genotype RY, while wrinkled green parents make gametes with the genotype ry. When these gametes get together to make the F_1, the resulting genotype of the F_1 plants is RrYy.

- The self-cross of the F1 generation produced all possible combinations of the two traits in a ratio of 9:3:3:1 of round yellow peas:round green peas:wrinkled yellow peas:wrinkled green peas.

  a. The genes for pea color and pea shape don't have to stay together when gametes are made. The round yellow parental plant (RRYY) made gametes that contained the dominant alleles (RY). The wrinkled green parental plant (rryy) made gametes that contained the recessive alleles (ry). If the genes for pea color and seed shape were tied together somehow, they would stay tied together in the F_1. When the F_1 plants made gametes, they also could make only gametes that were either RY or ry. Thus, the F_2 generation would have only round yellow and wrinkled green plants. (You can draw a Punnett square like those in Figure 15-1 to show how these two types of gametes would combine to produce this type of F_2. Remember to put both types of gametes on each side of the Punnett square.)

  b. Because all possible combinations of traits were seen in the F_2, the gene for pea color must be segregating independently of the gene for pea shape. The F_2 generation had all possible combinations of the two traits. Thus, the F_1 generation must be able to make gametes with all possible combinations of alleles: RY, rY, Ry, and ry. When the gametes of the F_1 plants combine, all possible combinations of traits are seen in the F_2 offspring. The details of the self-cross of the F_1 are best seen using a Punnett square.

When you're working with crosses that involve two traits at the same time, an easy way to make sure that you include all possible types of gametes that an individual could make is to remember the word FOIL, which stands for First, Outer, Inner, and Last. Look at the genotype of the individual that is making gametes and use FOIL to figure out what types of gametes they would make. For example, if the individual has the genotype AaBb, you take the First letter of each pair of alleles (AB), then the Outer letter of each pair Ab, then the Inner letter of each pair (aB), and then the Last letter of each pair (ab). The four types of gametes an individual with genotype AaBb can make are AB, Ab, aB, and ab. FOIL works on any individual with two gene pairs, even if they have a different combination of alleles. For example, an individual with the genotype AaBB would make gametes with the combinations AB, AB, aB, and aB. In this case, only two types of gametes are produced, but the FOIL method makes sure that you don't miss anything.

Puzzling over the Punnett

When Mendel crossed round yellow parentals (RRYY) with wrinkled green parentals (rryy), the resulting F_1 generation was heterozygous for both traits (RrYy). Individuals that are heterozygous for two traits are called *dihybrids.* Individuals from the F_1 generation were mated with each other in a *dihybrid cross.* The best way to track what is happening in a dihybrid cross is to use a Punnett square (see Figure 15-3):

1. **Figure out the types of gametes that will be produced by the individuals that are being mated.**

   The round yellow dihybrids of the F_1 generation have the genotype RrYy. These individuals can make four types of gametes: RY, Ry, rY, and ry.

2. **Make a Punnett square that has enough rows and columns for the types of gametes produced by the individuals that are being mated.**

   In this case, both dihybrids that are being mated will produce four types of gametes, so the Punnett square must have four rows and four columns.

3. **Write the gamete types made by each parent along the top and sides of the Punnett square.**

   In this case, write RY, Ry, rY, and ry along the top and also along the sides of the square.

4. **Figure out all possible offspring by combining the gametes in each box of the Punnett square.**

   Each box in the Punnett square is an intersection point between two gamete types. In the box, write the four alleles that the F_2 offspring would have if the two gametes joined together. For example, the upper-left corner of the Punnett square in Figure 15-3 is the intersection point between gametes RY and RY. When these gametes combine, they produce an offspring with genotype RRYY.

5. **Write down all the different genotypes that are produced and count up how many there are of each kind.**

   A dihybrid cross produces nine different genotypes: RRYY, RRYy, RRyy, RrYY, RrYy, rrYY, Rryy, rrYy, rryy. The ratios of genotypes are one RRYY, two RRYy, one RRyy, two RrYY, four RrYy, one rrYY, two Rryy, two rrYy and one rryy. Thus, the genotypic ratio for a dihybrid cross is 1:2:1:2:4:1:2:2:1.

6. **Figure out the phenotype of each offspring in the box by remembering which alleles are dominant and which are recessive.**

   In this case, round is dominant to wrinkled, and yellow is dominant to green. You can go through your Punnett square and color code the squares or write in the phenotypes.

7. **Write down all the different phenotypes that are produced and count how many there are of each kind.**

   In the dihybrid cross of the round yellow F_1, four phenotypes are produced: round yellow, round green, wrinkled yellow, and wrinkled green. The ratios of phenotypes are nine round yellow offspring, three round green offspring, three wrinkled yellow offspring, and one wrinkled green offspring. Thus, the phenotypic ratio for the dihybrid cross is 9:3:3:1.

A phenotypic ratio of 9:3:3:1 tells you that two genes are assorting independently, just like a 3:1 phenotypic ratio tells you that a single gene is segregating.

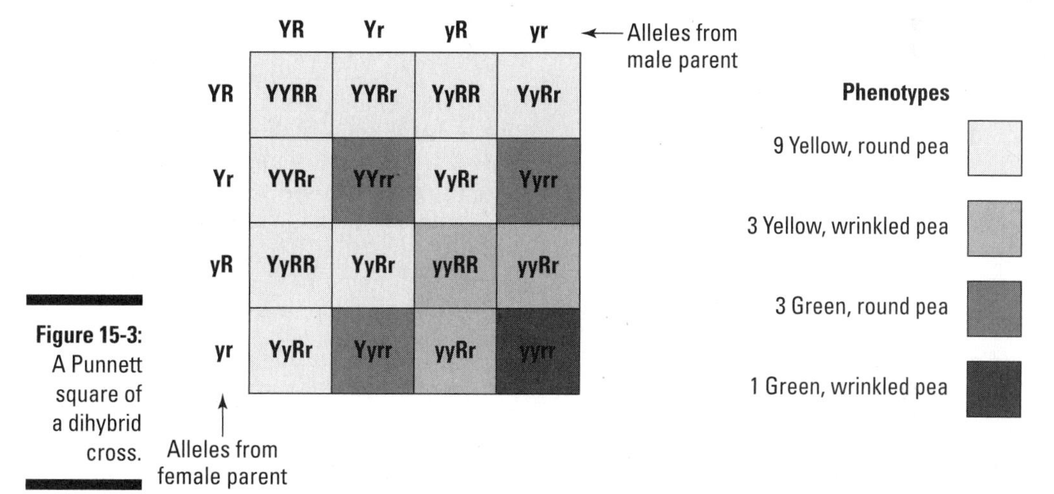

Figure 15-3: A Punnett square of a dihybrid cross.

Remembering meiosis

Genes separate from each other during anaphase I of meiosis when the homologous chromosomes move away from each other toward opposite sides of the cell. (See Chapter 14 for more on meiosis.) The separation of one pair of homologous chromosomes has no effect on the separation of other pairs of homologous chromosomes. So, if two genes are located on different chromosomes, their movement during meiosis is independent of each other. For example, imagine two pairs of homologous chromosomes inside a dihybrid pea plant:

- **The genes for pea shape and pea color are carried on different chromosomes.** One pair of homologous chromosomes has the gene for pea shape. In the pair, one chromosome carries the dominant allele (R), while the other carries the recessive allele (r). The other pair of homologous chromosomes has the gene for pea color. In this pair, one chromosome carries the dominant allele (Y), and the other carries the recessive allele (y).

- **In anaphase I, the two alleles for each gene separate as the homologous chromosomes go to opposite sides of the cell.** When the pairs of homologous chromosomes separate in anaphase I (see Chapter 14), one of each pair will go to opposite sides of the cell and end up in a different gamete. The chromosomes with the dominant shape allele (R) could go to the same side of the cell as the dominant color allele (Y), making a gamete with the genotype RY. The chromosome with the recessive shape allele (r) could go to the same side as the recessive color allele (y), making a gamete with the genotype ry. If meiosis happens this way in one cell of the dihybrid plant, the gametes produced will have the genotypes RY and ry.
- **Each pair of homologous chromosomes separates independently of the other homologous chromosomes, so segregation of alleles in different genes can vary from meiosis in one cell to another.** Another cell in the same dihybrid that is making gametes could separate the alleles differently. The chromosome with the dominant shape allele (R) could go to the same side of the cell as the chromosome with the recessive color allele (y), making a gamete that has the genotype Ry. The chromosome with the recessive shape allele (r) could go to the side of the cell with the dominant color allele (Y), making a gamete with the genotype rY. If meiosis happens this way in one cell of the dihybrid plant, the gametes produced will have the genotypes Ry and rY.
- **When you're predicting the outcome of a cross, you need to consider all the possible gamete types an individual can make.** When you consider all the possible ways that these gene pairs could separate during meiosis in different cells in the dihybrid plant, you come up with four possibilities for all the gametes produced by the dihybrid plant: RY, Ry, rY, and ry. These four gamete types are possible because the pairs of homologous chromosomes assort independently from each other during anaphase I.

Independent assortment of gene pairs occurs when homologous chromosomes segregate independently from each other during anaphase I of meiosis.

Chapter 16

Expect the Unexpected: Non-Mendelian Patterns of Inheritance

In This Chapter

- Discovering new patterns of inheritance
- Following linked genes
- Understanding sex-linked inheritance

Many traits don't behave exactly in the ways that Mendel described from his experiments with pea plants. (For more on the basics of genetics, see Chapter 15.) Unexpected patterns of inheritance led geneticists to probe deeper into the mysteries of genes and revealed many factors that contribute to inheritance. In this chapter, I present the most common variations on strict Mendelian inheritance.

It's News to Mendel: Inheritance Beyond Simple Dominance

Sometimes, the inheritance of a trait behaves differently from what you would expect based on Mendel's laws (see Chapter 15). Does that mean Mendel got it wrong? Not really. The laws that Mendel developed based on his observations are still believed to be correct. It's just that Mendel didn't see everything there was to see in his experiments. Inheritance involves a bit more than just single gene traits with a simple relationship to each other of dominant and recessive.

Since Mendel's time, scientists have discovered several types of inheritance that are complicated by other factors:

- In *incomplete dominance,* the effects of two different alleles combine to create a phenotype that is somewhere in between the dominant phenotype and the recessive phenotype.

- In *codominance,* the effects of two different alleles both contribute equally to the phenotype.
- *Pleiotropy* occurs when one gene has an effect on many different phenotypic characteristics.
- In *polygenic traits*, many different genes affect the phenotype.
- *Linked genes* are very close together on a chromosome and break the law of independent assortment.
- *Sex-linked traits* are located on the sex chromosomes, X or Y.

Mixing it up: Incomplete dominance

In incomplete dominance, the dominant allele only partially masks the presence of the recessive allele. A good example of incomplete dominance is seen in the color of some flowers, such as snapdragons:

- The red allele (R) is incompletely dominant to the white allele (W).
- Homozygous plants (RR or WW) have either red flowers or white flowers (see Figure 16-1).
- When the two alleles are present together in a heterozygote (RW), the result is a plant with pink flowers.

If the red allele was completely dominant to the white allele, heterozygous plants would have red flowers. But in the case of incomplete dominance, you can see the presence of the white allele because of the lighter color of the flowers. For the case of flower color, you can think of incomplete dominance like mixing paints — red paint mixed with white paint makes pink paint. Incomplete dominance shows the same kind of blending in the phenotype of the heterozygote.

Sharing the power: Codominance

Codominance is very similar to incomplete dominance (see preceding section) because heterozygotes can also show the effect of two different alleles. However, when two alleles are codominant to each other, they both show up distinctly in the heterozygote — the alleles don't blend their effects like they do in incomplete dominance. For example, if there were a case of codominant alleles for flower color, with one allele for red color and the other for white color, a heterozygote would have flowers that were red and white.

The ABO markers for human blood type are an excellent example of codominance. Three alleles for blood type exist in the human population: allele *IA,* allele *IB,* and allele *i*. The *IA, IB,* and *i* alleles are the blueprints for enzymes that attach carbohydrates to the surfaces of your red blood cells:

- If you have allele *IA*, then you make an enzyme that attaches carbohydrate A to your blood cells.
- If you have allele *IB*, then you make an enzyme that attaches carbohydrate B to your blood cells.
- Allele *i* is nonfunctional and won't result in either carbohydrate A or carbohydrate B being attached to your blood cells.

Every person has two copies of the gene for the ABO markers, which are called ABO antigens. A person's blood type depends on which two alleles they have (see Figure 16-2):

- People who have the genotype *IAIA* or *IAi* have Type A blood. Their red blood cells have the A antigen on their surfaces.
- People who have the genotype *IBIB* or *IBi* have Type B blood. Their red blood cells have the B antigen on their surfaces.
- People who have the genotype *IAIB* have Type AB blood. Their red blood cells have both the A antigen and the B antigen on their surfaces. The A and B alleles are codominant to each other because both alleles are visible, yet distinct, in the phenotype of the heterozygote.
- People who have the genotype *ii* have Type O blood. Their red blood cells have neither the A nor the B antigen on their surfaces.

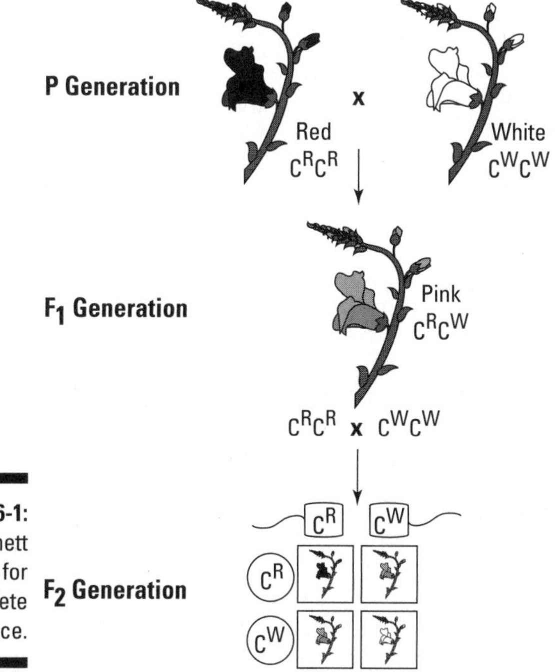

Figure 16-1: A Punnett square for incomplete dominance.

Knowing your blood type is very important when you're giving blood to or receiving blood from other people. Your body will attack any red blood cells that have ABO antigens that are different from your own by producing antibodies against the foreign ABO antigens. Thus, you can receive only blood that is compatible with your own:

- **People with Type AB blood can receive any type of blood.** The immune systems of people with AB blood are trained to accept both the A and B antigen as normal and won't make antibodies against any red blood cells (Figure 16-2). Thus, people with Type AB blood are called *universal receivers.*
- **People with Type A blood can receive Type A or Type O blood.** If any blood with the B antigen is introduced into the body, the immune system of Type A people will make antibodies against the B antigen. Type O blood cells do not have any ABO antigens on their surface, so they're acceptable to people of any blood type. Thus, people with Type O blood are called *universal donors.*
- **People with Type B blood can receive Type B or Type O blood.** If any blood with A antigen is introduced into the body, the immune system of Type B people will make antibodies against the A antigen.
- **People with Type O blood can receive only Type O blood**. If any blood with B or A antigen is introduced into the body, the immune system of Type O people will make antibodies against these antigens.

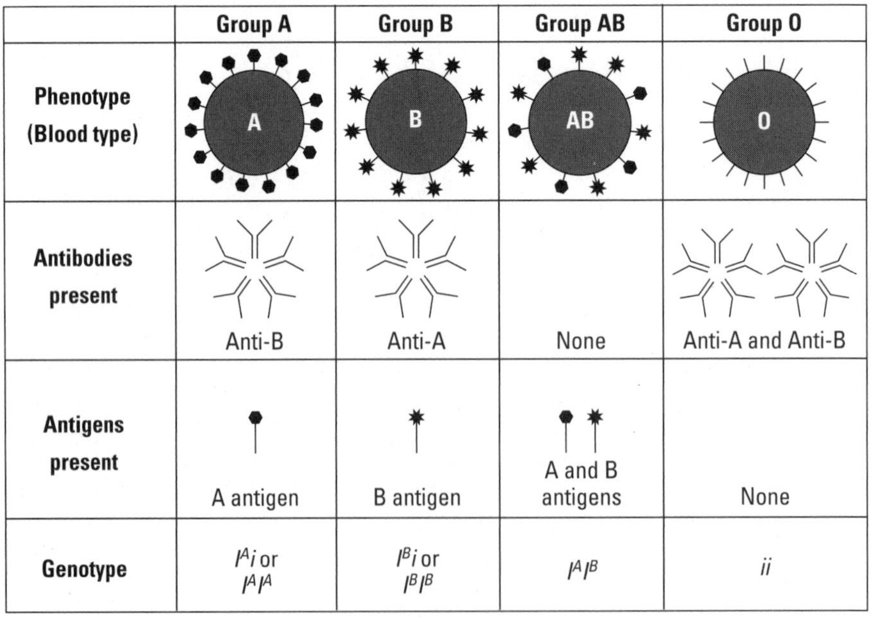

	Group A	Group B	Group AB	Group O
Phenotype (Blood type)	A	B	AB	O
Antibodies present	Anti-B	Anti-A	None	Anti-A and Anti-B
Antigens present	A antigen	B antigen	A and B antigens	None
Genotype	I^Ai or I^AI^A	I^Bi or I^BI^B	I^AI^B	ii

Figure 16-2: Human blood type.

Making an impact: Pleiotropic genes

Some genes affect a trait that is connected to many different characteristics. Genes that affect more than one trait are called pleiotropic genes. For example, people who have sickle cell anemia have two copies of a recessive allele for defective hemoglobin. The defective hemoglobin changes shape when oxygen levels in the cell decrease, causing the red blood cells to collapse, or sickle. On the one hand, the hemoglobin gene controls the structure and function of hemoglobin. However, when the defective alleles are present and the blood cells sickle, the sickling blood cells cause many other effects:

- Anemia and extreme fatigue result from the destruction of blood cells and lack of oxygen circulating around the body.
- The sickled cells plug capillaries, blocking blood flow to tissues and causing intense pain and muscle damage.
- The kidneys and spleen are damaged and can eventually fail.

These effects on the body represent changes to multiple aspects of a person's phenotype, all deriving from one pleiotropic gene.

It's not that simple: Polygenic traits

Human eye color is often used as an example of dominant and recessive alleles. Alleles for blue eyes (b) are recessive to alleles for brown eyes (B), the story goes, so if you have two alleles for blue eyes (bb), your eyes are blue. If you have one or two alleles for brown eyes (Bb or BB), your eyes are brown.

There's just one problem with this simple scenario — human eye color is much more diverse than just blue or brown. People have eyes ranging from blue to green to almost purple (think Elizabeth Taylor) to brown to almost black. If you walk up to ten people you know who have brown eyes and really look at their eyes, I'm guessing you'll see ten different variations on the color brown. The same is true for ten people you know who have blue eyes. So, just what is the story with eye color?

Well, the first part of the story about human eye color is partially correct — alleles for dark eyes are dominant to alleles for light eyes. Alleles for dark eyes lead to the production of the pigment melanin — the same pigment that darkens skin or makes freckles — and the deposit of melanin into your eyes. People have so many variations in eye color because several genes, not just one, are involved in the production and deposit of melanin in the eye. The number of dominant alleles you have determines the amount of melanin you have deposited in your eyes and thus your eye color:

- People with very dark, chocolate brown eyes have lots of melanin deposited in their eyes. They're homozygous dominant for all the genes that control melanin production and deposition in the eye.
- People with grey or light blue eyes have very little melanin deposited in their eyes. They have very few dominant alleles in the genes that control melanin production and deposition in the eye. (People who have no melanin deposited in their eyes at all have pink eyes.)
- People with eye colors in between dark brown and grey or light blue have different amounts of melanin deposited in their eyes.

When a trait is controlled by many genes, it's called a polygenic trait (see Figure 16-3).

Because people have many possible combinations of alleles in the genotypes for polygenic traits, a wide variety of phenotypes usually occurs. Human traits that are polygenic include eye color, skin color, and height. If you graph the number of people who have each variety of phenotype for a polygenic trait, the graph looks like a curve because of all the possible variations.

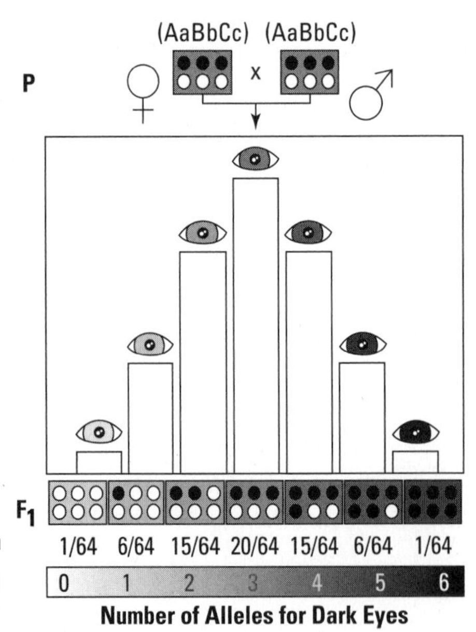

Figure 16-3: Distribution of polygenic trait.

Almost Inseparable: Linked Genes

One of the biggest surprises for scientists studying genetics after Mendel is that some traits break the Law of Independent Assortment (see Chapter 15). When an individual who is heterozygous for two traits makes gametes, the prediction is that all possible combinations of alleles will occur in the gametes. However, some traits seem to want to travel together as if they are linked. When a person is heterozygous for two linked traits, they don't make gametes with all combinations of alleles. Instead, some alleles behave as if they are stuck together and always show up together in gametes. A person who is heterozygous for two traits, controlled by genes A and B, will make very different gametes depending on whether these genes are unlinked or linked:

- If genes A and B assort independently, a heterozygote (AaBb) will make gametes with all possible combinations of alleles, AB, Ab, aB, and ab.
- If genes A and B are linked, a heterozygote (AaBb) will make only two types of gametes, such as AB and ab (see Figure 16-4).

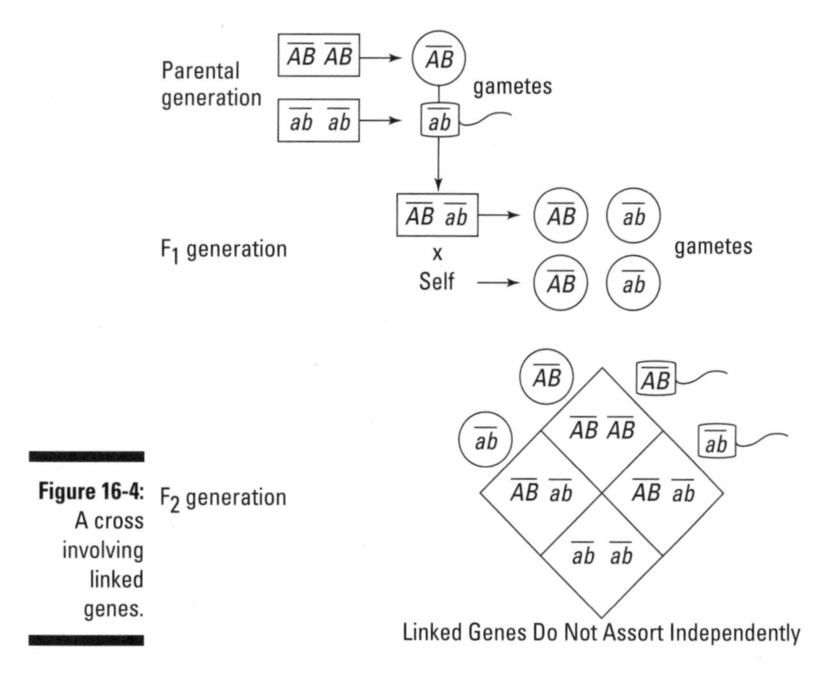

Figure 16-4: A cross involving linked genes.

Traveling together because of linkage

Traits are *linked* because the genes for those traits are on the same chromosome. If two genes are close together on the same chromosome, they'll travel together during anaphase I of meiosis when homologous chromosomes separate from each other. (See Chapter 14 for more on meiosis.) The strength of the tendency to stick together varies for different genes depending on how close together the genes are on the chromosome.

Genes that are very close together on a chromosome are *completely linked.* The alleles for any two genes that are completely linked will always travel together.

If a person is heterozygous for two genes, A and B, that are completely linked, they will make only gametes with two genotypes instead of four. For example, if alleles A and B are on one homologous chromosome and the alleles a and b are on the other homologous chromosome, then they'll make gametes only with the genotypes AB and ab, as shown in Figure 16-4. During meiosis, the A and B alleles will always travel together, and the a and b alleles will always travel together. Fewer gamete types are produced when genes are completely linked than when genes assort independently. Thus, a mating between two organisms that are heterozygous for linked genes will produce less diversity in offspring than will a mating between two organisms that are heterozygous for unlinked genes:

- If genes A and B assort independently, a mating between two heterozygotes (AaBb) is predicted to produce offspring with a phenotypic ratio of 9:3:3:1. (See Chapter 15 for more information on predicting ratios.)
- If genes A and B are completely linked, a mating between two heterozygotes (AaBb) is predicted to produce offspring with a phenotypic ration of 3:1.

Slipping away through recombination

Genes that are linked because they're on the same chromosome may be separated by crossing-over (see Chapter 14). A person who is heterozygous for two traits (AaBb) may have alleles A and B on one chromosome and alleles a and b on the other chromosome. This heterozygote (AaBb) will produce different gametes types depending on whether the genes are completely linked or if they're linked but recombination occurs between them:

- If the genes are completely linked, then the alleles will always travel together and gametes will have only the genotypes AB and ab. Because the combination of alleles in the gametes is the same combination as in the heterozygous parent, these combinations are called *parental types.*

- If crossing-over occurs between the linked genes, then alleles may be switched between the homologous chromosomes (a for A, and b for B). This switching creates new combinations of alleles (aB and Ab) in the gametes. Because these combinations of alleles are different than the combinations seen in the heterozygous parent, they're called *recombinant types.*

Crossing-over between linked genes allows the heterozygote to make all four types of gametes: AB and ab if no crossing-over occurs, and aB and Ab if crossing-over occurs. If two linked genes are very far apart, crossing-over will almost always occur between them very frequently and the genes will behave as if they are sorting independently. In other words, heterozygotes for the two genes will produce all four types of gametes in equal numbers, just as is predicted for unlinked genes (Chapter 15).

Building a map of a chromosome

The farther apart two linked genes are on a chromosome, the greater the chance that recombination will occur between the two genes. Thus, by performing crosses and observing the amount of recombination, you can figure out how far apart two genes are on a chromosome. If you do this process over and over again for different pairs of genes on the same chromosome, you can make a map of the chromosome showing where each gene is located. For each pair of traits, do the following:

1. **Mate parental organisms that are homozygous for both traits.**

   For example, cross a parental that has the genotype AABB with one that has the genotype aabb (see Figure 16-4).

2. **Mate two F1 heterozygotes.**

3. **Count the number of F2 offspring that have parental phenotypes.**

   The parental phenotypes are the combinations that were present in the parental generation. For example, if the parental with the genotype AABB in Figure 16-4 has a phenotype of tall (AA) and hairy (BB), then tall and hairy represents one parental combination of phenotypes. If the parental with the genotype aabb has a phenotype of short (aa) and hairless (bb), then short and hairless is the other parental combination of phenotypes. The total number of F_2 offspring that were either tall and hairy or short and hairless would represent the number of parental combinations in the F_2.

   Parental combinations in the F_2 result when no crossing-over occurs between the linked genes.

4. **Count the number of F2 offspring that have recombinant phenotypes.**

   Recombinant phenotypes are combinations that were not present in the parental generation. If the parental combinations of traits were tall and hairy versus short and hairless, then recombinant phenotypes would be tall and hairless or short and hairy. The total number of F_2 offspring that were either tall and hairless or short and hairy would represent the number of recombinants in the F_2.

   Recombinants in the F_2 result when crossing-over occurs between the linked genes.

5. **Calculate the percentage of recombinants in the F2.**

   Divide the total number of offspring with recombinant phenotypes by the total number offspring in the F_2 and then multiply the answer by 100. This percentage is called the *recombination frequency.* In Figure 16-4, no recombinant offspring existed, so the recombination frequency is 0.

6. **Calculate the distance between the genes in map units.**

   A *map unit* is defined as 1 percent recombination frequency. So, if you had 14 percent total recombinant phenotypes in the F_2, that amount would tell you that the two genes were 14 map units apart on the chromosome. In Figure 16-4, the recombinant frequency is 0, so the genes are 0 map units apart. In other words, they're completely linked.

Mama's Boy: Sex-Linked Inheritance

Some traits break Mendel's laws because they aren't inherited equally by males and females. Traits that are inherited differently in males and females are called *sex-linked traits.*

The disease hemophilia, which results from a defect in blood-clotting proteins, is an example of a sex-linked trait. People who have hemophilia can't clot their blood and are at risk from dying of blood loss whenever they're injured. The disease was passed from Queen Victoria of England to her descendants and then spread throughout the royal families of Europe as Queen Victoria's children married other European royals. A curious pattern occurred as the disease spread through these families: All the affected individuals — those actually having the disease hemophilia — were male.

Analyzing the pedigree

The pedigree of Queen Victoria's family, shown in Figure 16-5, provides information regarding the inheritance of hemophilia. The pedigree contains four important clues:

- **The disease can skip generations.** In other words, parents who don't have the disease can produce children who do. Thus, hemophilia must be a recessive trait.
- **The disease is much more common in males than in females.** The difference between males and females is that males have X and Y for their sex chromosomes, whereas females have two X chromosomes. For a boy to be born with sex-linked trait like hemophilia, only his mom needs to give him the defective allele. For a girl to be born with a sex-linked trait, both parents need to give a defective allele.
- **The disease is never passed from father to son.** Sons get their Y chromosomes from their fathers, so the disease isn't carried on the Y chromosome.
- **Affected sons are born to unaffected mothers.** Sons get their X chromosomes from their mothers, so the disease is carried on the X chromosome.

The clues from Queen Victoria's pedigree reveal that hemophilia is an *X-linked recessive trait.* In other words, the disease is caused by a recessive allele that is carried on the X-chromosome.

Males get their X chromosomes from their mother and their Y chromosomes from their father. Females get an X chromosome from each parent.

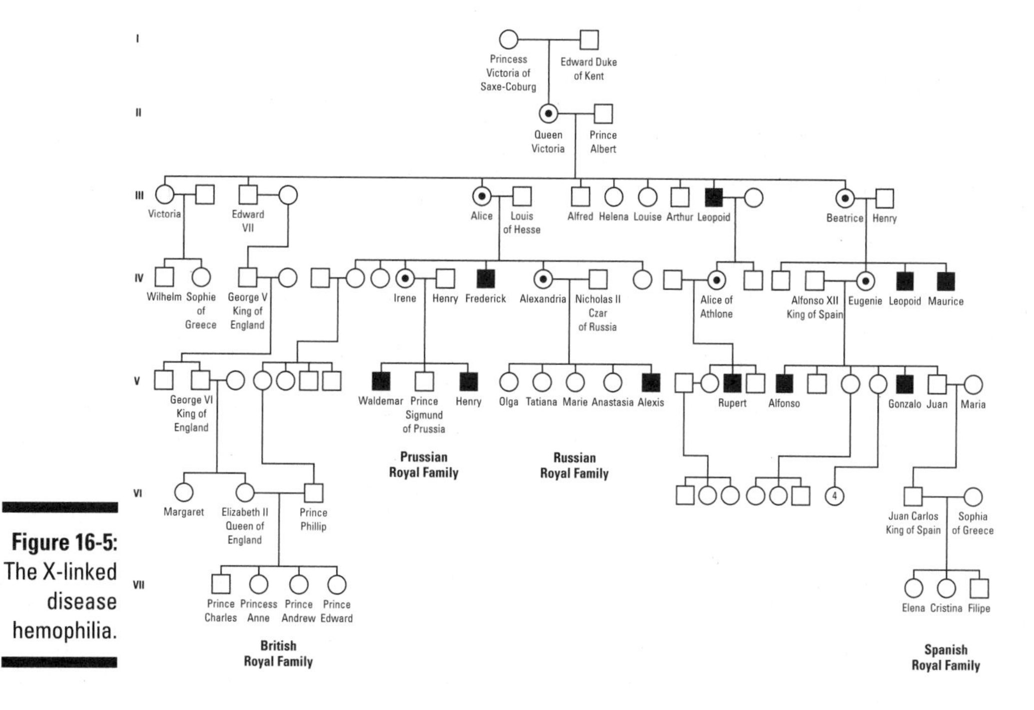

Figure 16-5: The X-linked disease hemophilia.

You can link sex-linked traits to the X chromosome or the Y chromosome, and they can be dominant or recessive. Traits that are linked to the Y chromosome, called *Y-linked traits,* are much less common than X-linked traits because the Y chromosome doesn't have a lot of genes on it. Sex-linked traits show distinctive patterns of inheritance in pedigrees:

- **X-linked dominant traits** affect both males and females.
  - Because the traits are dominant, they don't skip generations.
  - Affected mothers can have both affected sons and daughters.
  - All daughters of an affected father are affected.
- **Y-linked traits** affect only men.
  - Affected men pass the trait to all their sons.
  - The traits do not skip generations.

Explaining the differences

Scientists can explain the differences in the patterns of inheritance for sex-linked traits by taking a closer look at the X and Y chromosomes. Although X and Y pair up with each other during meiosis, they're actually quite different from each other:

- The human X chromosome contains more than a thousand genes that affect normal body functions. Some genes on the human X chromosome include genes for color vision and blood-clotting proteins, as well as the ability to make normal muscle proteins.
- The human Y chromosome is much smaller than the X and contains about 200 genes. The genes on the Y chromosome are completely different from the ones on the X chromosome. One of these genes, called SRY, turns on the development of the testes in male fetuses.

Because the X and Y chromosomes don't have any genes in common, males have only one copy of each gene on the X chromosome (see Figure 16-6). Because males have only one copy of genes on the X, they can't be homozygous or heterozygous for these genes. Instead, they're considered *hemizygous.* The difference in the number of X chromosomes causes X-linked recessive traits to be more common in males:

- If males have a defective allele on their X chromosome, they'll show the defect in their phenotype.
- If females have a defective allele on their X chromosome, they may have a normal allele on their X chromosome. The normal allele will hide the effect of the defective allele.

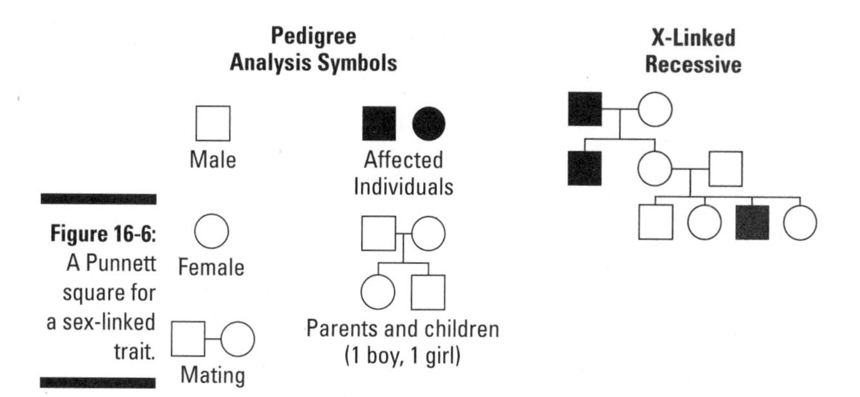

Figure 16-6: A Punnett square for a sex-linked trait.

Queen Victoria must have been heterozygous for hemophilia, because although she passed the disease on to her son Leopold, she herself didn't have the disease. One of her X chromosomes had the disease allele, but the other X had a normal allele for blood-clotting protein, so she was able to clot her blood. Because Queen Victoria had the disease allele but didn't show it, she is called a *carrier* of the disease.

You can predict the inheritance of sex-linked traits by using Punnett squares. (See Chapter 15 for more on Punnett squares.) Sex-linked traits are passed to gametes along with the sex chromosomes. When setting up a Punnett square for a sex-linked trait, remember to keep the alleles linked to the sex chromosomes:

- Male parents have only one allele for sex-linked traits.
  - If the trait is X-linked, the allele is passed to gametes that receive the X chromosome from the male parent.
  - If the trait is Y-linked, the allele is passed to gametes that receive the Y chromosome.
- Female parents have two alleles for X-linked traits and no alleles for Y-linked traits.

No boys allowed

Did you know that calico cats are almost always female? The gene for orange versus black coat color in cats is carried on the X chromosome. The Y chromosome doesn't carry a gene for coat color. Male cats are XY, so they're either orange (X chromosome carries orange allele) or black (X chromosome carries black allele). Female cats who are heterozygous for the two color alleles have orange and black fur. The only way a male cat can be a calico is if he has too many chromosomes and is XXY. Male calico cats do occasionally occur due to nondisjunction in meiosis (see Chapter 14), but they're very rare and can't produce offspring.

The predicted offspring of a cross involving a sex-linked trait are created by combining gametes in the boxes of the Punnett square.

To keep track of alleles for sex-linked traits, make sure that you always write them as attached to an X or Y chromosome. For example, the dominant allele for hemophilia is written X^H and the recessive allele is written X^h. By writing the alleles attached to an X or a Y, you'll be less likely to forget that the allele is on only one type of sex chromosome.

Part V
Molecular Genetics: Reading the Book of Life

In this part . . .

The DNA code is the underlying program for how cells function. DNA determines the behavior of cells largely because it contains the blueprints for proteins, the dominant worker molecules in cells. In order to build a protein, cells must copy the code from DNA into RNA in the process of transcription and then use the RNA as a template for protein construction in the process of translation.

How and when the DNA code is used is just as important as the code itself. Cells control access to the DNA code in order to regulate function and development of cells. In this part, I explain how DNA is copied, how it is used to make proteins, and how cells control the information encoded in DNA.

Chapter 17

DNA Synthesis: Doubling Your Genetic Stuff

In This Chapter

- Meeting the enzymes that copy DNA
- Exploring the steps of DNA replication
- Finding out about leading and lagging strands

Before a cell can reproduce itself, it must make a copy of its DNA, a process called DNA replication. Many enzymes play a role in DNA replication, and each one has a specific job. In this chapter, I examine the roles of the enzymes that copy DNA and explore the process of how the DNA is copied.

DNA Replication: An Overview

Cells copy their DNA by a process called *DNA replication*. DNA replication occurs whenever a cell is going to reproduce itself, either by mitosis or binary fission. (For more on these processes, see Chapter 13.) DNA replication also occurs before cells in reproductive tissue make sperm or eggs by meiosis (see Chapter 14).

If a cell is going to divide to make more cells, then the parent cell needs to copy all of its parts, including the DNA. Of course, because the DNA contains the blueprints for the functioning of the cell, it's especially important that the DNA is copied very carefully. Any mistake in the DNA code, called a *mutation,* can affect a cell's ability to function correctly. Thus, cells have evolved a process that carefully unwinds the DNA and uses the existing code as a pattern for copying the code into new DNA molecules. (For more on the structure of DNA, see Chapter 7.)

In order to make an exact copy of each DNA molecule, cells use *semiconservative replication.* The two halves of the double helix are separated from each other, and each half is used as a template, or pattern, to direct the synthesis of the other half. Each old half gets a new partner, resulting in two DNA molecules from one.

Everybody Lend a Hand: Enzymes Involved in DNA Replication

You've probably seen construction crews hard at work putting up a new building. Many people are members of the crew, and each person is a specialist in one particular part of the job. The same is true for DNA replication — many enzymes are part of the crew that copies the DNA, and each enzyme has a particular job. The enzymes involved in DNA replication are as follows:

- **Helicase** breaks, or melts, the hydrogen bonds that hold the two halves of the double helix together.
- **Topoisomerase** breaks and reseals the DNA backbone to allow the backbone to release tension caused by twisting.
- **Primase**, also called RNA polymerase, puts down small pieces of RNA called *primers*. The primers serve as starting points for copying the DNA.
- **DNA polymerase** makes new strands of DNA. Two different DNA polymerases are involved in DNA replication:
    - *DNA polymerase III* copies the DNA by reading the code on the existing strands and building new strands that are complementary to the originals. The enzyme follows the base pairing rules described in Chapter 7. For example, if an A appears in the original strand, it brings in a nucleotide with a T.
    - *DNA polymerase I* removes RNA nucleotides from the RNA primers and replaces them with DNA nucleotides so that no RNA is remaining in the DNA when the job is done.
- **DNA ligase** seals up breaks in the DNA backbone by forming covalent bonds between nucleotides.

The names of the enzymes involved in DNA replication give clues to their jobs. Helicase unzips the helix. Primase gets things started, or primes the pump, by putting down primers. DNA polymerase makes polymers of DNA. Ligate means to tie, so DNA ligase ties up the broken bits of DNA by forming covalent bonds along the sugar-phosphate backbone.

It Takes a Village: Events at the Replication Fork

The enzymes involved in DNA replication stay together in a large enzyme work crew, called an *enzyme complex,* that slides along the DNA, copying as it goes. All the enzymes are working simultaneously, each doing their part in the sequence of events that needs to occur to copy the DNA. DNA replication is easiest to understand if you look at each of the events of DNA replication in order from what must happen first to what happens last.

Start at the very beginning: Origins of replication

DNA replication begins at specific locations within the chromosome. These locations are marked by particular DNA sequences called *origins of replication* (ORI), shown in Figure 17-1.

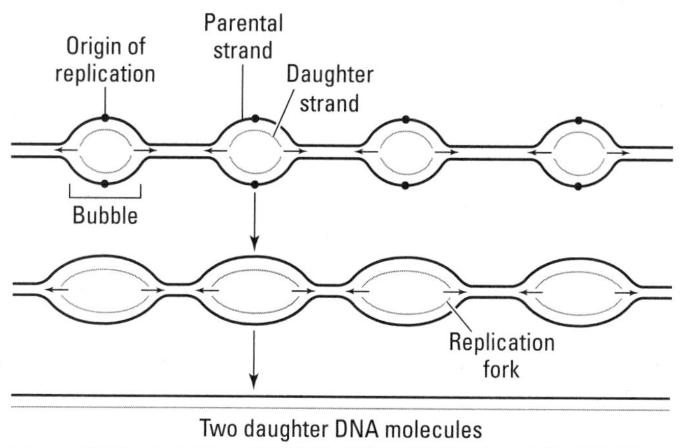

Figure 17-1: Origins of replication.

A specific set of proteins recognizes this sequence and opens the double helix at each origin of replication. Once the double helix has been opened, the enzyme complex that copies the DNA can begin. Replication occurs in *both* directions away from the origin, conducted by two enzyme complexes that begin at each origin and move away from each other along the DNA. Bacterial chromosomes have one origin of replication, so two enzyme complexes move around the circular chromosome away from the origin.

Eukaryotic chromosomes, including those of humans, have multiple origins of replication (see Figure 17-1), so several enzyme complexes work on each chromosome once replication begins.

As the enzymes unwind the DNA to begin copying, a visible loop called a *replication bubble* is created in the DNA. At both sides of the bubble is a V-shaped area where the DNA that is being copied meets up with the DNA that hasn't been unwound yet. These V-shaped structures are called *replication forks.* Replication forks are the places where DNA replication (see Figure 17-2) is actively happening — in other words, they're where the enzymes are doing their jobs.

Learning to unwind with helicase

Once the DNA has been opened at the origin, the enzyme helicase, shown in Figure 17-2, catalyzes the breaking of the hydrogen bonds between the base pairs. You can think of helicase like the metal piece that slides along a zipper and causes it to unzip. Helicase unzips the hydrogen bonds, and the two halves of the DNA helix separate. Once the strands are separated, proteins called *single-strand DNA-binding proteins* (SSBPs) act like thumbtacks to hold the DNA open by binding to the separated strands and stopping them from rejoining each other.

One difference between the double helix and a zipper is that a zipper is flat and the double helix is twisted. As helicase unzips the double helix, these twists get pushed together, creating tension in the DNA backbone. The DNA backbone would become kinked as it is unzipped if it weren't for the action of another enzyme, called *topoisomerase*. As the twists in the DNA get tight, topoisomerase catalyzes the breaking of covalent bonds in the backbone of the DNA. Once the backbone is cut, it can untwist and release the tension. Then, topoisomerase forms a new covalent bond to reseal the DNA.

To help you imagine how DNA molecules may be stressed as they unwind, try taking a rubber band and twisting it around and around until it starts to kink. Kinks like these would form in the DNA if it weren't for topoisomerase.

Putting down some primer

After helicase has opened the double helix, the genetic code is available to be copied. However, there is a small problem — DNA polymerase III, the enzyme that reads the code and builds new strands of DNA, can't start a strand of DNA on its own. DNA polymerase III can only add nucleotides to an existing strand. So, before DNA polymerase can do its job, it needs a helper enzyme to get the new strands started.

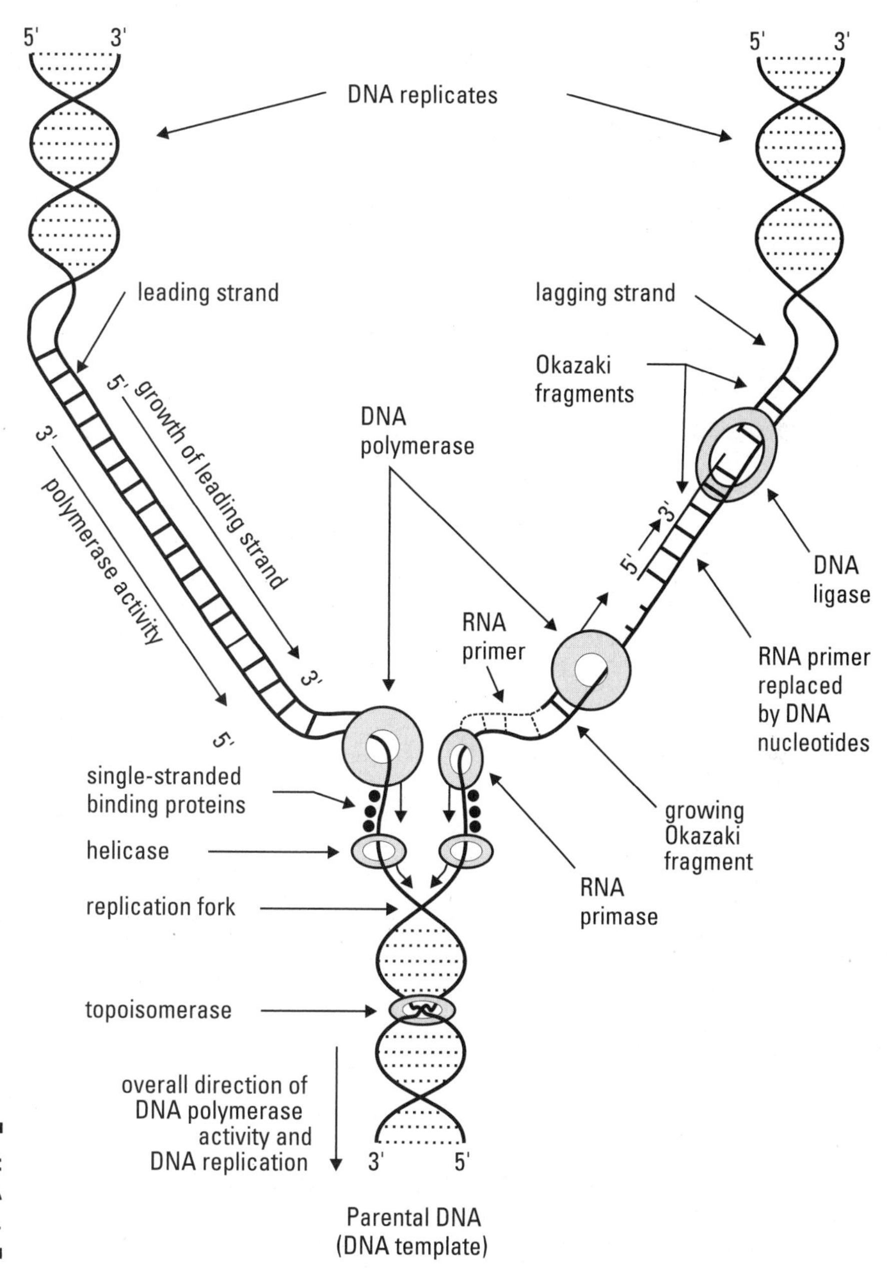

Figure 17-2: DNA replication.

The helper enzyme is Primase, an RNA polymerase that builds strands of RNA. Primase can start a strand on its own and it makes short pieces of RNA, called primers (Figure 17-2), that are complementary to the DNA. Once the primers have been made by Primase, DNA polymerase III can start making DNA by attaching DNA nucleotides to the 3' ends of the primers. (See Chapter 7 for more on 5' and 3' ends of DNA.)

Rolling down the line

Once the primers are in place, DNA polymerase III can begin to synthesize new DNA, as shown in Figure 17-2. It uses the original DNA strands as a template and follows the base-pairing rules described in Chapter 7 when bringing in nucleotides. If the original strand has adenine (A), DNA polymerase III brings in thymine (T) and attaches the 5' end of the thymine to the 3' end of the growing new strand. Thus, the new DNA strands always get longer at their 3' ends. Scientists say that the DNA "grows in the 5' to 3' direction."

DNA polymerase III has certain quirks about the way it functions:

- It must have a template to match nucleotides against.
- It can't start a new strand on its own.
- It always adds new nucleotides to the 3' end of an existing strand.

Replacing some tiles

When your cells divide and your DNA gets copied, I'm guessing you want a good job to be done — everything copied correctly, nice and perfect DNA, right? You probably don't want a bunch of RNA primers cluttering up your chromosomes. That is why cells have the fix-it enzyme, DNA polymerase I. DNA polymerase I slides along the DNA that has been copied and, wherever it finds a piece of RNA, it removes the RNA nucleotides and replaces them with their DNA counterpart. So, if it removes the RNA version of adenine, it replaces it with the DNA version of adenine, and so on. (RNA has uracil instead of thymine, so uracil RNA nucleotides get replaced with thymine DNA nucleotides.) When DNA polymerase I is done with its job, all the RNA has been removed, and only DNA is left, which I'm sure is a big relief to you and your cells.

Reactions that build complex molecules from smaller pieces, like the synthesis of DNA, are *endergonic* — in other words, they require energy. Cells get around this problem by bringing in the nucleotides in a very energy-rich form as *deoxyribonucleoside triphosphates* (dNTPs). dNTPs are essentially nucleotides with three phosphates attached; for example, ATP is a dNTP. The removal of two phosphates from a dNTP is so energetically favorable — in other words, so exergonic — that it makes the process of DNA replication exergonic overall.

Tying up loose ends

When DNA polymerase I is done replacing RNA primers with DNA, covalent bonds are still missing along the sugar-phosphate backbone. Every time a new stretch of DNA is begun with a primer, the DNA is copied until DNA polymerase III bumps into the next primer. Then, DNA polymerase III detaches and moves to the next primer without forming a covalent bond between its last nucleotide and the 5' end. Even after the primers are replaced with DNA by DNA polymerase I, these gaps remain. Sealing up these gaps is the job of the enzyme DNA ligase.

DNA ligase slides along the new DNA and closes gaps between DNA fragments by catalyzing the formation of covalent bonds in the sugar-phosphate backbone.

Finishing the job

The enzyme complexes that replicate the DNA begin at the origin of replication and move in both directions away from the origin. They continue moving along the DNA until all the DNA has been copied. This moves the replication forks along the chromosomes. When two replication forks meet, then all the DNA between the origins has been copied, and the enzyme complex lets go of the DNA.

Figure 17-2, earlier in this chapter, shows all the enzymes in the replication complex doing their jobs simultaneously. It represents a moment in time and can be confusing because everything is happening at once. The best way to understand replication is to draw several pictures yourself, beginning with an unwound double helix, and moving through each step chronologically. After you draw a double helix, ask yourself, "What would happen next?" It should be fairly obvious that you need to open up the DNA to get in there and copy it. So, then draw a replication bubble with some helicase and single-strand DNA-binding proteins. Again, ask yourself, "What would happen next?" Keep doing this until you've sketched out the whole process. One of your drawings should look just like Figure 17-2, but it will mean so much more when you see it as part of a sequence of events.

Keeping It Together: Leading and Lagging Strands

The two strands of the double helix are antiparallel to each other, which means that their 5' and 3' ends are pointing in opposite directions. Likewise, when the helix is unwound and each original strand gets a new partner strand, the partner strands will be built antiparallel to the originals. For example, the two new strands that are being built at the replication fork in Figure 17-2, earlier in this chapter, are pointing in opposite directions. The 3' ends of the new strands are labeled and also indicated by an arrowhead. On one side of the replication fork, the 3' ends of the new DNA point away from the fork; on the other side, they point toward the replication fork.

When DNA polymerase III is adding nucleotides to the new strands of DNA, it always adds them to their 3' ends. Thus, one new strand of DNA is growing away from the replication fork, while the other is growing toward the fork. The opposite polarity of the DNA strands creates something of a problem because all the enzymes involved in DNA replication stay together as they copy the DNA on *both* sides of the replication fork. Figure 17-2 shows all the enzymes as if they're separate from each other so that you can see each one, but really they're all joined in one large work crew, or enzyme complex. The enzymes work crew keeps moving toward the fork as more DNA is unwound, but one DNA polymerase III keeps trying to copy DNA in the opposite direction! To stay with the work crew, this DNA polymerase keeps making short pieces of DNA and then skipping along the DNA back toward the fork and starting again on a new primer. Thus, one new strand of DNA at a replication fork is made in pieces called *Okasaki fragments,* which you can see in Figures 17-2 and 17-3.

The new strand that is made in fragments is called the *lagging strand,* shown in Figure 17-3. The 3' end of the primer that begins the lagging strand points *away* from the fork and the direction that the enzyme crew is moving, so the lagging strand must be restarted over and over again as the enzyme crew moves away. RNA polymerase keeps putting down new primers on the side of the fork where the lagging strand is being built so that DNA polymerase III can keep working. Thus, DNA polymerase I has lots of work to do on the lagging strand removing all those primers. In addition, DNA ligase has to form covalent bonds between all the fragments of DNA.

On the opposite side of the replication fork from the lagging strand, there is no problem — the 3' of the primer on that side points toward the fork, in the direction that the enzyme crew is moving. Thus, once DNA polymerase III gets started making new DNA on that side of the fork, the enzyme can just keep going and the new strand is made in one long continuous piece known as the *leading strand.*

No matter how a replication fork is drawn, you can always tell where the leading and lagging strands will be built. Just do the following:

1. **Mark the 5' and 3' ends of the original two DNA strands at the replication fork.**
2. **Draw a primer on each side of the fork and label its 5' and 3' ends.**

   Remember that it will be antiparallel, or opposite, to the original DNA.
3. **Look at the 3' ends of the primers.**

   The 3' end that points toward the fork is where the leading strand will get started.

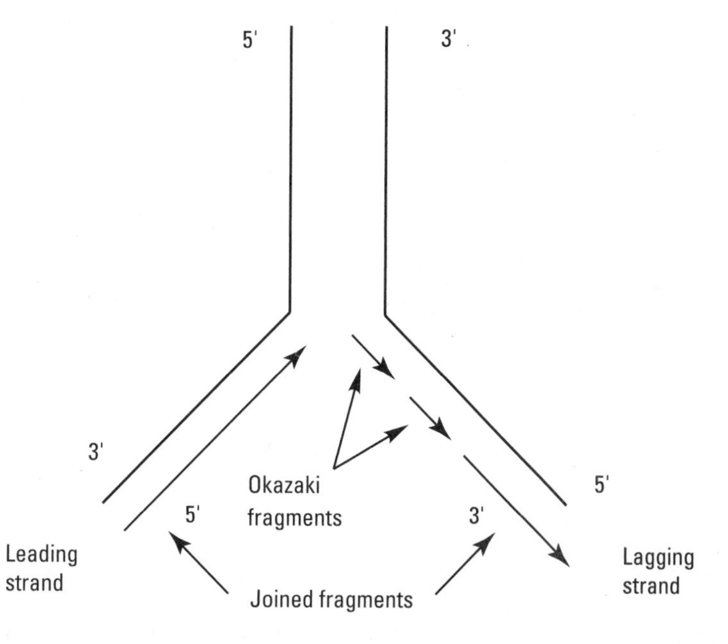

Figure 17-3: Leading and lagging strands.

Sometimes, people think that the terms *leading* and *lagging* refer to the original strands of DNA at the replication fork, but this idea is wrong. The terms leading and lagging refer to the new strands of DNA that are made as new partners for the original strands. So, if you're looking at a replication fork, you must find the new strands, or draw them in, in order to figure out which is the leading strand and which is the lagging.

In cells that have linear chromosomes, including those of humans and other eukaryotes, the lagging strand gets a little bit shorter every time the DNA is copied. The shortening occurs at the very ends of the chromosomes. Once the primer is removed, there is no room for a new primer, and DNA polymerase III can't copy the DNA without one. Fortunately, the ends of the chromosomes are capped with very repetitive sequences of DNA called *telomeres*. The telomeres don't contain any code that is needed by the cell, so they can be shortened without affecting cell function. However, once the telomeres become critically short, the cell can't divide anymore because it can't copy its chromosomes without losing important information.

Achieving immortality

Most cell types can divide only a certain number of times before their telomeres become too short and the cells can't copy their DNA. However, some cells in the body, like those that give rise to blood cells, have to divide forever. How do they do it? These cells activate a particular blueprint in the DNA to make a protein called *telomerase* that can fix telomeres. Telomerase carries around a little template, made of RNA, that lets it build new telomeres onto the ends of chromosomes, keeping the chromosomes long. This telomere repair lets the cells continue dividing indefinitely.

On a scary note, many cancer cells have figured out a way to turn on the telomerase blueprint, too. One reason tumors can keep growing and growing is because cancer cells within them have the ability to keep their telomeres long. By using telomerase to repair their telomeres, cancer cells can keep copying their DNA and thus keep dividing!

Chapter 18

Transcription and Translation: What's in a Gene?

In This Chapter

- Copying the genetic code to make RNA
- Reading the mRNA code
- Translating the mRNA code to protein
- Making mistakes in the DNA code

DNA contains the blueprints for important worker molecules in the cell, such as RNA and protein. The information in DNA is copied by transcription to make RNA. One type of RNA, called messenger RNA, carries the information for proteins to the ribosome. Ribosomes and molecules of tRNA work together to translate the mRNA message and build proteins. In this chapter, I present the details of transcription and translation and explore the effect of mutation on protein and cell function.

File It Under Genes: The Blueprints for RNA and Proteins

DNA contains the *genetic code* of the cell, a code that controls the cell's structure and function. However, the DNA itself is just the information, a detailed blueprint, not the active agents that carry out the plan. The plan is carried out by the molecules that are coded for by the DNA. So, the code in the DNA leads to the creation of worker molecules, and these worker molecules act according to the plan in the DNA and based on signals received by the cell. The actions of the worker molecules lead to the particular structure and function of the cell. Thus, the code in the DNA ultimately controls the structure and function of the cell.

Two types of worker molecules are encoded by the DNA:

- **RNA molecules** of several types are involved in the synthesis of proteins. Some RNA molecules act as information carriers, while others have enzymatic activity and catalyze reactions during protein synthesis. RNA molecules that act as enzymes are called *ribozymes.*
- **Proteins** are the main worker molecules of the cell. (For more on proteins, see Chapter 6.) They perform a wide range of cellular functions including catalysis (enzymes), structural support (cytoskeleton), signaling (receptors), transport (membrane proteins and cytoskeleton), and cellular identity (membrane proteins).

Defining a gene

A *gene* is a single blueprint for a worker molecule. If you think of your entire genome as a big file cabinet, each drawer would be a chromosome, and the files within the drawers would be the genes. Humans have 46 chromosomes, so your file cabinet would have 46 drawers. Based on the Human Genome Project, which determined the DNA sequence of the entire human genome, humans have about 22,500 genes scattered among those chromosomes. So, that would be 22,500 files in your file cabinet. Each file would contain the blueprint for the construction of a different worker molecule, either an RNA or a polypeptide (protein).

Going with the flow

The main worker molecules of the cell are proteins, and the genes for proteins are contained within the DNA. However, in order to build the protein, the information in the gene must first be copied into a messenger molecule, called *messenger RNA* (mRNA). The messenger RNA takes the copy of the blueprint to the ribosome where it's used to build the protein. DNA is copied into RNA by the process of *transcription*, and then an mRNA copy is taken to the ribosome where it's *translated* into a protein sequence.

The flow of information to build a protein in a cell starts with the DNA, then flows to mRNA, and then is used to build the protein. This pathway of information from DNA to RNA to protein is called the *central dogma of molecular biology.*

Make a Copy, Please: Transcription

The code in DNA is copied into RNA by a process called transcription. During transcription, the double helix of DNA is opened, and one strand of DNA is used as a pattern for the construction of an RNA molecule. Several types of RNA molecules perform different jobs in the cell (see Chapter 7). The RNA molecules either do a job directly for the cell, or, as in the case of mRNA, the mRNA molecules serve as a copy of important information.

Because protein synthesis is so important to the life of a cell, it's usually a big topic in molecular and cellular biology. Biology professors place lots of emphasis on the flow of information from DNA to mRNA to protein. With all this emphasis on the central dogma, you can easily forget that other kinds of RNA exist besides mRNA and that transcription makes all kinds of RNA. So, when you think about transcription, remember that this process can make several different types of RNA , but that only mRNA goes on to serve as the blueprint for a protein during the process of translation.

Locating the file

Genes may make up only a small percentage of the total DNA of a cell. In humans, for example, only about 2 percent of DNA contains the codes for protein. Scientists are still figuring out the functions of the other 98 percent, although much of it seems to be involved in regulating how the information in the DNA is used. (See Chapters 19 and 21 for more on gene regulation and the human genome.)

When a cell needs to build a protein, it recognizes the genes for proteins among all that other DNA. Cells recognize genes for proteins by their *promoters,* unique sequences of DNA that are located at the beginning of genes (see Figure 18-1).

If you think of genes as the files in your genetic file cabinet, then promoters are like the tabs that stick up off the files and label the files with their contents. In bacteria, promoters are 40 to 50 base pairs long and are located right next to the genes they label. In eukaryotes, promoters are more variable.

Proteins that turn on transcription recognize certain DNA sequences in promoters. For example, all bacterial promoters have two very similar sequences located in key spots within the promoter. These sequences are named for their distance from the spot in the gene where transcription begins, called the +1 ("plus one") site:

- The **–10 box** has the sequence TATAAT and is located ten bases away from the transcription start site.
- The **–35box** has the sequence TTGACA and is located 35 bases away from the transcription start site.

Although eukaryotic promoters are more variable, about 20 percent of eukaryotic promoters contain a common sequence TATAAA, called a TATA box ("ta-ta" box), located near the +1 site of the gene.

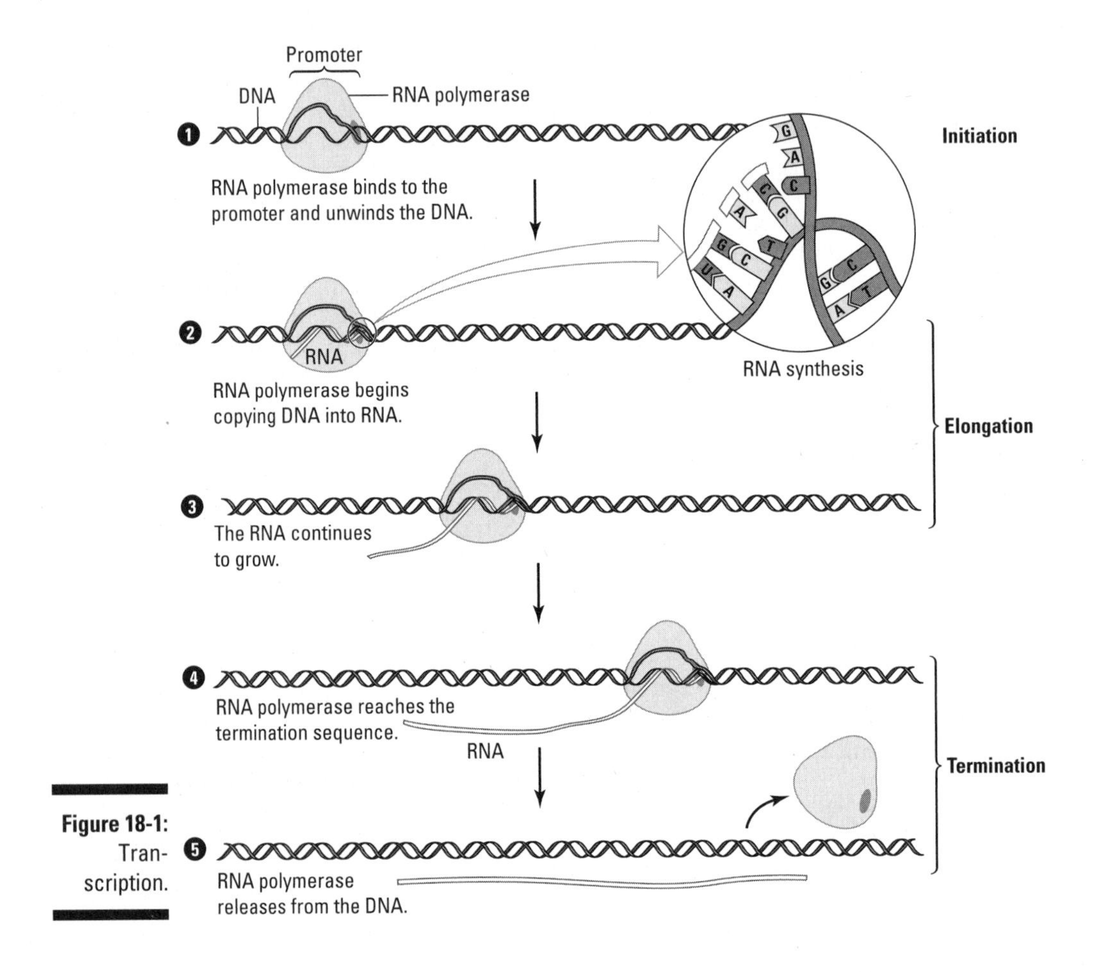

Figure 18-1: Transcription.

Hiring a worker

The enzyme that reads genes and copies the information into RNA is called *RNA polymerase* (see Figure 18-1). RNA polymerase opens the double helix and uses one of the strands of DNA as a template for the construction of an RNA molecule. In order to copy the code, RNA polymerase follows the base pairing rules (see Chapter 7) as it brings in RNA nucleotides to match up with the DNA nucleotides.

RNA polymerase catalyzes the formation of covalent bonds between the nucleotides in a growing RNA molecule.

By following the base pairing rules, RNA polymerase reads one strand of DNA and builds a complementary molecule of RNA. The strand that is read is called the *template strand* because it provides the pattern. The strand of DNA that isn't used by RNA polymerase is called the *nontemplate strand.*

The nontemplate strand of DNA is sometimes called the *coding strand.* This terminology can be confusing because people think the coding strand must be the one that RNA polymerase uses to get the code. But this assumption isn't correct — RNA polymerase uses the template strand. Both the nontemplate (coding) strand and the RNA molecule are complementary to the template strand, so they essentially contain the same code. The similarity between the code in the RNA and the nontemplate strand is the reason the nontemplate strand got the name coding strand.

When RNA polymerase pairs RNA nucleotides with the DNA nucleotides of the template strand, the nucleotides must be turned antiparallel to each other. So, RNA polymerase reads the template strand of DNA in the 3' to 5' direction in order to build the new RNA molecule in the 5' to 3' direction: In other words, the 5' end of incoming RNA nucleotides are joined to the 3' of the growing RNA chain.

RNA polymerase needs helpers in order to recognize promoters and begin transcription. These helpers are regulatory proteins that control the process of transcription (see Chapter 19). In bacteria, *sigma proteins* attach to RNA polymerase, forming a complex called a *holoenzyme.* The holoenzyme then binds to promoters and begins transcription. In eukaryotes, regulatory proteins called *transcription factors* (see Chapter 19) help RNA polymerase bind to promoters. Once RNA polymerase binds to promoters, it can begin reading and copying the DNA message into an RNA molecule.

Marking the end

Just like promoter sequences mark the beginning of genes, *transcription terminators* mark the ends of genes.

Transcription terminators are sequences in the DNA that trigger RNA polymerase to release the DNA and stop transcribing.

A transcription terminator may contain a code that creates a piece of RNA that folds back on itself, forming a hairpin loop. This lumpy piece of RNA essentially knocks RNA polymerase loose from the DNA, ending transcription of this gene. Of course, RNA polymerase will go on to transcribe many more genes and so can say "I'll be back!" to the terminator.

The synthesis of RNA molecules during transcription is an example of a chemical reaction that builds complex molecules from smaller pieces. Synthesis reactions like this one are typically endergonic (see Chapter 10) — in other words, requiring energy. In order to make transcription energetically favorable, cells bring in the RNA nucleotides as *ribonucleoside triphosphates* (NTPs), a very energy-rich form of nucleotides that contains three phosphate groups. The removal of two phosphates from each NTP as the nucleotide joined to the growing RNA molecule is so exergonic that it makes the overall process of transcription exergonic, too.

Finishing Touches: RNA Processing in Eukaryotes

After eukaryotic genes for proteins are transcribed, they're not quite ready to be translated. In fact, the RNA that is made from protein-encoding genes in eukaryotes is called a *pre-mRNA* or *primary RNA transcript* to indicate that the transcript isn't yet finished.

One issue that needs to be dealt with before translation is that eukaryotic genes for proteins don't just contain the code for the protein. The code for the protein contains short stretches of nucleotides, called *introns,* that break up the code (see Figure 18-2). The introns must be removed from the pre-mRNA before the mRNA is translated. The portions of the pre-mRNA that do contain the information for the protein are called *exons*. The exons remain in the finished mRNA and are expressed in the final protein.

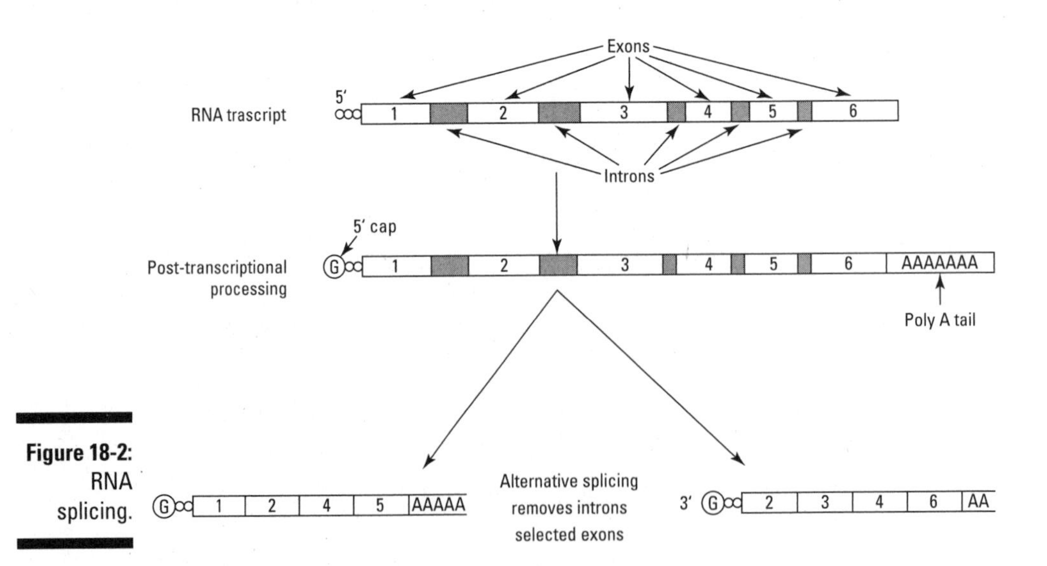

Figure 18-2: RNA splicing.

To remember the difference between exons and introns, think "Exons are expressed; introns interrupt."

Small particles made of RNA and protein remove introns from the pre-mRNA. The particles are called *snRPs* (pronounced "snurps") for *small nuclear ribonucleoproteins.* Several snRPs gather together to form *spliceosomes* that clip the pre-mRNA at the boundaries between exons and introns, remove the introns, and then form bonds between the exons. This method of removing introns is called *splicing* because the removal of introns is similar to the process of clipping out pieces of movie film and then resealing the remaining pieces to make an edited version of the film.

The enzymes in the spliceosome that cut and reseal the pre-mRNA seem to be part of the *small nuclear RNA* (snRNA) that makes up the snRPs. Most enzymes are proteins, but the snRNA in the snRPs is another example in a growing list of cases in which RNA molecules act as enzymes.

Along with splicing, the cell makes two more changes to the pre-mRNA to get it ready for translation:

- A protective cap, called the *5′ cap,* is added to the 5' end of the pre-mRNA. The 5' cap identifies the finished mRNA as an RNA that should be translated.
- Enzymes clip off the 3' end of the pre-mRNA and replace it with a piece of RNA that has between 100 and 250 adenine nucleotides. This *poly-A tail* protects the finished mRNA from being broken down by the cell.

Making a Protein: Translation

When a cell needs to build a protein, RNA polymerase reads the code for the protein in the DNA and copies it into a molecule of mRNA. The mRNA is then shipped out of the nucleus and into the cytoplasm where ribosomes and tRNAs use the code in the mRNA to build the protein. When the protein is finished, it will be available to do its job for the cell.

The code in mRNA is written in the pattern of the four nucleotides — C, G, A, and U — that occur in the strand. The length of the strand, the proportion of each nucleotide in the strand, and the order of the nucleotides can all be different from one mRNA molecule to another. For example, the mRNA that contains the code for the protein insulin would be quite different from the mRNA that contains the code for the protein collagen. Insulin and collagen are unique in shape and function — insulin is a globular signaling protein and collagen is a fibrous structural protein, so their primary structure, the sequence of amino acids in the protein, is also unique. The differences in the code in the mRNA molecules is what determines the differences in the primary structure of the protein.

Reading the code

The nucleotide code in mRNA molecules determines the primary structure of a protein. In other words, the pattern of nucleotides decides the type and ordering of the amino acids in the polypeptide chain. The cell "reads" the mRNA code in blocks of three nucleotides called *codons.* Each codon tells the cell which amino acid should be added to the polypeptide chain. Twenty different amino acids are found in cells, but you can make 64 three-letter codons out of four nucleotides. Thus, the genetic code is *redundant,* and some amino acids are represented by more than one codon.

Scientists have figured out what all 64 codons represent to the cell and have organized this information into a codon dictionary, shown in Figure 18-3. If you're given the nucleotide sequence of a piece of mRNA, you can use this dictionary to determine the type and order of amino acids represented by the mRNA by following these steps:

1. **Find the large row on the left of the table that is marked with the first letter of the codon you want to look up.**

   For example, to determine which amino acid is represented by the codon CAG, start with the second large row in the table because that row is marked with the letter C.

2. **Find the column in the table that is marked with the second letter of the codon you want to look up and then put your finger on the area of the table where the row you looked up in Step 1 intersects with the column.**

   For CAG, you find the third column because this column is marked with an A. Then, you slide your fingers along the table and find the section of the table where the row and column intersect. At the intersection, the names of two amino acids, histidine and glutamine, are each written twice.

3. **For the area of the table you identified in Step 2, find the small row at the right of the table that is marked with the third letter of the codon you want to look up.**

   Follow the line of the small row across to the intersection you identified in Step 2 to determine the correct amino acid. For CAG, the third letter is G, which marks the bottom small row of the area you identified in #2. Thus, the codon CAG represents the bottom amino acid, glutamine.

First Letter	Second Letter: U	C	A	G	Third Letter
U	phenylalanine	serine	tyrosine	cysteine	U
U	phenylalanine	serine	tyrosine	cysteine	C
U	leucine	serine	STOP	STOP	A
U	leucine	serine	STOP	tryptophan	G
C	leucine	proline	histidine	arginine	U
C	leucine	proline	histidine	arginine	C
C	leucine	proline	glutamine	arginine	A
C	leucine	proline	glutamine	arginine	G
A	isoleucine	threonine	asparagine	serine	U
A	isoleucine	threonine	asparagine	serine	C
A	isoleucine	threonine	lysine	arginine	A
A	methionine & START	threonine	lysine	arginine	G
G	valine	alanine	aspartate	glycine	U
G	valine	alanine	aspartate	glycine	C
G	valine	alanine	glutamate	glycine	A
G	valine	alanine	glutamate	glycine	G

Figure 18-3: The codon dictionary.

When you're decoding an mRNA molecule, you have to begin and end at the right place in the sequence. Each mRNA molecule has a section near the 5' cap that is recognized by the ribosome. This *ribosomal binding site* isn't part of the code for the protein. The protein code begins at the *start codon,* which has the sequence AUG.

Translation begins at the start codon closest to the 5' end of the mRNA molecule. The protein code is then decoded in the 5' to 3' direction until another special codon, a *stop codon,* is reached. Three stop codons — UAA, UGA, and UAG — mark the end of translation.

After you identify the start codon in the protein code, you can figure out all the codons from the start codon to the stop codon simply by marking off the nucleotides in groups of three. The grouping of the nucleotides into threes determines the *reading frame* for the protein code. In other words, by grouping the nucleotides into threes, you determine how the protein code will be read. For example, imagine that you want to decode a short piece of mRNA with the following sequence:

5' CUAGAUGACCUUGUGA3'

If you just start at the left end of the sequence and break the nucleotides up into threes, you'd get CUA, GAU, GAC, and so on. If, however, you start at the AUG closest to the 5' end, you get AUG, ACC, UUG, and UGA. These two groupings represent two different reading frames. If you look up the amino acids determined by these two different arrangements of codons, you see very different results! Only the second reading frame, which began with the start codon, gives you the correct amino acid sequence: methionine, threonine, leucine. Notice that the start codon, AUG, also represents an amino acid, methionine, whereas the stop codons don't specify an amino acid.

One way to remember the rules for decoding mRNA is to compare them to the rules for reading English, described in Table 18-1.

Table 18-1 Relating Rules for English to Rules for mRNA

English Rule	*mRNA Rule*
Read left to right.	Read 5' to 3'.
Look for words like "The" that begin sentences.	Find the start codon closest to the 5' end.
Break sentences up into words.	Break the code into threes.
Stop at ending punctuation marks.	Stop at stop codons.

The decoder: tRNA

Cells follow the same rules for decoding mRNA as people do. Cells, however, don't have a codon dictionary like the one shown in Figure 18-3. Instead, cells rely upon a decoder molecule called transfer RNA, or tRNA, shown in Figure 18-4. tRNA molecules carry in the amino acids for the protein and figure out where to put them in the growing polypeptide chain.

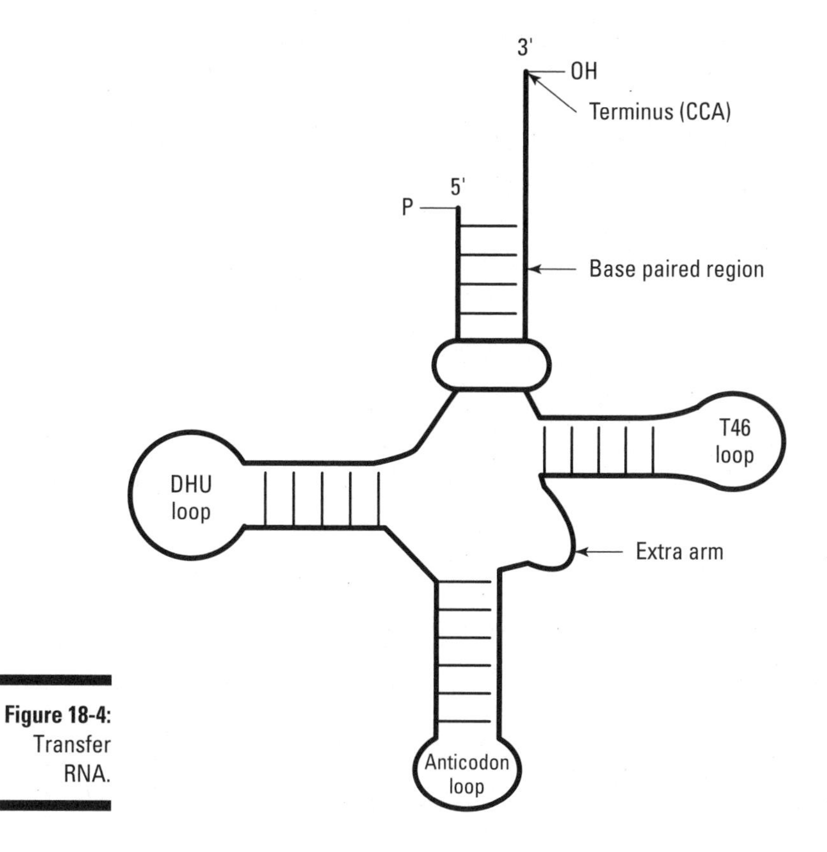

Figure 18-4: Transfer RNA.

To figure out where to put the amino acids, tRNA molecules have a special structure called an *anticodon.* Three nucleotides at one end of the tRNA form the anticodon. Anticodons fit up against codons in the mRNA, binding with hydrogen bonds just like the two halves of the DNA double helix.

In order for anticodons and codons to pair, they must have complementary sequences. For example, the start codon 5'AUG3' only fits correctly with the tRNA that has the anticodon 3'UAC5'. Just like all nucleic acids, the codons and anticodons pair with each other in an antiparallel arrangement, with opposite 5' and 3' ends.

In a cell, tRNA molecules with anticodons carry the amino acids. Sometimes people get confused and think that the codon dictionary shows which anticodons go with each amino acid, but that thought process is incorrect. People wrote the codon dictionary to serve as a reference for decoding mRNA. It totally skips over the role of the tRNA. So, the codon dictionary shows codons in the mRNA. The Table of the Genetic Code is for you to use when you're decoding the mRNA and has nothing to do with anticodons.

The matching of mRNA codons and tRNA anticodons correctly arranges the amino acids in the growing protein chain. The 20 amino acids that are found in proteins are carried by about 40 different tRNAs. tRNAs that carry amino acids are called *aminoacyl tRNAs.* Each aminoacyl tRNA carries its amino acid at its 3' end and has a unique anticodon in its anticodon loop that can pair only with its complementary mRNA codon. Thus, the mRNA code determines the order in which the different tRNAs deposit their amino acid cargo.

Master craftsman: The ribosome

The ribosome organizes the meeting of mRNA and tRNA molecules in order to translate the mRNA code into a sequence of amino acids. When the large and small subunits of the ribosome join to form a completed ribosome, pockets are formed inside the ribosome (see Figure 18-5). These pockets, called the *A site, P site,* and *E site,* organize the arrival and departure of the tRNAs as they bring their amino acids to the growing protein chain. During translation, tRNAs enter the ribosome through the A site, then move to the P site, and finally exit from the E site.

The ribosome catalyzes peptide bond formation between the amino acids as they're brought together by the tRNAs.

You can think of the ribosome as the craftsman that drives in the nail to hold together the building blocks brought in by the tRNAs. The part of the ribosome that catalyzes the peptide bond is made of rRNA. Thus, the ribosome acts as a ribozyme (say that ten times fast!).

Wobble power

Remember how 61 codons for amino acids represent just 20 amino acids? (The three stop codons don't represent amino acids.) And only 40 tRNAs carry those amino acids? So, why are there 61 codons if only 40 anticodons exist? How can 40 tRNAs recognize all 61 codons?

The answer is that the third nucleotide in the anticodon isn't very picky about which nucleotide it pairs up with. For example, if you look at the table of the genetic code shown in Figure 18-3, you can see that the amino acid proline is represented by the codons 5'CCU3', 5'CCC3', 5'CCA3', and 5'CCG3'. A single tRNA can match all four of these codons as long as its anticodon starts with 3'GG5'. The nucleotide in the third position in the anticodon can be any nucleotide. This variability in the third position of the anticodon is called *wobble* in the genetic code. Because of wobble, 40 tRNAs can recognize 61 codons!

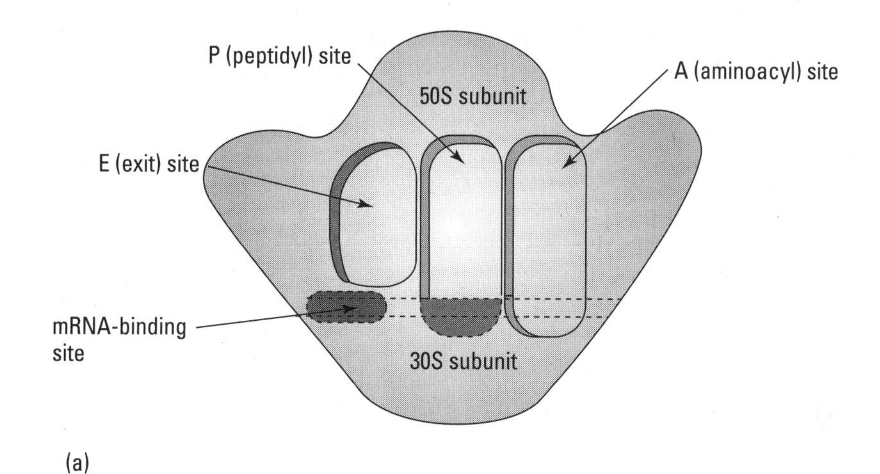

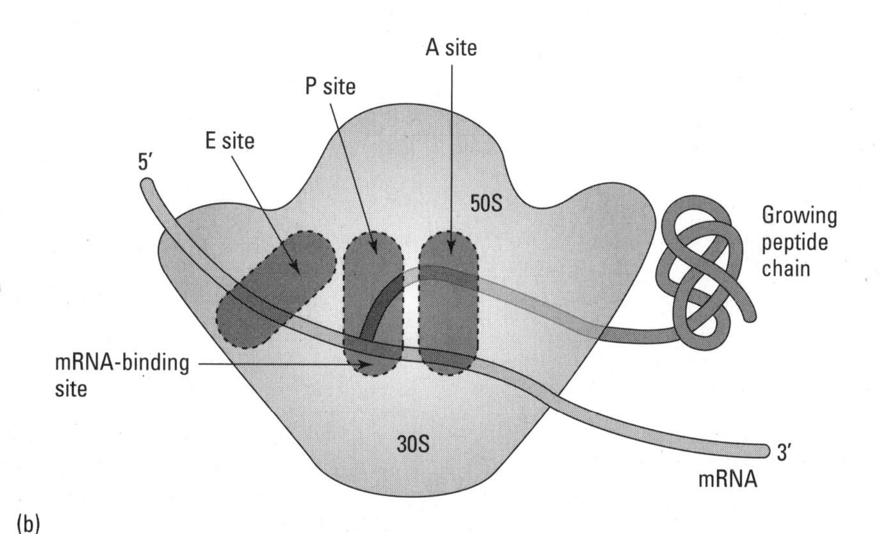

Figure 18-5: The ribosome.

The steps of translation

Translation occurs in three phases, shown in Figures 18-6 and 18-7.

- **Initiation of translation** begins when the small subunit of the ribosome binds to the ribosomal binding site on the mRNA. Then, the aminoacyl tRNA that carries methionine binds by its anticodon to the start codon. Finally, the large subunit of the ribosome binds to the small subunit and mRNA. At the end of initiation, the ribosome is complete, and the first tRNA is positioned in the P site of the ribosome. The second codon in the mRNA is positioned in the A site of the ribosome.

- **Elongation of translation** occurs as amino acids are brought together and joined to form a polypeptide chain. Elongation has several steps, which repeat until the polypeptide chain is complete:

  a. *An aminoacyl tRNA enters the A site of the ribosome.* The tRNA that enters the A site has the complementary anticodon to the codon in the A site. As the two tRNA molecules are side by side in the P and A sites of the ribosome, their amino acids are next to each other.

  b. *The ribosome catalyzes bond formation between the two adjacent amino acids.* The amino acid carried by the tRNA in the P site is attached to the amino acid carried by the tRNA in the A site. The growing protein chain is temporarily held by the tRNA in the A site.

  c. *The ribosome and mRNA slide relative to each other.* You can think of this step, called *translocation,* like the movement of fabric through a sewing machine. The tRNA that was in the P site is pushed into the E site, whereas the tRNA that was in the A site is pushed into the P site. This places a new codon in the A site and the growing polypeptide chain in the P site. The tRNA in the E site exits the ribosome, and the A site is available for a new aminoacyl tRNA to enter, and the steps of elongation repeat.

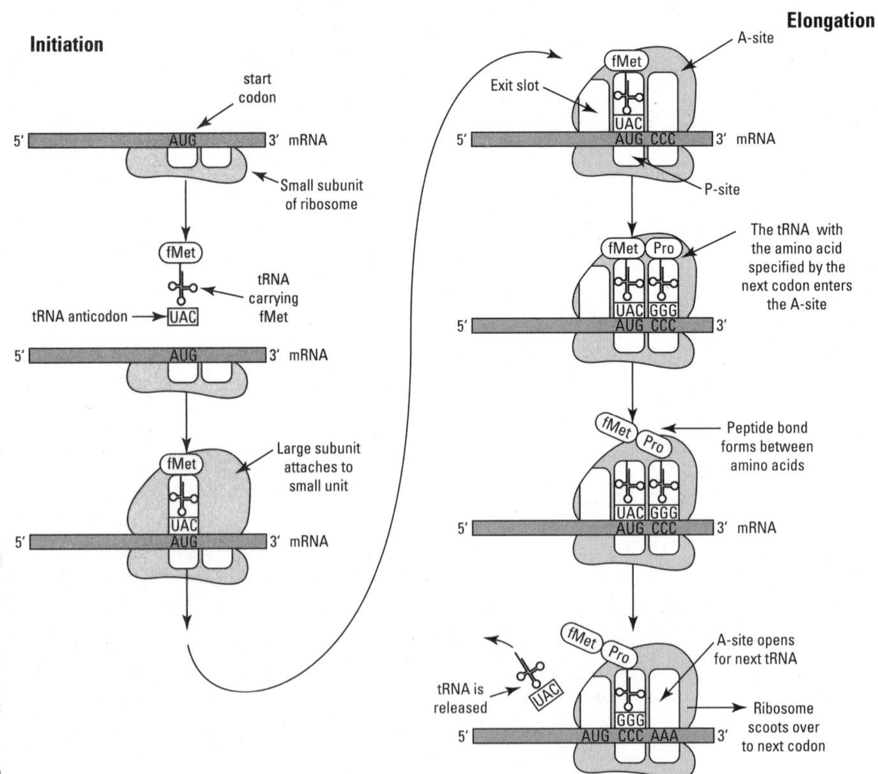

Figure 18-6: Initiation and elongation of translation.

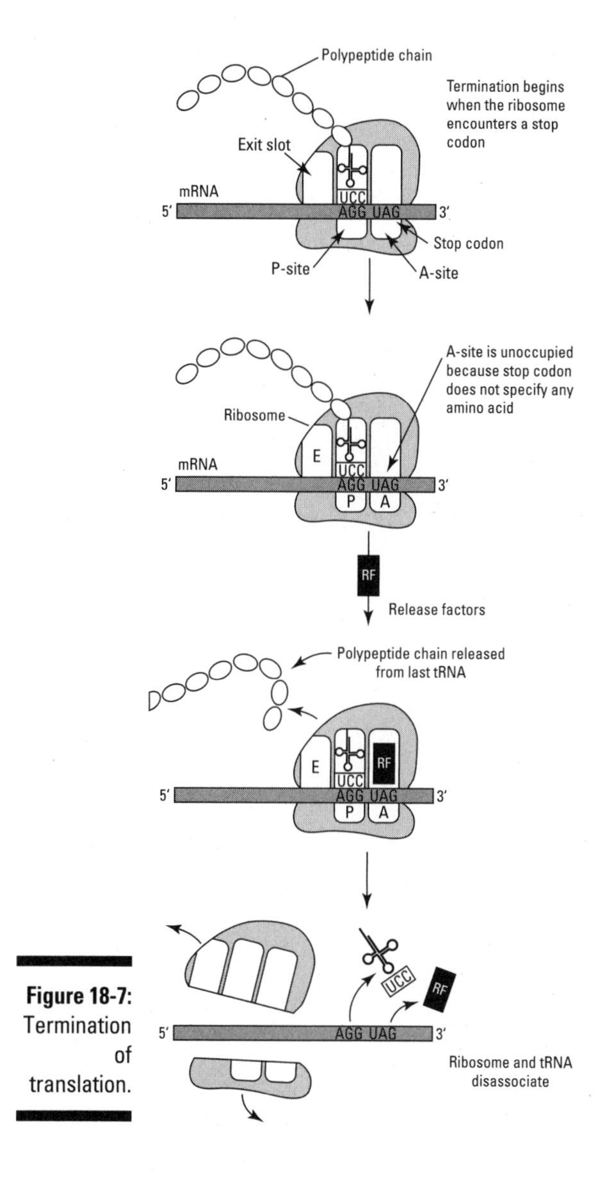

Figure 18-7: Termination of translation.

- **Termination of translation** begins when a stop codon enters the A site of the ribosome. Stop codons aren't recognized by aminoacyl tRNAs. Instead, the stop codon is recognized by an enzyme called *release factor.* Release factor enters the ribosome and catalyzes the breaking of the bond between the growing polypeptide chain and the tRNA that holds it. Release factor releases the polypeptide chain from the ribosome so that it can go out into the cell and be converted into a finished protein.

The synthesis of a polypeptide is an endergonic process (see Chapter 10). Cells make translation energetically favorable by coupling translation with the hydrolysis of guanosine triphosphate (GTP). Just like ATP, GTP has three phosphate groups. The hydrolysis of the bonds between the phosphate groups is highly exergonic. When GTP hydrolysis is coupled with the formation of peptide bonds, the overall process is exergonic and will occur spontaneously in cells.

Don't Drink and Drive: Mutation

The genetic information in DNA is transcribed into RNA, which is then translated into protein. If the DNA code is changed, then RNA molecules will also change, which may lead to a change in proteins. If proteins are changed so that they can't do their job for the cell, then cell function may be altered. For humans, changes in protein function can lead to disease. For example, cancer, hemophilia, Tay-Sachs, and cystic fibrosis are all human diseases that result from changes in the DNA. Changes in DNA are called *mutations.*

Everybody makes mistakes

DNA polymerase, the enzyme that copies DNA when cells are about to divide, is a darn good enzyme. It doesn't make many mistakes (as compared to other enzymes that copy nucleic acids), and it proofreads its own work. Still, when all is said and done, DNA polymerase makes approximately one uncorrected mistake for every billion base pairs of DNA that it copies. Mutations that "just happen" because of the way the enzyme functions are called *spontaneous mutations*.

DNA polymerase's *mutation rate* of one error for every billion base pairs copied does sound pretty good, doesn't it? There's just one problem — cells can have a lot of DNA. You, for example, have about 3 billion base pairs of DNA in every one of your cells. That means that every time one of your cells divides, and DNA polymerase copies the DNA, the resulting cells will have about six mutations. When you remember that your DNA is really valuable to your health, the picture begins to look a little more scary.

In addition, certain chemicals and types of radiation in the environment of the cell can increase the mutation rate of DNA polymerase. Agents that increase the mutation rate are called *mutagens.* Mutations that result from the action of mutagens are called *induced mutations.* If cells are exposed to mutagens, they will accumulate more mutations than normal, which increases the chances that cell function will be altered. In humans, higher rates of disease often follow exposure to environmental mutagens. For example, Japanese people who survived the nuclear bombings of Hiroshima and Nagasaki during World War II got cancer ten times more frequently than other Japanese people.

Dealing with the consequences

DNA polymerase makes mistakes all the time, yet cells continue to function just fine. So although the idea of mutation is scary, cells seem to be able to live with a certain level of it. One reason cells continue to function despite mutation is that not all mutations cause changes in proteins. Some mutations in DNA occur outside of genes. Even if a mutation does occur in a gene, the mutation may cause a change in mRNA that doesn't affect the final protein. For example, if the codon AAA were changed to AAG, the amino acid would not change; both of these codons represent the amino acid lysine. Mutations like this one that don't affect the final protein are called *silent mutations*.

Several other types of mutation, however, do affect proteins:

- **Missense mutations** result in changes in the amino acids of a protein. A single base pair change in the DNA can result in a missense mutation. For example, if an mRNA codon were changed from AAA to AGA, the amino acid would change from lysine to arginine. The effect of missense mutations on the protein depends on the differences between the amino acids and where in the protein the replacement occurs.
- **Nonsense mutations** occur when a stop codon is introduced into mRNA. For example, the codon UGG represents the amino acid tryptophan. If a mutation occurred to change UGG to UGA, then the protein would stop early. Nonsense mutations typically have severe consequences for protein function.
- **Frameshift mutations** cause the reading frame of the mRNA to be altered. For example, if DNA polymerase slips while copying the DNA, it may copy one nucleotide twice, which would cause an *insertion* of an extra nucleotide into the mRNA. All the codons after the insertion point would be off by one, and the resulting protein would be completely different from the original plan.

  Deletions of nucleotides can also cause frameshift mutations. Because frameshift mutations cause *extensive missense,* they typically have severe consequences for the protein.

One gene, one enzyme

Have you ever noticed the warning "Attention phenylketonurics, this product contains phenylalanine" on products that contain the artificial sweetener aspartame (Nutrasweet)? Did you ever wonder why this warning appeared?

The warning is there because of a mutation. Some people are born with mutations in the gene for the enzyme that converts the amino acid phenylalanine into the amino acid tyrosine. This mutation may not sound like a big deal, but it is. If you're born with this mutation and consume foods with phenylalanine in them, the phenylalanine will build up in your system. The accumulation of phenylalanine causes progressive mental deterioration and seizures. In fact, before this disorder was understood, affected babies would be born normal and then progressively decline and die. If you have children, you probably remember that your babies' heels were pricked at birth. Babies' blood is tested right away to make sure that the enzyme levels are normal and that the babies are free from phenylketonuria.

In 1908, a British doctor named Archibald Garrod made observations of some of his patients who were affected by a disorder very similar to phenylketonuria. Based on his observations, he concluded that some people were born with "inborn errors of metabolism." Scientists now know that these "inborn errors" are mutations in the genes for proteins like enzymes.

Chapter 19

Control of Gene Expression: It's How You Play Your Cards That Counts

In This Chapter

- Studying gene regulation in bacteria
- Finding the connection between gene regulation and cell differentiation
- Exploring opportunities for gene regulation in eukaryotes

Genes are expressed when cells use the blueprint in the gene to build a particular worker molecule in the cell. At any given time, cells use only a fraction of the genes in their genome. Powerful changes in a cell occur through control of which genes are turned on and off in a cell. In this chapter, I cover both the importance of gene regulation and also how cells control the expression of their genes.

Controlling the Situation: Gene Regulation and Information Flow

Gene regulation is about cells choosing which genes to use and which not to use. Most types of cells have a complete set of genes, yet cells use only a fraction of those genes at any given moment. Cells may change what genes they're using depending on signals or changes in the environment. The information in genes is copied into RNA and then proteins, so when cells change which genes they're using, it affects which RNA molecules and proteins are available in the cell.

For example, some bacteria can make a very long lived resting structure called an *endospore.* Endospores let these bacteria go into hibernation if conditions are bad and then come out of hibernation when conditions get better. Bacteria need the proteins to make endospores only if conditions get bad, so it would be a waste of energy if they made them all the time. So, most of the time, the bacteria doesn't use the genes for endospore-making proteins. Then, when conditions get bad, the bacteria access the information in the genes needed to make endospores.

Scientists use the following terms to describe gene regulation:

- Genes are *expressed* when cells are using the information in the gene to build a molecule called the *gene product* (usually a protein but could also be RNA). Another way to say that a gene is being expressed is to say that the gene is on.
- Genes aren't expressed when cells aren't using the information in the gene. The gene product isn't present in the cell, and the gene is off.

Becoming a specialist

Gene regulation is important when cells become specialized for certain tasks. Take you, for example. You started as a single cell that divided by mitosis (see Chapter 13) to give rise to the trillions of cells in your body. As the cells divided, they made copies of your DNA so that every cell got a complete copy. Your original single cell was a *stem cell* that had the ability to become any type of cell. However, as it divided by mitosis, its descendants were given signals to become specialists in certain tasks. For example, some of your cells became skin cells that protect you from invading microorganisms, or they became heart cells that contract to keep your blood flowing.

Once cells become specialists, they normally don't change jobs or become stem cells again.

Cells that are specialists are *differentiated* or programmed for a certain task. Cells become differentiated as they turn on certain genes and turn off others. Each type of differentiated cell in your body requires unique tools in order to do its job. So, differentiated cells build the correct tools, or proteins, that they need.

For example, skin cells build lots of a strong protein called keratin that helps toughen up your outer layer. Heart cells make the proteins actin and myosin so that they can contract. The blueprints for keratin, actin, and myosin are located in different genes. Skin cells and heart cells would waste energy and resources if they made every protein they could all the time, so instead, these differentiated cells use just the genes they need. You can think of a skin cell and a heart cell as both having the same set of files, but in skin cells, certain drawers of the cabinet are locked while others are open. Heart cells have a different pattern of locked and open drawers. Each cell type becomes different because it expresses a different set of genes. In other words, cell differentiation is achieved through *differential gene expression.*

Cells that are differentiated are specialized for certain functions. Differentiated cells have access only to the genes they need to survive and do their job for the organism.

Keeping house

Some proteins are essential to the survival of the cell in any conditions. For example, proteins needed for basic metabolism like the enzymes that catalyze reactions during glycolysis (see Chapter 11) are needed so that cells can transfer energy from food to ATP. Also, cells always need enzymes such as RNA polymerase so that cells can make proteins. Without the enzymes for glycolysis or RNA polymerase, cells wouldn't have the ATP and proteins they need to do anything.

Everything old is new again

Stem cells are unique because they have the ability to become any type of cell. In other words, they can access any of the genetic files in the cells genomic file cabinet. Stem cells are important to doctors and scientists who hope to be able to figure out how to use them to heal people.

For example, someone who has diabetes because his pancreas can't make insulin may be healed if stem cells can be transplanted into his pancreas and triggered to become new, functioning pancreatic cells. Or someone who has a paralyzing spinal cord injury may be able to walk again if stem cells can be persuaded to grow into new nervous tissue in the injured person's spine.

By studying stem cells and comparing them to differentiated cells, scientists hope to discover how to control the switch from stem cell to differentiated cell and back again. Apparently, two major events happen as a stem cell becomes a differentiated cell. First, certain genes that let the cell access all the genetic files are turned off. Second, another set of genes that are specific to the function of the differentiated cell are turned on. So, as cells differentiate, they lose access to some genetic files and activate the genes for their new function.

One roadblock to making differentiated cells go backward and become stem cells again has been regaining the lost access to all the genetic information of the cell. Recently, scientists discovered that a single gene, called G9a, controls the switch that restricts access to genetic information. With this knowledge, scientists may soon be able to figure out how to reverse the effects of G9a and regenerate stem cells from differentiated cells.

The genes for these essential *housekeeping proteins* are typically in use all the time in most cells. Other genes, for proteins involved in special functions, are turned on and off as needed. Thus, cells have two types of genes based on gene expression:

- **Constitutive genes** are genes that are expressed all the time. These genes contain the blueprints for proteins that do essential housekeeping functions for the cell.
- **Regulated genes** are turned on and off as they're needed by the cell. Regulated genes contain the blueprints for proteins that are needed by the cell in some situations but not others.

I Can Be Flexible: Gene Expression in Bacteria

Scientists first discovered the basic principles of gene regulation by studying how gene expression works in bacteria. Bacteria regulate their gene expression in order to respond to an ever-changing environment. For example, the availability of food and water changes constantly, and bacteria must be able to take advantage of their current situation in order to survive. Bacteria respond to environmental changes by turning on genes for proteins that will help them survive. Genes are turned on and off by DNA-binding proteins that bind to DNA and control transcription. (For more on transcription, see Chapter 18.)

The basic steps of gene regulation in bacteria are as follows:

1. **The cell receives an environmental signal.**
2. **The environmental signal either activates or inactivates a DNA-binding protein.**
3. **DNA-binding proteins either bind to or let go of regulatory sequences in the DNA.**
4. **Transcription is turned on or off by the DNA-binding proteins.**

Organizing bacterial genes

Bacteria organize multiple genes under the control of one promoter. The set of genes plus promoter is called an *operon.*

Typically, the genes in one operon contain the blueprints for proteins that work together in a single process. That way, transcription and translation of the operon make all the proteins needed for the process at the same time.

Following are the important elements of a bacterial operon:

- The promoter marks the beginning of the operon. RNA polymerase binds to the operator to begin transcription.
- *Structural genes* are genes for proteins within the operon. One operon contains several structural genes.
- The *operator* is a regulatory sequence of DNA located between the promoter and the structural genes. DNA-binding proteins bind to the operator to control transcription of the operon.

Taking E. coli to dinner

Escherichia coli, better known as *E. coli,* is best known to most people as the bringer of gastrointestinal distress. However, your view of *E. coli* is probably skewed from the bad rap the bacterium gets in the news media. Basically, a few strains of *E. coli* run around causing very dangerous diseases and give the whole species a bad name. Many nicer strains of *E. coli* exist, some of them happily living in your small intestines right now. Others are excellent bacterial lab rats, helping scientists learn about cells and how they function. In particular, scientists have discovered a lot about the structure and function of DNA, including gene regulation, from studying *E. coli.*

Like you, *E. coli* uses enzymes to break down food molecules. *E. coli* has hundreds of genes for all the different enzymes it makes. Some genes are constitutively expressed. In other words, *E. coli* makes them all the time. Other genes are regulated — *E. coli* turns them on when it needs them and turns them off when it doesn't. For example, the sugar glucose is usually available to *E. coli,* so *E. coli* always makes the enzymes it needs to break down glucose. Other food sources are more unpredictable, so *E. coli* turns on the enzymes to break down these food sources only when they become available.

Looking at lac

Lactose, the sugar found in milk, is occasionally available to *E. coli*. The genes for the enzymes to break down lactose are located within the *lactose operon*, or *lac operon* (see Figure 19-1). The *lac* operon has three components:

- The *lac promoter* is the binding site for RNA polymerase.
- The *lac operator* is the binding site for a DNA-binding protein called the *lac repressor protein.*
- Three structural genes contain the blueprints for the two proteins needed to breakdown lactose plus a blueprint for a protective enzyme.
  - The *lacZ* gene is the blueprint for the enzyme *beta-galactosidase* (β-galactosidase). Beta-galactosidase catalyzes the splitting of the disaccharide lactose into two monosaccharides, glucose and galactose, and then glycolysis breaks it down (see Chapter 11).
  - The *lacY* gene is the blueprint for the membrane protein *galactosidase permease,* which brings lactose into the cell.
  - The *lacA* gene is the blueprint for *transacetylase,* a protective enzyme that allows certain sugars to be removed from the cell when the sugars are too plentiful.

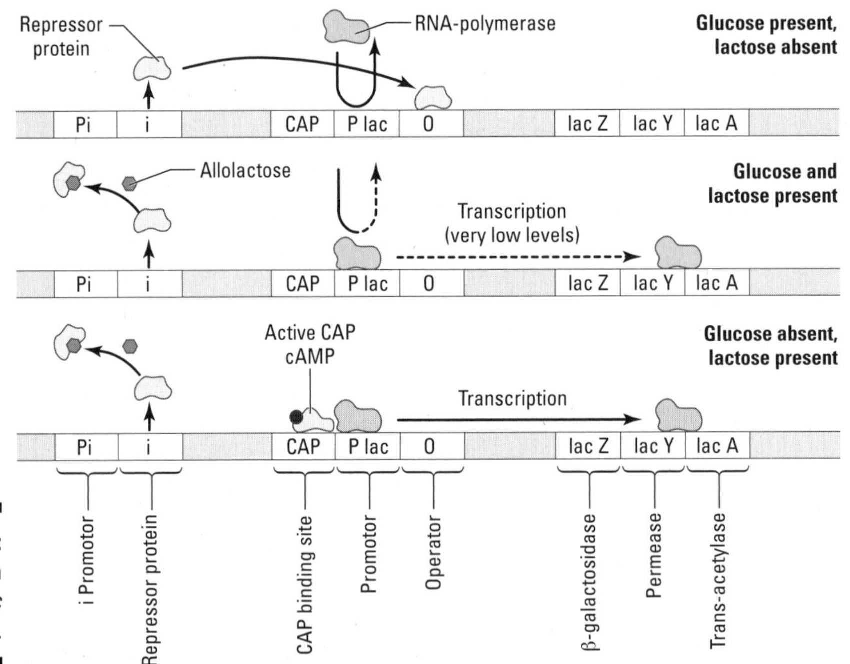

Figure 19-1: Regulation of the *lac* operon.

Feeling repressed

The DNA-binding protein that controls transcription of the *lac* operon is called the *lac* repressor protein.

The *lac* repressor is a DNA-binding protein that regulates the *lac* operon. When the *lac* repressor is active, it binds to the operator, and transcription is blocked.

The *lac* repressor has two binding sites that are critical to its function:

- A **DNA-binding site** that binds to the operator sequence of the *lac* operon
- An **allosteric site** that binds to an isomer of lactose (called allolactose)

The allosteric site of the *lac* repressor controls its activity. When the allosteric site is empty, the *lac* repressor is active. When lactose is available, an isomer of lactose binds to the allosteric site, inactivating the *lac* repressor. The operator of the *lac* operon is right next to the promoter, so when the *lac* repressor is active, it gets in the way of RNA polymerase. If RNA polymerase can't bind to the promoter, then transcription can't occur.

The blueprint for the *lac* repressor is contained in the *I* gene, which isn't part of the *lac* operon. The *I* gene is constitutively expressed, so the *lac* repressor is always available in the cell. When the *lac* repressor is active, it shuts down the *lac* operon. Because the active repressor shuts down the operon, regulation of the *lac* operon by the *lac* repressor is called *negative control.*

Game on: Inducing the lac operon

Lactose is the *inducer* of the lac operon: When lactose is available, it turns the *lac* operon on so that *E. coli* can make the enzymes needed to break down lactose.

Lactose regulates the *lac* operon by induction.

Induction, or turning on, of the *lac* operon allows *E. coli* to use the lactose as a food source, which makes sense if you think about things from the perspective of *E. coli*; when lactose is available, *E. coli* "wants" to eat it.

The tricky part about understanding gene regulation is to think about how lactose turns on the *lac* operon. After all, *E. coli* can't really think about what it wants; it's only a bacterium after all, so how does it "know" when lactose is available? The answer is the *lac* repressor. When lactose is available, an isomer of lactose (allolactose) binds to the *lac* repressor and inactivates it. The *lac* repressor lets go of the operator, and RNA polymerase can bind to the promoter (Figure 19-1). Transcription of the *lac* operon can occur and the enzymes to break down lactose are made.

Game over: Repressing the lac operon

The beauty of the regulation of the *lac* operon is its efficiency: when lactose is available, lactose turns on the *lac* operon. *E. coli* makes the enzymes to break down lactose and uses the lactose as a food source. As *E. coli* uses up the lactose, the allolactose isn't available to bind to the repressor protein. The repressor protein becomes active again, and binds to the operator. So, when the lactose is gone, the *lac* operon is turned off again. Again, this process makes sense from the perspective of *E. coli* — if it doesn't have any lactose, why make the enzymes to break it down? The tricky part is remembering how this process works: No lactose means no isomer (allolactose) bound to the allosteric site of the *lac* repressor, which means that the repressor protein binds to the operator and blocks transcription.

To help you remember the regulation of the *lac* operon, think like *E. coli*. If lactose is available, would you want to make the enzymes to break it down? Of course! So, when lactose is available, the *lac* operon is on. If lactose isn't available, do you want to make the enzymes to break it down? Why bother? So, when lactose isn't available, the *lac* operon is off.

Advancing to the next level: Catabolite repression of the lac operon

Lactose isn't the only sugar that has an effect on the regulation of the *lac* operon; glucose is a player in the game, too. *E. coli* always makes the enzymes to break down glucose so that the bacterium is always ready to use glucose when glucose is available. In fact, if glucose is available, *E. coli* won't even bother making the effort to break down other sugars like lactose. How can a simple bacterium make these kinds of choices? The answer is through the regulatory DNA-binding protein *catabolite activator protein* (CAP), which responds to the levels of glucose available to the cell.

CAP is an *activator* protein, which means it turns on transcription of genes (as compared to *repressor* proteins that turn off transcription).

When CAP is active, it turns on transcription of catabolic operons like the *lac* operon that produce enzymes to break down different food sources (other than glucose).

- **When glucose is available, CAP is inactive,** so *E. coli* doesn't make new enzymes and doesn't bother breaking down food molecules other than glucose.
- **When glucose runs out, CAP becomes active** and turns on catabolic operons, which allows *E. coli* to make enzymes for food sources other than glucose.

Reguation by CAP makes sense from the perspective of *E. coli; E. coli* always makes the enzymes to break down glucose, so if glucose is available the bacterium just eats the glucose. If glucose runs out, *E. coli* needs to find new food sources, so CAP turns on catabolic operons that produce enzymes for other food sources. That way, *E. coli* won't run out of food.

The details of how CAP responds to the presence of glucose are a bit complicated, but very similar to the workings of the *lac* repressor protein (see the preceding section "Feeling repressed"). Like the *lac* repressor, CAP is an allosteric protein with two important binding sites:

- A *DNA-binding site* that binds to a regulatory sequence of DNA called the CAP-binding site. The *CAP-binding site* is a sequence of DNA next to the promoters of catabolic operons. When CAP binds to the CAP-binding sites in the DNA, it enhances transcription of the adjacent operons.
- An *allosteric site* that binds to a signaling molecule named *cyclic AMP* (cAMP). When cAMP is bound to CAP, CAP is active. When cAMP isn't bound to CAP, CAP is inactive. In other words, CAP needs cAMP in order to function. (Just like I need a cup of tea in the morning!)

CAP and cAMP function as a team to monitor and respond to the presence of glucose. The amount of cAMP in the cell changes as the amount of glucose changes:

- When glucose levels are high, cAMP levels are low.
- When glucose levels are low, cAMP levels are high.

You can think of cAMP like an alarm signal. When glucose runs out, more cAMP molecules run around the cell saying, "Oh no! We're running out of food!" The cAMP molecules bind to CAP and activate it. Together, cAMP and CAP bind to the CAP binding site near the promoters of catabolic operons and turn them on, so *E. coli* makes enzymes to break down other food molecules besides glucose.

The levels of cAMP change as glucose levels change because glucose acts as a regulator of the enzyme adenylate cyclase, which makes cAMP out of ATP. When glucose is available, it binds to an allosteric site on adenylate cyclase and inactivates the enzyme so that no cAMP is made. When glucose isn't available, it can't bind to adenylate cyclase. Adenylate cyclase is active and produces cAMP from ATP.

To fully understand the regulation of the lac operon, you have to put the two regulatory systems together: negative control by the lac repressor combined with positive control by CAP. Remember that these two systems are independent of each other:

- Lactose is the environmental signal that interacts with the *lac* repressor.
- Glucose is the environmental signal that interacts with CAP.

So, when you think about the regulation of the *lac* operon, consider these two systems separately. You can think of lactose as the on/off switch for the *lac* operon. When lactose is present, the operon is turned on. When lactose is absent, the operon is turned off. Glucose acts as volume control. When glucose is present, the volume of transcription on catabolic operons, including the *lac* operon, is very low. When glucose is absent, the volume of transcription of the *lac* operon gets cranked way up. The combined effects of these two systems determine the level of transcription of the *lac* operon:

- If glucose is present and lactose is absent, the volume is low, and the switch is off. No transcription of the *lac* operon occurs.
- If glucose and lactose are present, the volume is low, and the switch is on. Very low levels of transcription of the *lac* operon occur.
- If glucose is absent and lactose is present, the volume is high, and the switch is on. Very high levels of transcription of the *lac* operon occur.

The hardest part of understanding gene regulation is getting the details straight on which molecules bind where and understanding their effects. The best strategy I've found for learning these details is to practice drawing the regulation of the *lac* operon in different conditions (different sugars available). In other words, practice re-creating Figure 19-1 for yourself on a blank piece of paper. Draw the *lac* operon and then say to yourself, "If lactose is present, what is the effect?" Draw all the regulatory molecules onto the *lac* operon, describing what is happening out loud. Repeat your drawings for all the conditions shown in Figure 19-1. The combination of drawing, seeing, and hearing will help reinforce the information. Practice until you can re-create Figure 19-1 with ease, narrating as you go.

The Master Plan: Gene Expression in Eukaryotes

The regulation of gene expression in eukaryotes is more complicated than the regulation of gene expression in bacteria. However, the regulation of gene expression in eukaryotes and bacteria have some similarities:

- Gene regulation at the level of transcription, called *transcriptional control,* occurs in both bacteria and eukaryotes. Regulation of the *lac* operon in *E. coli* is an example of transcriptional control in bacteria. In eukaryotes, other types of gene regulation occur, but transcriptional control is the most common mechanism.
- DNA-binding proteins regulate the activity of genes by binding to regulatory sequences in the DNA. Repressor proteins and CAP are both examples of important regulatory proteins in bacteria. In eukaryotes, many different types of regulatory proteins interact with each other and bind to regulatory sequences in the DNA to control transcription.

Seizing the opportunity

The pathway from gene to protein is more complex in eukaryotes than the pathway is in bacteria (see Chapter 18), which creates many more opportunities for regulation of gene expression in eukaryotes. Genes are expressed when their gene product is present in the cell, so any step in the pathway from gene to protein represents an opportunity for control. In eukaryotes, the major opportunities for control are as follows (see Figure 19-2):

- **Access** to genes is affected by the compaction of DNA into chromosomes.
- Transcriptional control regulates whether RNA polymerase binds to promoters and transcribes the DNA.
- Control of RNA processing determines whether the pre-mRNA is converted into a finished mRNA molecule.
- Control of mRNA stability determines how long mRNA remains in the cell available to be translated.
- Translational control regulates whether mRNA is translated into protein.
- Control of protein modification determines whether the polypeptide made by translation is modified into a fully functional protein.

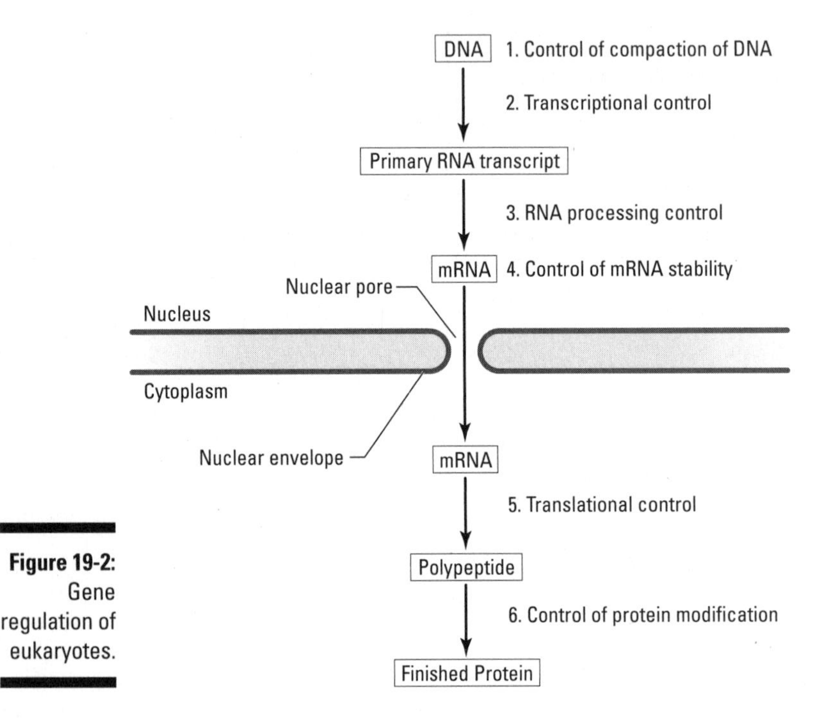

Figure 19-2: Gene regulation of eukaryotes.

Unpacking the plan

Cells contain a lot of DNA, much more than would fit inside the cell if the cells didn't do a little creative packaging. For example, if you took the DNA out of just one of your cells and laid all the DNA molecules (each of your 46 chromosomes) end to end, the DNA molecule would be about 2 to 3 meters long. That's about 6 feet in length, which may even be taller than your height! Of course, the string of DNA molecules would be extremely thin, so thin that you couldn't even see it with your naked eyes. Still, that's a long piece of DNA to pack into each and every cell.

Eukaryotic cells solve the DNA packing problem by packaging the DNA into *chromatin.* Chromatin is a complex of DNA wound around proteins.

To package the DNA into chromatin, cells do the following process, shown in Figure 19-3:

1. **The DNA molecules are wound around positively charged proteins called histones that gather together in groups of eight proteins.**

   The negatively charged DNA is attracted to the positively charged histones. When the DNA is wrapped around these clusters of histones, it looks like beads on a string. The beads of DNA plus histones are called *nucleosomes.*

2. **The string of nucleosomes and DNA twist until they form coiled fibers that are 30 nanometers wide.**

   These fibers are called *30-nanometer fibers.*

3. **Continued twisting of the DNA and protein fibers packs the DNA even more tightly until the DNA is very condensed.**

   To imagine how this compaction works, take a rubber band and keep twisting it in one direction until it beings to form knots. In its most condensed form, which occurs during mitosis and meiosis, DNA becomes visible under the microscope as chromosomes. (See Chapters 13 and 14 for more on mitosis and meiosis, respectively.)

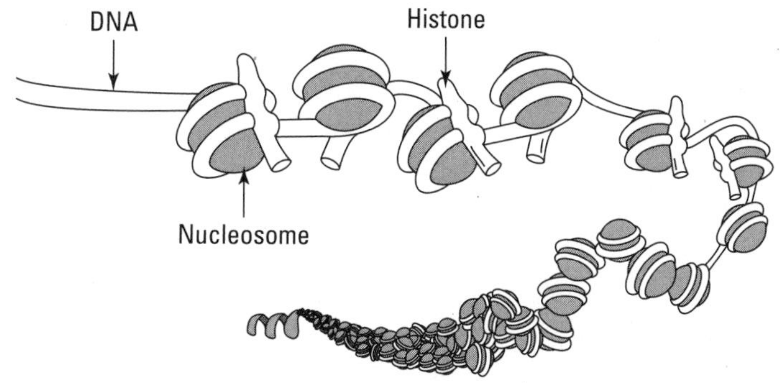

Figure 19-3: Compaction of DNA.

Gene expression is blocked by the condensation of DNA. When the DNA is tightly coiled up, regulatory proteins and RNA polymerase just can't get access to the genes. Before transcription of a gene can occur, the chromatin around the gene must be *decondensed.* One control of chromatin condensation is through proteins called *histone acetyl transferases* (HATs). HAT proteins attach acetyl groups to histone proteins, which neutralize their positive charge. When the histones lose their positive charge, DNA is less attracted to them, and the bond between DNA and histones is weakened. When the bonds within nucleosomes loosen, the DNA decondenses.

Controlling transcription

Transcription in eukaryotes is controlled by DNA-binding proteins, called *regulatory proteins,* which bind to *regulatory sequences* of DNA located near promoters. Regulation by DNA-binding proteins is the same fundamental mechanism for transcriptional control in bacteria. However, the types of regulatory proteins and regulatory sequences are more diverse in eukaryotes.

Several important regulatory sequences regulate gene expression in eukaryotic cells:

- **Promoter proximal elements** are regulatory sequences located very close to promoters. When regulatory proteins bind to promoter proximal elements, transcription is turned on. The exact sequence of promoter proximal elements is unique for each gene so that eukaryotic cells can identify which genes they're turning on.
- **Enhancers** are regulatory sequences that are located far away from the genes they regulate. When regulatory proteins bind to enhancers, transcription is turned on.
- **Silencers** are regulatory sequences that are located far away from the genes they regulate. When regulatory proteins bind to silencers, transcription is blocked.

Groups of regulatory proteins interact with each other and regulatory sequences in order to control the binding of RNA polymerase to promoters. The regulatory proteins that bind to regulatory sequences in order to turn on transcription are called *transcription factors*. Two types of transcription factors control transcription in eukaryotic cells:

- **General transcription factors,** also called basal transcription factors, are required for transcription of any gene in all cell types. They are part of the *transcription initiation complex,* a group of proteins that work together to attract RNA polymerase and help it bind to promoters.
- **Regulatory transcription factors** are more targeted than basal transcription factors. Regulatory transcription factors bind to the regulatory sequences of specific genes in order to turn them on or off. In other words, regulatory transcription factors are the regulatory proteins that control which genes in eukaryotes are being expressed. For example, the regulatory transcription factors in a skin cell would be different from the ones in the muscle cell. In a skin cell, the regulatory transcription factors would turn on the genes for tools needed for the skin cell function, like the protein keratin. In a muscle cell, the regulatory transcription factors would turn on the genes for tools needed muscle cell function, such as the proteins actin and myosin.

In addition to transcription factors, proteins called *coactivators* also play a role in the initiation of transcription. Coactivators bind to general and regulatory transcription factors, bringing them together to form the transcription initiation complex (see Figure 19-4). Transcription factors and coactivators work together to initiate transcription:

1. **Regulatory transcription factors bind to enhancers for the gene to be transcribed.**
2. **Regulatory transcription factors gather coactivators and general transcription factors together.**

   The proteins bind to each other and the promoter for the gene to be transcribed.
3. **RNA polymerase binds to the gathered proteins, completing the transcription initiation complex.**
4. **Transcription begins.**

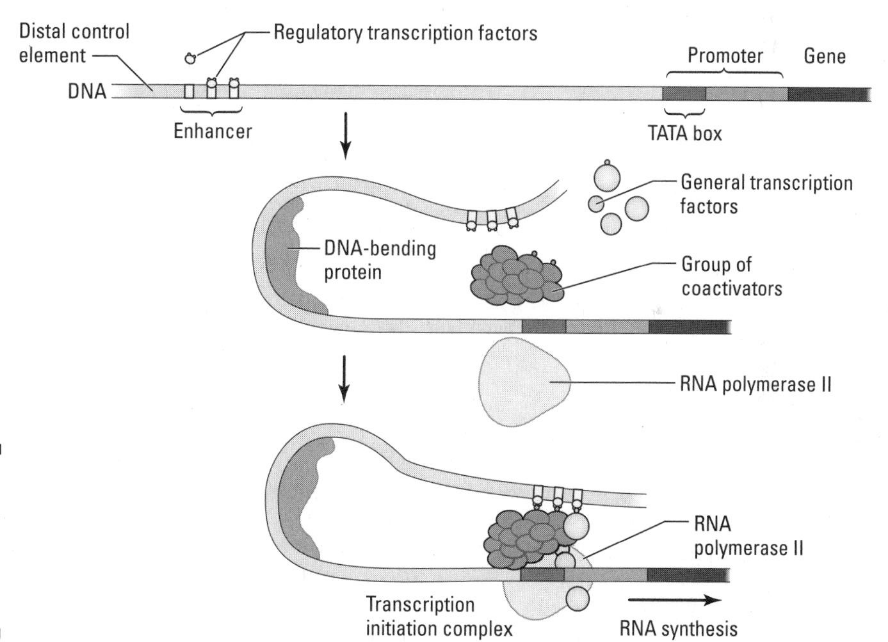

Figure 19-4: Transcription factors in eukaryotic cells.

Transcription factors bind to regulatory sequences in order to control gene expression in eukaryotes. To turn genes on, transcription factors bind to promoter proximal sequences and enhancers. To turn genes off, transcription factors bind to silencers. Regulatory transcription factors are unique to different cell types and turn on the genes that cause cell differentiation.

Controlling events between transcription and translation

Two opportunities exist for regulation of gene expression between transcription and translation. Control of mRNA processing (see Chapter 18) can affect the conversion of a pre-mRNA into a finished mRNA molecule. Also, the cellular lifespan of an mRNA determines how many times it can be translated before the mRNA is broken down by the cell, thus affecting the amount of gene product in the cell.

The mechanisms for controlling mRNA processing and stability are as follows:

- mRNA processing is regulated by proteins that interact with the spliceosomes. These proteins can affect the way a pre-mRNA is spliced to become a finished mRNA. For example, these proteins can determine which introns are removed from the pre-mRNA. Through *alternative splicing*, spliceosomes can splice one type of pre-mRNA different ways, leading to several different finished mRNAs. The differences in the mRNAs leads to differences in the proteins. Thus, alternative splicing produces several different gene products from one original pre-mRNA.
- The lifespan of an mRNA in the cell, or *mRNA stability,* is controlled by *RNA interference.* Small RNA molecules, called *microRNA* (miRNA), bind to complementary sequences on mRNA molecules, marking them for destruction by the cell. The release of miRNA molecules from the nucleus controls the timing of destruction of specific mRNAs.

The real difference between men and women

Men are from Mars, and women are from Venus, right? Well, at least according to author John Gray. But it turns out that to get from Venus to Mars all you need is one gene that codes for a single transcription factor.

In terms of DNA, men and women are almost identical. In fact, women have all the genes they need to become men scattered among their 46 chromosomes — all the genes except one, that is. Located on the Y chromosome is a single gene, SRY, which contains the blueprint for a transcription factor called testes determining factor (tdf).

Human fetuses are identical until about six weeks of life when transcription of SRY produces testes determining factor. TDF binds near the promoters of the genes necessary for the formation of testes and activates these genes. Testes are formed, which produce testosterone, which signals more "male" genes to turn on and voilà — a boy baby is born!

In female fetuses, there is no SRY gene, so although all the other necessary male genes are present, they never get turned on. The default developmental pathway is to become female, but all it takes is one very powerful transcription factor to travel from Venus to Mars!

Controlling translation and beyond

Eukaryotic cells often control gene expression by regulating translation. A cell can use several mechanisms to block translation of an mRNA:

- Regulatory proteins can bind to mRNA, blocking their translation.
- The cell may modify the 5' cap or poly-A tail of the mRNA in order to block translation.
- The addition of a phosphate group to a protein in the ribosome, called *phosphorylation,* may stop or slow translation.

Prevention of translation of an mRNA prevents the gene product from being made, blocking expression of the gene from which the mRNA was made.

Once translation occurs, the protein is available in the cell. However, cells can still control the effect of that protein by regulating its activity. Enzymes are commonly regulated by competitive and noncompetitive inhibition (see Chapter 6). The activity of other proteins, particularly those involved in signaling pathways, are often controlled by the addition or removal of a phosphate group. Phosphorylation and dephosphorylation are particularly important in signal transduction pathways (see Chapter 9).

Part VI

Tools of Molecular Biology: Harnessing the Power of DNA

"...finally, those researchers working after hours should limit their investigation to the behavior of protons and electrons, and hereafter refrain from putting eggs in the particle accelerator."

In this part . . .

Powerful tools enable scientists to explore and manipulate the DNA code, opening a new frontier in biological science. Today, scientists have many tools to study DNA, how it governs cells, and the effects of changes to the DNA code. DNA is cut with enzymes, stored in genomic libraries, copied rapidly, and sequenced so that its code can be read. Tools like these are greatly enhancing scientists' understanding of life on Earth and the ability to solve some of the problems of human societies.

Scientists now have the ability to study not just pieces of DNA but the entire genome of an organism. In this part, I explain how scientists use the tools of molecular biology to explore genomes and apply biological knowledge to solve current world problems.

Chapter 20

Recombinant DNA Technology: Power Tools at the Cellular Level

In This Chapter

- Working with DNA
- Bringing genes together
- Reading the genetic code
- Making a difference in the world

Recombinant DNA technology enables scientists to study and experiment with DNA. Scientists copy genes, read their sequences, and introduce them into new cells. Scientists are using powerful techniques that act at the molecular level to engineer crop plants and diagnose and treat diseases. In this chapter, I explain the basics of the techniques used to work with DNA and introduce some possible applications to solve today's problems.

Piecing It Together: Recombinant DNA Technology

DNA is so small that you can barely see it with an electron microscope — and yet, people have figured out how to read it, copy it, cut it into pieces, sort it, and put it back together in new combinations. When DNA from two different sources is combined together, the patchwork DNA molecule is called *recombinant DNA*. For example, scientists have combined human genes and DNA from *E. coli* and then placed the recombinant DNA into *E. coli* so that the bacterium makes human proteins. Doctors can use these proteins, such as insulin, to treat human diseases. The tools that let scientists manipulate DNA in these amazing ways are called *recombinant DNA technology*.

Cutting DNA with restriction enzymes

Bacteria make enzymes called *restriction endonucleases,* or more commonly, *restriction enzymes,* that cut strands of DNA into smaller pieces (see Figure 20-1). Bacteria use restriction enzymes to fight off attacking viruses, chopping up the viral DNA so that the virus can't destroy the bacterial cell. Scientists use restriction enzymes in the lab to cut DNA into smaller pieces so that they can analyze and manipulate DNA more easily.

Each restriction enzyme recognizes and cuts DNA at a specific sequence called a *restriction site.*

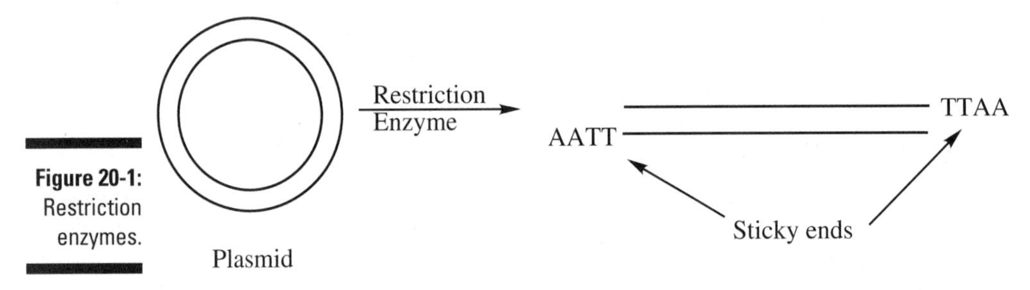

Figure 20-1: Restriction enzymes.

The restriction enzyme called EcoR1 cuts DNA at the sequence 5'GAATTC3'. If you mix DNA and a restriction enzyme, the enzyme will find all the restriction sites it recognizes and cut the DNA at those locations.

Restriction enzymes make cutting and combining pieces of DNA easy. For example, if you wanted to put a human gene into a bacterial plasmid, you'd follow these steps:

1. **Choose a restriction enzymes that forms sticky ends when it cuts DNA.**

   Sticky ends are pieces of single-stranded DNA that are complementary and can form hydrogen bonds. Restriction enzymes that form sticky ends cut the DNA backbone asymmetrically so that a piece of single-stranded DNA hangs off each end. For example, the sticky ends shown in Figure 20-1 have the sequences 5'AATT3' and 3'TTAA5'. A and T are complementary base pairs, so these ends could form hydrogen bonds and thus stick to each other.

2. **Cut the human DNA and bacterial plasmids with the restriction enzyme.**

   If you cut a plasmid DNA and human DNA with the same restriction enzyme, all the DNA fragments will have the same sticky ends.

3. **Combine human DNA and bacterial plasmids.**

   The two types of DNA have the same sticky ends, so some pieces of plasmid DNA and human DNA will stick together. Thus, some plasmids will end up with a human gene inserted into the plasmid.

4. **Use DNA ligase to seal the backbone of the DNA.**

   DNA ligase will form covalent bonds at the cut sites in the DNA, sealing together any pieces of DNA that combined together.

Any plasmids that contain human DNA are recombinant. These plasmids could now be inserted into bacterial cells.

Sorting molecules using gel electrophoresis

Gel electrophoresis separates molecules based on their size and electrical charge. The "gel" part of gel electrophoresis is a slab of a gelatin-like substance made from polysaccharides. Molecules of different sizes are placed into pockets in the gel, and then electricity is used to move the molecules through the gel, as shown in Figure 20-2. The types of molecules most commonly separated by gel electrophoresis are DNA and proteins. Electrophoresis means to carry something using electricity, so in this case molecules are being carried through the gel:

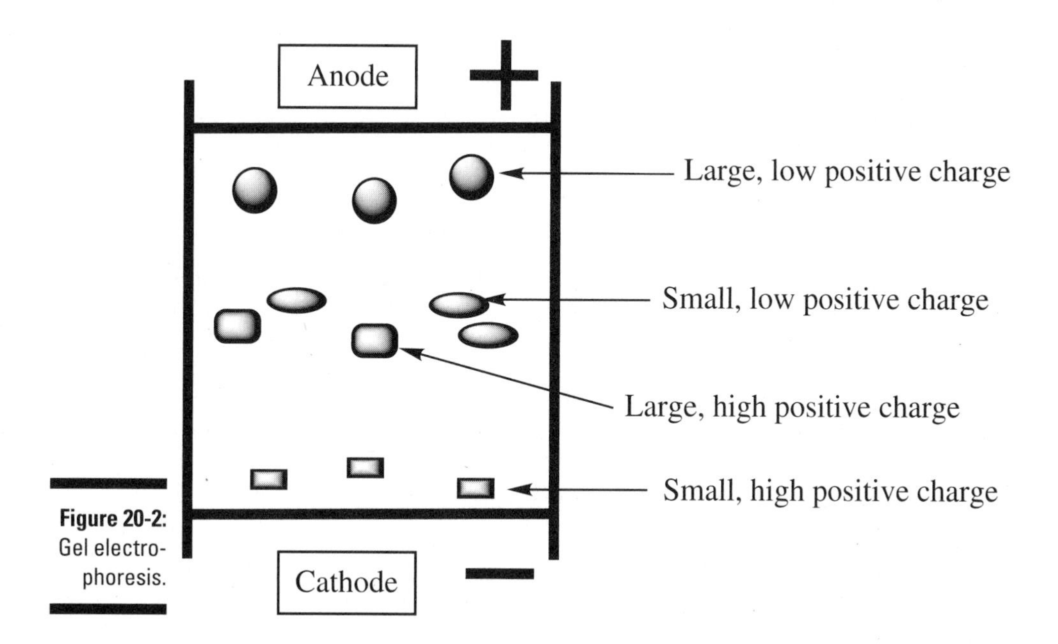

Figure 20-2: Gel electrophoresis.

- **Molecules are attracted to the electrode with opposite charge.** Positively charged molecules will move through the gel toward the negatively charged electrode, while negatively charged molecules will move through the gel toward the positively charged electrode.
- **Small molecules can move more quickly through the gel than larger molecules, so molecules of different sizes will start to spread apart from each other in the gel.** You can think of gel electrophoresis as a race of molecules through an obstacle course. The long fibers of the polysaccharide in the gel are the obstacles that the molecules must get through. When you turn on the electricity and run a current through your gel, the molecules begin the race. Smaller molecules can wiggle around the obstacles more easily, so they run faster in the race. When you turn off the electricity, the race is over, and all the molecules stop where they are. Bigger molecules will be closer to the starting line, and smaller molecules will be closer to the finish line.

Gel electrophoresis is often used to separate molecules of DNA (see Figure 20-3). Electricity pulls DNA through a gel because DNA is negatively charged due to its many phosphate groups. When an electrical current is run through a gel, the DNA molecules are attracted to the positive electrode.

The positive electrode is colored red on the electrophoresis chamber, so scientists say that "DNA runs to the red."

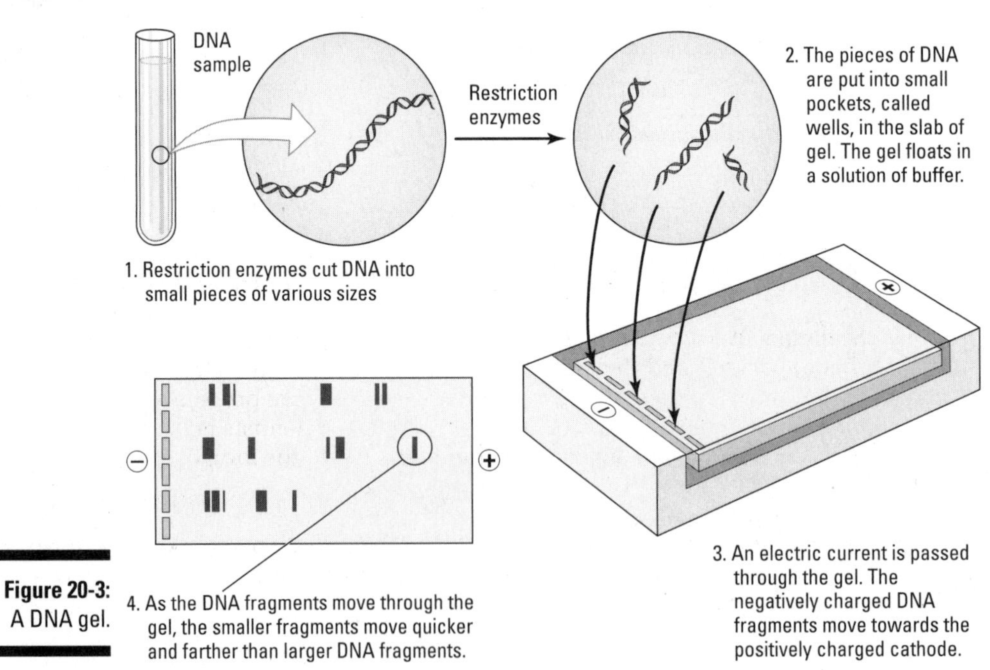

Figure 20-3: A DNA gel.

The steps for doing gel electrophoresis on DNA samples are very similar to the steps of other types of gel electrophoresis:

1. **Pour the liquid gel and let it solidify.**

   A spacer called a *comb* fits into a *gel platform,* which is sealed with tape. The comb creates little pockets called *wells* in the gel.

2. **Place the gel into an electrophoresis chamber and fill the chamber with buffer.**

   The buffer solution conducts electricity throughout the electrophoresis chamber.

3. **Cut DNA samples with restriction enzymes and mix samples with loading buffer.**

   The restriction enzymes cut the DNA into pieces of different sizes. The loading buffer contains a dense molecule to help settle the DNA into the wells and also has marker dyes so that you can see the approximate position of your samples as they move through the gel.

4. **Load the samples into the wells using a micropipettor.**

   Micropipettors allow precise measurements of very small volumes. The DNA samples should sink to the bottom of the wells.

5. **Seal the box and start the current.**

   The electrical current will separate the DNA samples in the gel.

6. **Stop the current and stain the gel.**

   The stain will stick to the DNA in the gel, creating stripes called *bands.*

Each band on a gel represents a collection of DNA molecules that are the same size and therefore stop in the same place.

Making cDNA with reverse transcriptase

Scientists use recombinant DNA technology to combine eukaryotic DNA with that of bacteria and then introduce eukaryotic genes into bacterial cells. However, bacteria can't use eukaryotic genes to make proteins unless the introns are removed from the eukaryotic genes. Scientists get around this problem by creating intron-free eukaryotic genes in the form of *complementary DNA* (cDNA).

cDNA is made from eukaryotic mRNA that has already been spliced to remove the introns. (For more on this process, see Chapter 18.)

Figure 20-4 shows the steps for making cDNA:

1. **Isolate mRNA for the protein you're interested in.**
2. **Use the enzyme reverse transcriptase to make a single-stranded DNA molecule that is complementary to the mRNA.**

   Reverse transcriptase is a viral enzyme that uses RNA as a template to make DNA.
3. **Use reverse transcriptase or DNA polymerase to make a partner strand for the DNA molecule, creating a finished double-stranded molecule of cDNA.**

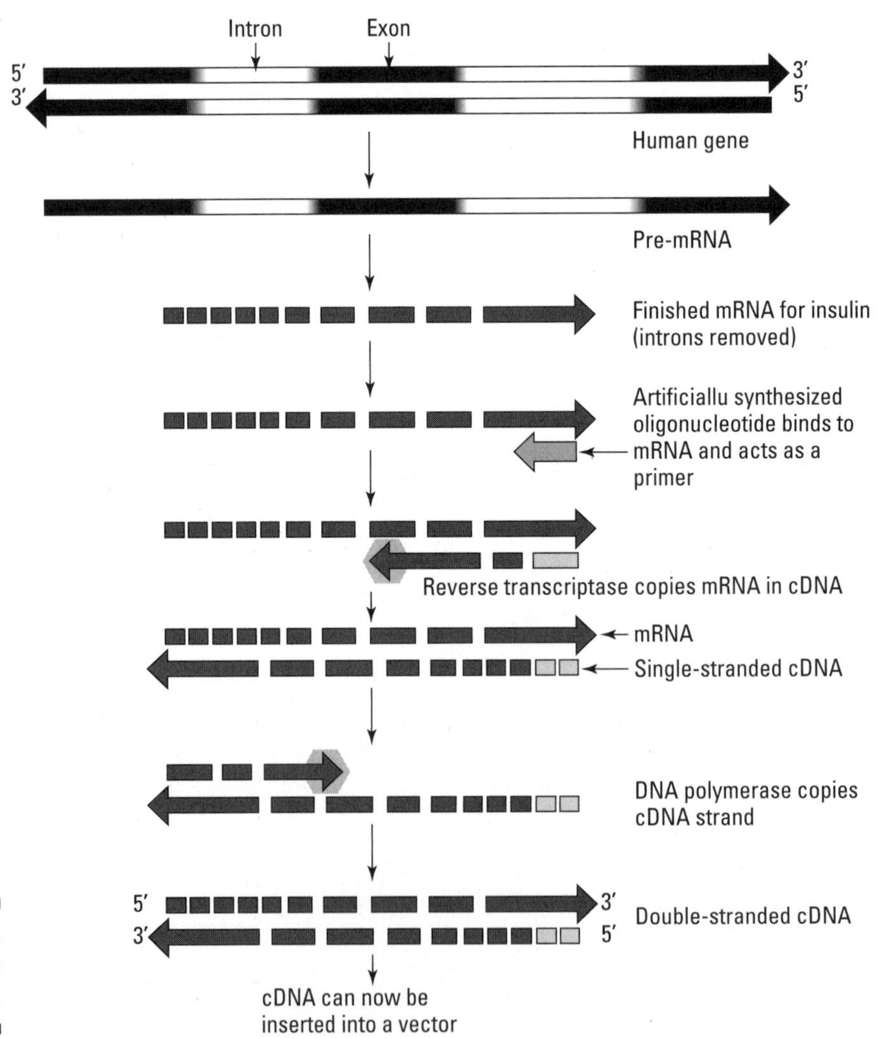

Figure 20-4: Making cDNA.

Cloning genes into a library

Scientists store the DNA they're working with in *DNA libraries,* recombinant DNA molecules that contain the gene of interest. Once a gene is put into a DNA library, *DNA cloning* makes many identical copies of the gene. To clone a gene into a library, you first need to put the gene into a *vector.* A vector, such as a plasmid or virus, helps carry DNA into a cell. Figure 20-5 illustrates the process for introducing a gene into a vector and then cloning the gene into a library.

1. **Using the same restriction enzyme, cut the vector and the DNA containing the gene to be cloned.**

   That way, the vector and the DNA to be cloned will have the same sticky ends.

2. **Mix the vector and DNA to be cloned together and add DNA ligase.**

   Some vectors will pick up the genes to be cloned. DNA ligase will form covalent bonds, sealing the genes into the vectors. The vectors that pick up genes are recombinant.

3. **Introduce the vector into a population of cells.**

   The vector will be reproduced inside the cells. Once the vector is reproduced, the gene has been cloned.

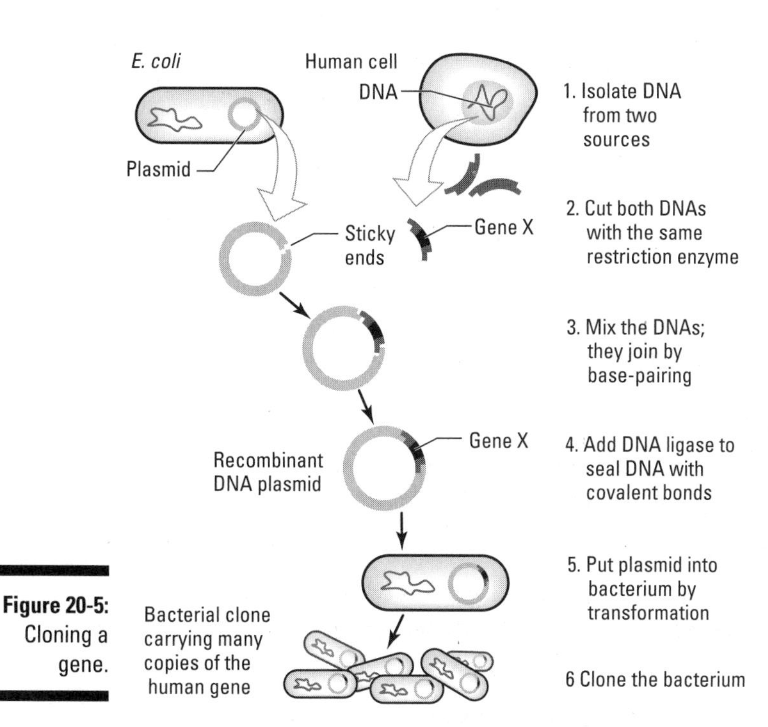

Figure 20-5: Cloning a gene.

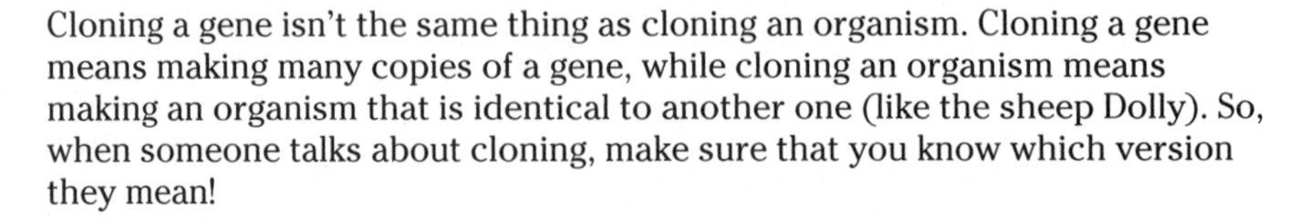

Cloning a gene isn't the same thing as cloning an organism. Cloning a gene means making many copies of a gene, while cloning an organism means making an organism that is identical to another one (like the sheep Dolly). So, when someone talks about cloning, make sure that you know which version they mean!

DNA libraries are recombinant vectors that store genes and keep them handy for scientists.

DNA libraries make it easier for scientists to work with DNA they're interested in, such as DNA from a particular type of cell or organism. Scientists use several types of DNA libraries:

- DNA libraries contain fragments of DNA inserted into a vector.
- *cDNA libraries* contain fragments of cDNA inserted into vectors.
- *Genomic libraries* contain DNA fragments that represent the entire genome of an organism.

Finding a gene with DNA probes

Once genes are cloned into a library, scientists use *DNA probes* to find the vectors that contain specific genes of interest. Probes are pieces of single-stranded DNA that are used to locate a particular DNA sequence (see Figure 20-6).

Probes are made with a sequence that's complementary to the sequence you're looking for. Using a probe is like going fishing — you use the right bait (a complementary sequence) to catch something you want (a certain gene). Probes will attach with hydrogen bonds to their complementary sequence. For example, if you were looking for a gene that contained the sequence 5'TAGGCT3', you'd make a probe with the sequence 3'ATCCGA5'.

Probes are also labeled with a fluorescent or radioactive marker so that you can locate them in a DNA sample. To use a probe to locate DNA, complete the following steps:

1. **Prepare a DNA sample to be probed for the gene of interest.**

   You can look at DNA in many different forms — DNA in a gel, DNA attached to a microscope slide, and even DNA in colonies on a plate. To prepare any of these samples to be probed, you must treat the DNA with heat or chemicals to make it single-stranded and ready to pair with another strand of DNA.

2. **Wash your DNA probe over the surface of your DNA sample.**

   The probe will attach to its complementary sequence in the sample.

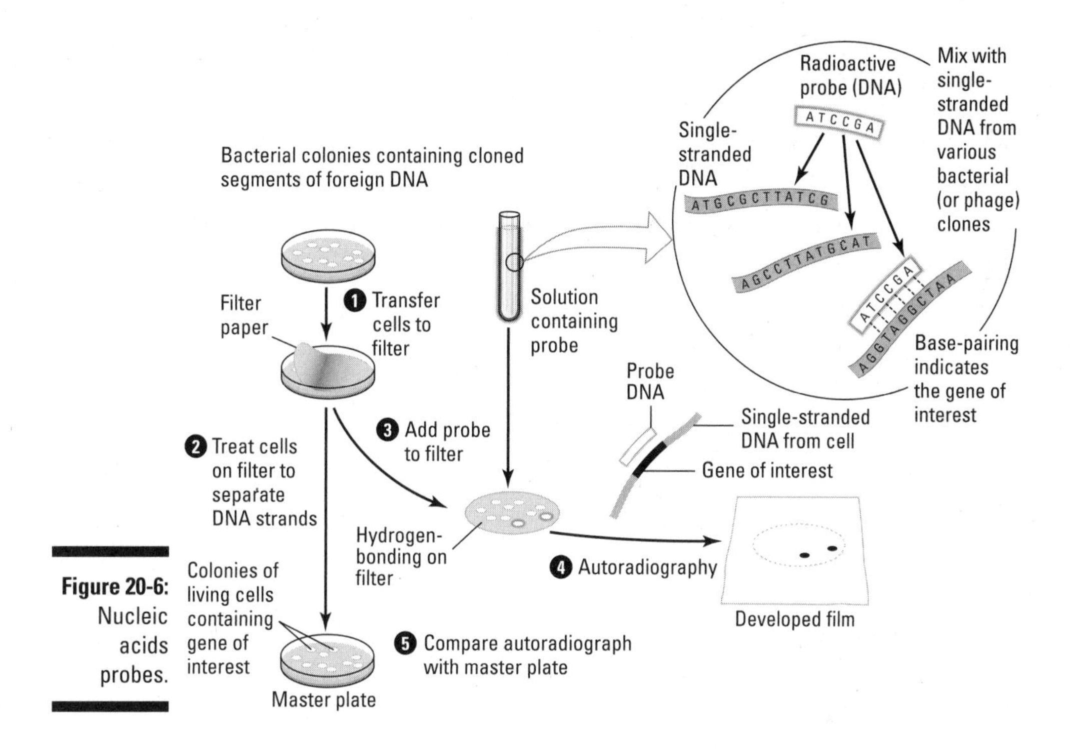

Figure 20-6: Nucleic acids probes.

3. **Locate your probe to find the gene of interest.**

   A certain wavelength of light activates fluorescent probes. Radioactive probes are located by using the treated DNA to expose photographic film.

Copying a gene with PCR

The *polymerase chain reaction* (PCR) is one of the most important techniques in molecular biology because it allows you to make many copies of a gene that you're interested in. For example, you may need more copies of a gene in order to read its sequence, or you may be a forensic scientist that has only a few cells to work with to solve a crime. PCR makes this kind of work possible.

PCR can turn a single copy of a gene into more than a billion copies in just a few hours. Making many copies of a gene using PCR is called *gene amplification.*

PCR targets the gene to be copied with *primers*. Primers are single-stranded sequences of DNA that are complementary to sequences next to the gene to be copied (see Figure 20-7).

DNA polymerase uses the primers to begin DNA replication, so it starts copying DNA next to the gene you want and then keeps going and copies the gene. To begin PCR, the DNA sample that contains the gene to be amplified is combined with thousands of copies of the primers. Then, this basic cycle of steps, shown in Figure 20-7, is repeated over and over:

1. **Heat the sample for 1 minute to disrupt the hydrogen bonds in the DNA and make it single stranded.**

   Temperatures around 95° C (203° F) are needed to "melt" the hydrogen bonds.

2. **Cool the sample for 1 minute so that hydrogen bonds can reform.**

   Because there is so much primer, primers will stick with hydrogen bonds to the single-stranded target DNA.

   Temperatures around 65° C (149° F) are cool enough for the primers to stick, or *anneal,* to the target DNA.

3. **Heat the sample to the optimum temperature for the DNA polymerase for 1 minute.**

   DNA polymerase will copy the target gene. The heat-stable enzyme, *Taq polymerase,* that is used for PCR works best at temperatures around 75° C (167° F).

By repeating the preceding steps just 30 times, one copy of a gene can be amplified into more than 1 billion copies. This multiplication works like compound interest — every time you make copies, the copies are copied, then those copies are copied, and so on. Using PCR, the DNA from just a single cell can provide enough DNA for scientists to work with in a very short amount of time.

Reading a gene with DNA sequencing

DNA sequencing determines the order of nucleotides in a DNA strand. DNA sequencing techniques use a special kind of nucleotide, called a dideoxyribonucleotide triphosphate (ddNTP). Dideoxyribonucleotides are identical in structure to typical nucleotides (deoxyribonucleotide triphosphates, or dNTPs) except that, instead of having a hydroxyl group attached to their 3' carbon, they just have a hydrogen atom (see Figure 20-8). This small change is very significant to DNA polymerase, which requires a 3' hydroxyl in order to add nucleotides to a growing chain of DN.A (See Chapter 17 for more on DNA replication.)

If DNA polymerase is copying DNA and a dideoxyribonucleotide becomes part of a growing chain, it will be the last nucleotide that gets added; DNA polymerase will not be able to add any more nucleotides. DNA sequencing uses this chain interruption to determine the order of nucleotides in a strand of DNA.

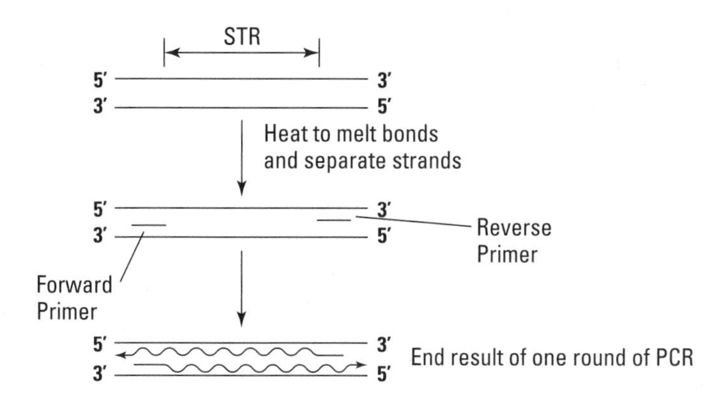

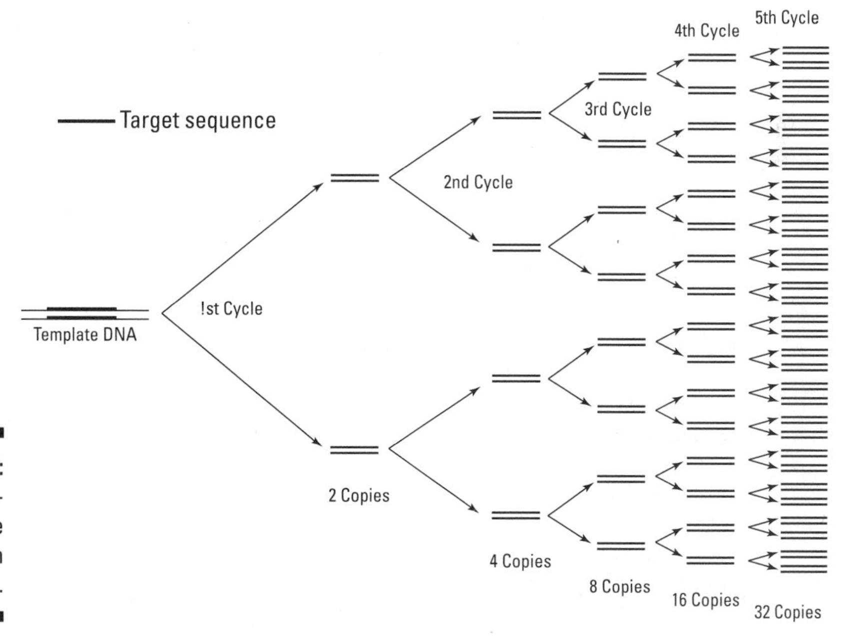

Figure 20-7: The polymerase chain reaction.

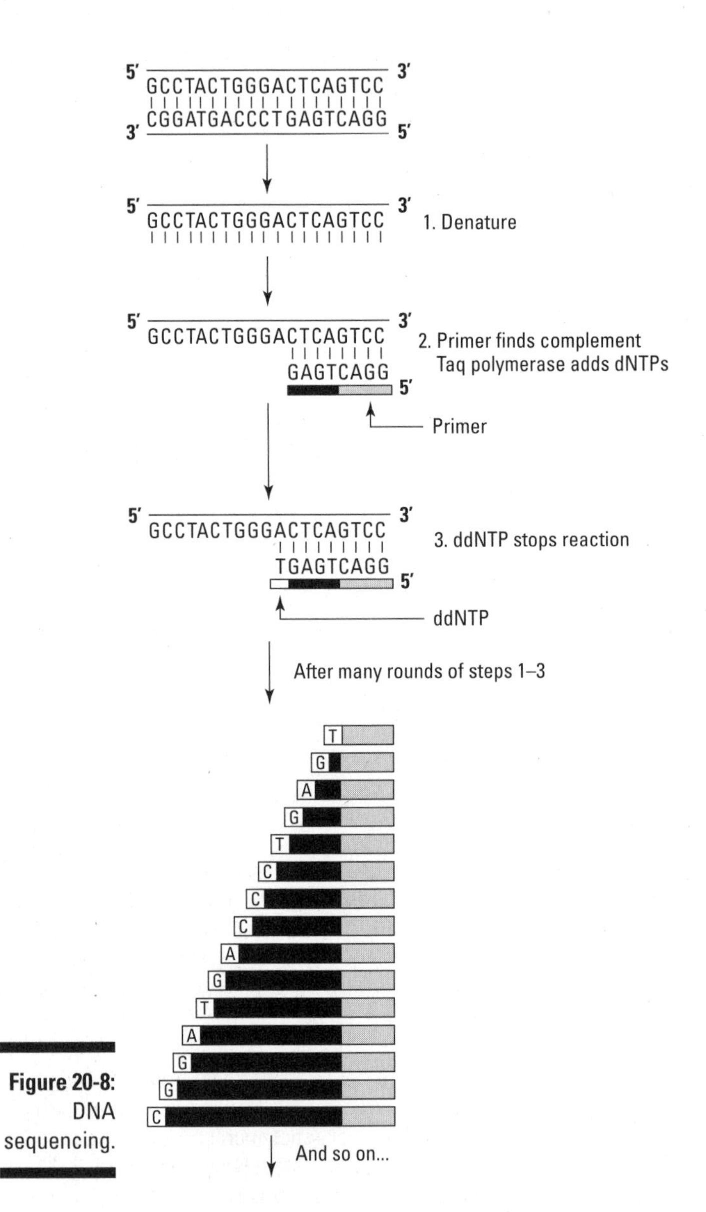

Figure 20-8: DNA sequencing.

Most DNA sequencing today is *cycle sequencing,* which combines the original methods of DNA sequencing with the heating and cooling cycle of PCR. (See the section "Copying a gene with PCR," earlier in this chapter, for more on this topic.) Like PCR, cycle sequencing uses primers that target the gene to be sequenced. In order to begin cycle sequencing, you'd combine the DNA to be sequenced, primers that target the DNA to be sequenced, Taq polymerase (a DNA polymerase), lots of typical nucleotides (dNTPs), and lots of dideoxyribonucleotides (ddNTPs). Then, the basic cycle of PCR is repeated over and over, producing many copies of the gene to be sequenced.

Cycle sequencing is different from PCR in several ways:

- During cycle sequencing, ddNTPs get randomly incorporated into the DNA as it's being copied.
- Whenever a ddNTP is incorporated into a growing chain of DNA, the chain is stopped.
- Instead of producing many exact copies of a DNA sequence like PCR, cycle sequencing produces many partial copies of the sequence, each one stopped at a different place.
- The partial copies are loaded into a machine that uses gel electrophoresis to put the copies into order by size (Figure 20-8).
- As the partial sequences pass through the machine, a laser is used to read the fluorescent tag on each ddNTP. Each type of ddNTP is tagged with a different color fluorescent marker, so a computer can read the DNA sequence as the fluorescent ddNTPs flow past the laser.

Some like it hot

Human beings — and human enzymes — like life right around 37° Celsius (or 98.6° Fahrenheit). If your body gets much hotter than that temperature, your enzymes denature, and you die.

However, just because humans can't take the heat doesn't mean that nobody can. Happily perking along in those hot springs are bacteria and archaea called *thermophiles*. Thermophiles like it hot — in fact, put them at your body temperature, and they can't grow because it's just too cold and slow for thermophiles.

One of these thermophiles, a bacterium called *Thermus aquaticus*, is famous to biologists all around the world. *T. aquaticus* is the source of the enzyme Taq polymerase, the DNA polymerase that is used in PCR. Like the cells that it came from, Taq polymerase likes it hot, which means the enzyme isn't destroyed during the heating part of the PCR reaction. Before biologists discovered Taq polymerase, they used a DNA polymerase that denatured at the high PCR temperatures. PCR was a lot more work because new DNA polymerase had to be added after every heating step! Now, thanks to a "hot" bacterium, the process is completely automated and much easier to do.

Changing the Plan: Using Molecular Biology to Solve Problems

Recombinant DNA technology can be controversial. People, including scientists, worry about the ethical, legal, and environmental consequences of altering the DNA code of organisms:

- *Genetically modified organisms* (GMOs) that contain genes from a different organism are currently used in agriculture, but some people are concerned about the following potential impacts on wild organisms and on small farms:
  - Genetically modified plants may interbreed with wild species, transferring genes for pesticide resistance to weeds.
  - Crop plants that are engineered to make toxins intended to kill agricultural pests can also impact populations of other insects.
  - Small farmers may not be able to afford genetically modified crop plants, putting them at a disadvantage to larger corporate farms.
- Genetic testing of fetuses allows the early detection of genetic disease, but some people worry that genetic testing will be taken to extremes, leading to a society where only "perfect" people are allowed to survive.
- Genetic testing of adults allows people to learn whether they have inherited diseases that run in their family, but some people worry that some day insurance companies will use genetic profiles of people to make decisions about who to insure.
- Parents of children with life-threatening diseases that can be treated with bone marrow transplants are using genetic testing to conceive children that can provide stem cells for their sick siblings. The umbilical cord is an excellent source of these stem cells, so the new babies aren't harmed, but people worry that this may lead to an extreme future scenario where babies are born to serve as bone marrow or organ donors for existing people.
- Human hormones like insulin and human growth hormone are produced by bacteria through recombinant DNA technology and used to treat diseases like diabetes and pituitary dwarfism. However, some people seek hormones like human growth hormone for cosmetic reasons (for example, so that their children can be a little taller). People question whether it's ethical for parents to make these choices for their children and whether too much emphasis is being placed on certain physical traits in society.

Making useful proteins through genetic engineering

The bacterium *E. coli* is used as a little cellular factory to produce human proteins for treatment of diseases. To get *E.* coli to produce human proteins, cDNA copies of human genes are put into plasmid vectors and then the vectors are introduced into *E. coli* (see Figure 20-9).

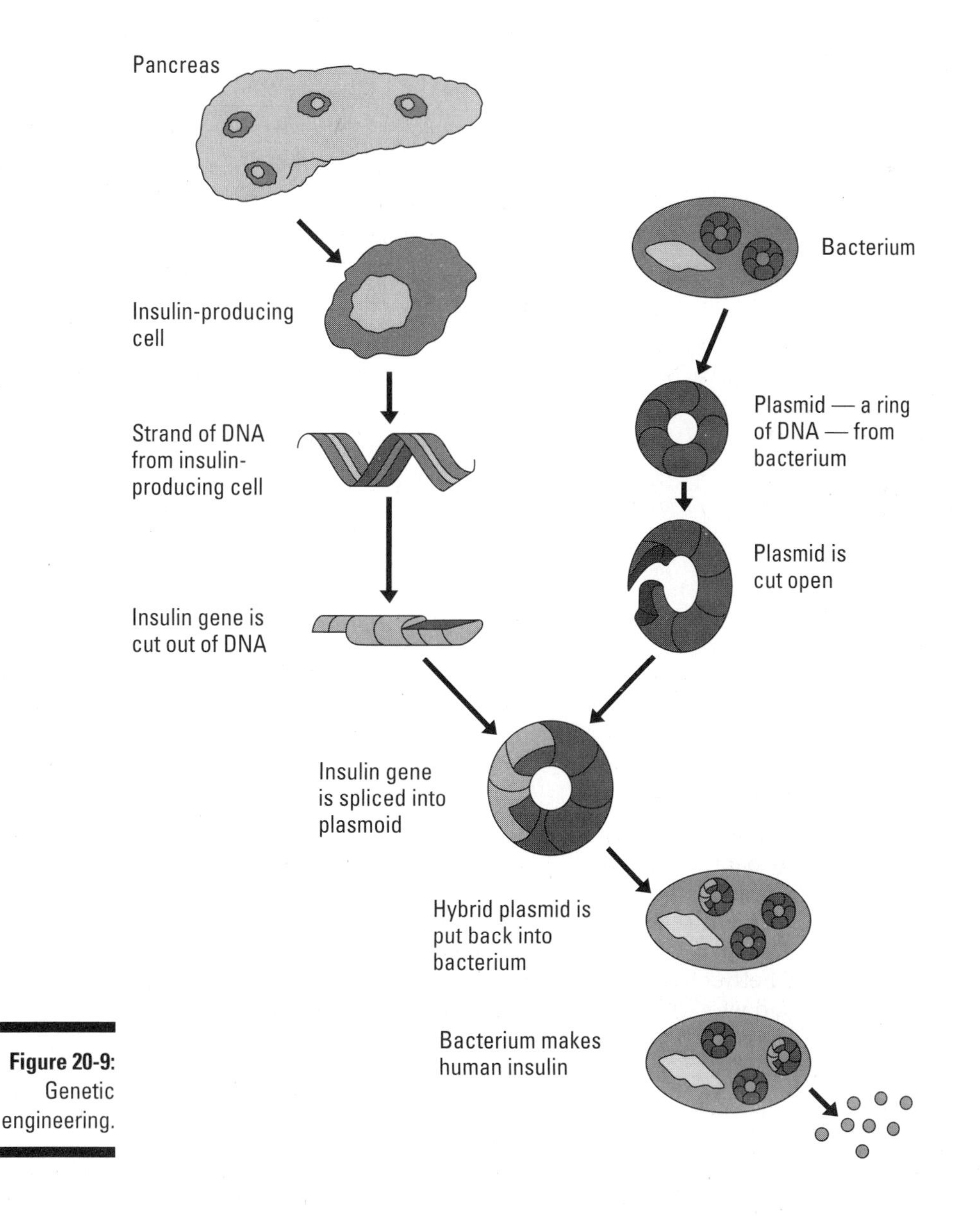

Figure 20-9: Genetic engineering.

The bacterium transcribes and translates the human gene, producing a human protein that is identical to the protein made by healthy human cells. Several human proteins are currently produced by this method, including the following:

- Human insulin for treatment of diabetes
- Human growth hormone for treatment of pituitary dwarfism
- Tumor necrosis factor, taxol, and interleukin-2 for treatment of cancer
- Epidermal growth factor for treatment of burns and ulcers

Searching for disease genes

Some people carry the potential for future disease in their genes. *Genetic screening* allows people to discover whether they're carrying recessive alleles for genetic diseases, allowing them to choose whether or not to have children. Also, diseases that show up later in life, such as Alzheimer's and Huntington's disease, can be detected early, to seek the earliest possible treatment.

In order to screen for a particular genetic disease, scientists must first discover the gene that causes the disease and study the normal and disease-causing sequences. Scientists have identified the genes for several genetic diseases, including cystic fibrosis, sickle-cell anemia, Huntington's disease, an inherited form of Alzheimer's, and an inherited form of breast cancer.

Once the gene for a genetic disease has been identified, doctors can screen people to determine whether they have normal or disease-causing alleles.

Looking before you leap

In my opinion, people should continue to debate the issues associated with altering the DNA code of organisms as science leads to new applications for recombinant DNA technology. As a society, people should openly consider the ethical implications of what they do and choose carefully to try to reap the benefits of this technology without creating even larger problems. Many scientists agree with my view and are proceeding carefully into controversial areas with advice from people who specialize in *bioethics*. Scientists have given testimony before lawmaking bodies, such as the United States Congress, to help craft the laws that will guide and protect societies as new applications for recombinant DNA technology emerge. This technology has great power to solve problems and cure diseases, and I don't think people should turn their backs on it for fear of the future. Instead, they should weigh the pros and cons of each application and try to balance personal freedom and the common good.

In order to screen a person for a particular gene, scientists amplify the genes linked to the disease using PCR. (See the section "Copying a gene with PCR," earlier in this chapter.) Then, scientists use molecular techniques to screen the genes for the disease allele:

- **Restriction enzymes can reveal the presence of normal and disease alleles in a genotype.** Normal alleles and disease-causing alleles may have different restriction sites for restriction enzymes. Thus, when a person's DNA is cut with restriction enzymes and separated with gel electrophoresis, normal alleles and disease-causing alleles will produce a different pattern of bands on a gel (see Figure 20-10). By examining the pattern of bands, you can figure out a person's genotype.
- **Probes for normal alleles and disease-causing alleles detect the presence of these alleles in DNA.** Once the difference in sequence for normal versus disease-causing alleles is known, scientists search the DNA for matching sequences using probes specific to each sequence (Figure 20-6).

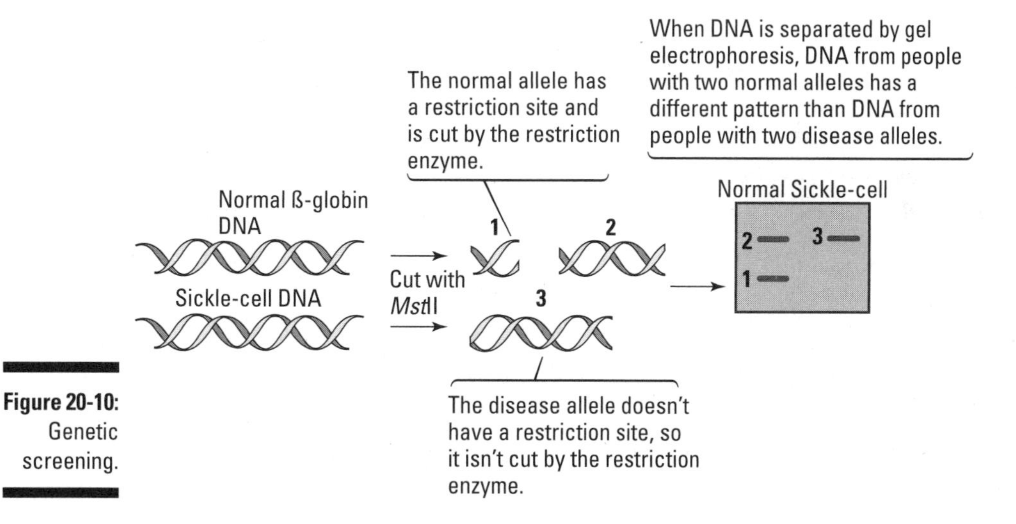

Figure 20-10: Genetic screening.

Building a "better" plant with genetic engineering

Many important crop plants contain recombinant genes. These *transgenic plants,* which are a type of genetically modified organism (GMO), provide labor-saving advantages to farmers who can afford them:

- **Transgenic plants that contain genes for herbicide resistance require less physical weed control.** Farmers can spray crop plants that are resistant to a particular herbicide with that herbicide to control weeds. Weed plants will be killed, but the modified crop plants will not.

- **Transgenic plants that contain genes for insect toxins will be less damaged by grazing insects.** The crop plants use the introduced gene to produce insect toxins that kill insects that graze on the plants.

Another potential benefit of transgenic plants is that certain crop plants may be altered to become more nutritious. For example, scientists are currently working on developing a strain of golden rice that may help combat Vitamin A deficiency in people around the world. Vitamin A deficiency can cause blindness and increase susceptibility to infectious diseases. Golden rice is being engineered to contain the genes necessary for the rice plants to produce beta-carotene. When people eat golden rice, their bodies will use beta-carotene to make Vitamin A. Rice is a staple food for half of the world's people, so golden rice has great potential for fighting Vitamin A deficiency!

Fixing a broken gene with gene therapy

Recombinant DNA technology offers the tantalizing potential of a cure for genetic diseases. If scientists can transfer genes successfully into bacteria and plants, perhaps they can also transfer them into people that have defective disease-causing alleles (see Figure 20-11). By introducing a copy of the normal allele into affected cells, the cells could be made to function normally, eliminating the effects of the disease.

The introduction of a gene in order to cure a genetic disease is called *gene therapy*.

Gene therapy for humans is being studied, and clinical trials have occurred for some diseases, but this type of treatment is far from being perfected. Many barriers to successful human gene therapy still need to be overcome:

- **Scientists must discover safe vectors that can transfer genes into human cells.** One possible vector is viruses that naturally attack human cells and introduce their DNA. Viral DNA is removed and replaced with *therapeutic genes* that contain the normal allele sequence. The viruses are allowed to infect human cells, thus introducing the therapeutic genes. Following are several safety issues associated with the use of viruses as vectors in gene therapy:
    - Viruses that have been altered may recombine with existing viruses to recreate a disease-causing strain.
    - Viruses that have been altered so that they can't directly cause disease may still cause a severe allergic reaction that is potentially life threatening.
    - Viruses that introduce genes into human cells may interrupt the function of normal genes.

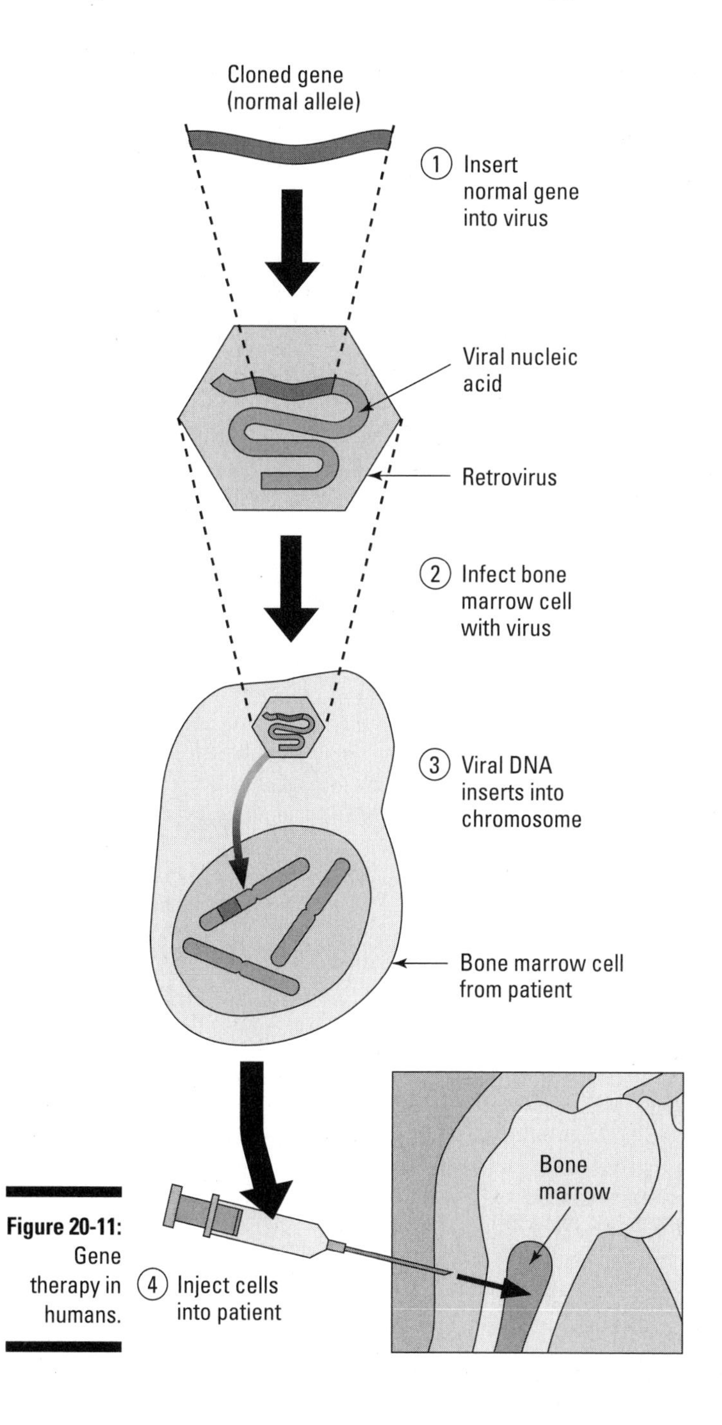

Figure 20-11: Gene therapy in humans.

- **Scientists must develop methods for introducing therapeutic genes into populations of target cells.** Humans are multicellular and have complex tissues. Genetic diseases can affect entire populations of cells. If gene therapy is to cure these diseases, the therapeutic genes must be introduced into all of the affected cells.
- **Stem cells that produce target populations of cells need to be identified.** If therapeutic genes are introduced into cells that have a limited lifespan in the body, then gene therapy will need to be repeated at regular intervals to maintain populations of healthy cells. On the other hand, if stem cells could be repaired with normal alleles, then they would continuously produce new populations of healthy cells, and the cure would be permanent.

Chapter 21

Genomics: The Big Picture

In This Chapter:

- Reading the genetic code
- Comparing genomes
- Discovering the function of DNA sequences
- Developing better treatments for disease

Genomics is the study of the entire DNA sequence of an organism. The Human Genome Project successfully tackled the task of sequencing the entire genome of humans and developed faster and less expensive methods for DNA sequencing that are now being used to determine the sequence of many other species. Comparisons between the genomes of different species are giving scientists insight into cellular function and the evolutionary history of life on Earth. In this chapter, I present some of the basic methods and goals of the science of genomics, as well as the impact of genomics on all of biology.

I Read the Whole Thing: Sequencing Genomes

You've probably heard of the *Human Genome Project* (HGP), but you may not know what an ambitious, exciting, and even controversial project it was. When HGP, whose primary goal was to sequence all 3 billion base pairs of DNA in the human genome, was first proposed in 1985, the pace of DNA sequencing was so slow that it would have taken 1,000 years to complete! Some scientists argued that HGP just couldn't be done, but others were inspired by the possibilities for advancement in human knowledge of disease, evolution, and human diversity.

HGP was officially launched in 1990 and brought together scientists from all around the world to work together to

- Determine the sequences of 3 billion base pairs that make up human DNA.
- Store the DNA sequence information in databases.
- Identify all the 20,000–25,000 genes in human DNA.
- Improve tools for data analysis.
- Address the ethical, legal, and social issues related to the project.

Unleashing the power of genomics

The Human Genome Project launched the science of *genomics,* the study of entire genomes. Prior to HGP, genes were identified and mapped to chromosomes one at a time (see Chapter 16). Genomics, on the other hand, takes a *systems approach*, looking at the entire genome at once. Since HGP, the genomes of hundreds more organisms have been sequenced. Because *all* the DNA sequence is available, genomics opens the door to many new types of information. Research in genomics is taking several approaches to make sense of all the new sequence data:

- **Genomics** is concerned with sequencing genomes of all different types of organisms.
- **Comparative genomics** compares genomes from different species in order to gain a better understanding of how species have evolved and to determine the function of genes and noncoding regions of the genome.
- **Functional genomics** tries to figure out the functions of genes, including when genes are expressed and how their products interact.
- **Pharmacogenomics** studies how a person's unique genome affects their response to drugs, with the goal of developing drugs that are highly effective for treating disease in each individual person.

Reading the book of life with shotgun sequencing

A type of rapid DNA sequencing, called *shotgun sequencing* (see Figure 21-1), makes it possible to sequence entire genomes in a reasonable amount of time.

Genomes are broken up into small fragments, which are sequenced and then reassembled like a puzzle to determine the complete genome. Figure 21-1 shows the steps for one type of shotgun sequencing, called *hierarchical shotgun sequencing*:

1. **Genomic DNA is cut into large fragments about 150,000,000 base pairs (150 Mb) long.**
2. **The large fragments are put into a vector called a bacterial artificial chromosome and then cloned into bacteria to be replicated.**
3. **The large fragments are mapped and organized to determine their relative order to each other.**
4. **Each large fragment is broken up into a set of smaller fragments that can be sequenced.**
5. **The small fragments are sequenced.**
6. **A computer lines up the overlapping sequences to determine the order of the small fragments.**

   Once all the small fragments are sequenced and aligned, the sequence of the genome is complete.

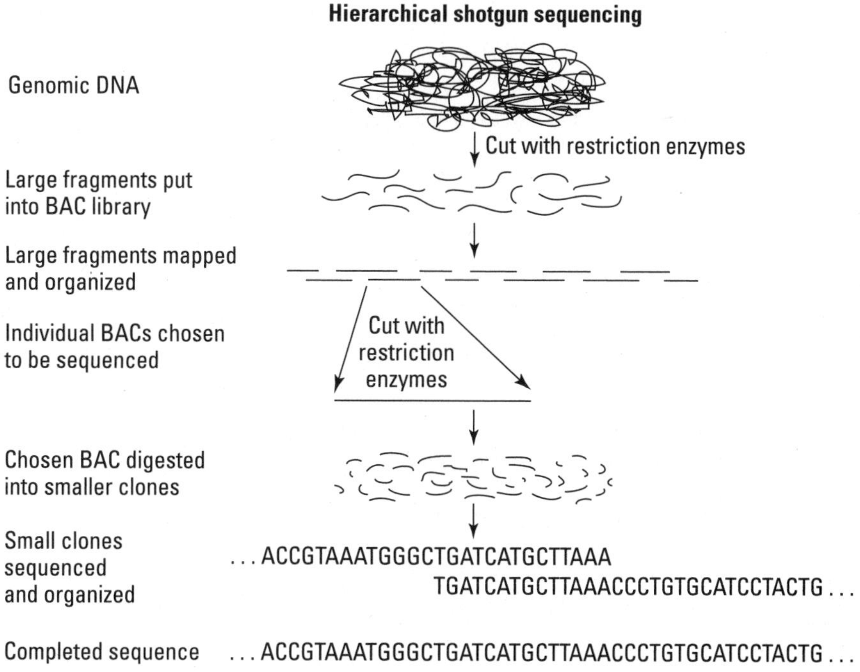

Hierarchical shotgun sequencing

Genomic DNA

Cut with restriction enzymes

Large fragments put into BAC library

Large fragments mapped and organized

Individual BACs chosen to be sequenced

Cut with restriction enzymes

Chosen BAC digested into smaller clones

Small clones sequenced and organized

. . . ACCGTAAATGGGCTGATCATGCTTAAA

TGATCATGCTTAAACCCTGTGCATCCTACTG . . .

Completed sequence . . . ACCGTAAATGGGCTGATCATGCTTAAACCCTGTGCATCCTACTG . . .

Figure 21-1: Shotgun sequencing.

Using shotgun sequencing, the genomes of hundreds of different species have been sequenced. Organisms are chosen for sequencing primarily based on their importance to humans:

- *Pathogens* are being sequenced in order to help develop new strategies to fight disease.
- *Extremophiles*, organisms that live in extreme conditions like high heat or high salt, are being sequenced to look for proteins with industrial applications.
- *Crop plants*, like corn and rice, are being sequenced to increase human understanding of the organisms that provide food to most of the people in the world.
- *Model organisms*, such as mice, fruit flies, and *E. coli*, are being sequenced so that their scientists can compare their genomes to those of humans.

Looking within the human genome

The human genome contains many unexpected patterns in the sequence. For example, humans have much fewer genes than were originally expected based on the size of the human genome — just 25,000 genes instead of the predicted 100,000! Other features of the human genome include

- The average gene size is about 3,000 base pairs long.
- The number of genes on each chromosome varies from 2,968 genes on chromosome 1 to just 231 genes on the Y chromosome.
- At least 50 percent of the human genome consists of small repeating sequences that do not code for protein or RNA.
- Genes are randomly clustered together on chromosomes and separated from other clusters of genes by noncoding DNA in between.
- The areas of DNA that contain genes contain lots of G-C base pairs, whereas the regions of noncoding DNA contain lots of A-T base pairs.
- Less than 2 percent of the genome contains the code for proteins and RNA molecules. The other 98 percent of the genome consists of various types of DNA and DNA whose function isn't yet known. The known functions of noncoding DNA include
    - Regulatory sequences that control how DNA is used, such as promoters and origins of replication.
    - Sequences that are involved in the structuring of DNA within the cell such as those that determine nucleosome positioning (see Chapter 19).

- Small repeated sequences, sometimes called *junk DNA,* that are scattered throughout the genome. The function of this DNA, which makes up 50 percent of the human genome, isn't yet known.
- Sequences, called *pseudogenes,* that are the mutated relics of formerly functioning genes.

The information in the human genome can be divided into two main categories: coding sequences that are found in genes and repetitive sequences that make up noncoding DNA.

We Have a Lot in Common: Comparative Genomics

Scientists who study comparative genomics use computer programs to search databases of genomic data to look for regions of similarity in the genomes of different species. Comparative genomics can answer several kinds of questions:

- **What is the evolutionary relationship between different species?** The greater the similarity between two genomes, the closer the relationship between the species.
- **What genes are common to different species?** Some genes exist in slightly different forms in many cells, forming gene families. Databases of genomic data are searched for sequences from known genes to see whether the known sequence is present in other organisms. If a known sequence is found in the genome of another organism, then the other organism probably has a gene with a similar function to the known gene.
- **Which sequences determine the unique features of organisms?** Comparisons of closely related species, such as chimpanzees and humans, reveal which sequences are most important in creating the unique features of each species.
- **Which sequences show the effects of natural selection?** By matching up the DNA sequences of related organisms, such as humans and mice, and then comparing the rates of mutation in different areas of the genome, scientists can identify areas of DNA that have undergone natural selection.

One of the big take-home messages of comparative genomics is that all life on Earth is very similar. Your DNA sequence shows the close relationship between you and your earthly neighbors. For example, your genome is 99.9 percent identical to that of other humans, 96 percent identical to that of chimpanzees, and about 40 percent identical to that of mice.

Monkey business

A recent comparison between the human and chimpanzee genomes shows that the genomes are 96 percent similar in sequence, which leaves people wondering how two species can be so similar in their DNA sequence, but so different in appearance and behavior. The leading hypothesis is that the development of organisms has as much to do with how they use their DNA sequence as what their DNA sequence actually says.

Noncoding DNA, the DNA that is outside of genes, is thought to play an important regulatory role in controlling when and how genes are used. The fact that the human genome has such a great amount (98 percent!) of noncoding DNA makes this idea seem quite possible. Also, genes for transcription factors (see Chapter 19) in humans seem to be evolving more quickly than similar genes in chimpanzees. The comparison between the human and chimpanzee genomes also shows differences in genes that control inflammation and a gene that protects chimpanzees from Alzheimer's. Understanding the effects of these differences between humans and chimps may help scientists better treat disease.

Another interesting discovery learned from comparing genomes is that a bigger genome and more genes don't necessarily mean a more complicated organism. And, organisms that are very similar to each other can have very different genome sizes (see Table 21-1).

Large genomes don't necessarily mean a lot of genes. Organisms with large genomes typically have more repetitive DNA in their genomes.

Table 21-1 A Comparison of Genomes of Different Types of Organisms

Organism	*Genome Size in Base Pairs (bp)*	*Number of Genes*
E. coli	4,600,000	3,200
Yeast	12,000,000	6,300
Mustard weed (*A. thaliana*)	120,000,000	25,500
Corn (*Z. mays*)	2,500,000,000	59,000
Nematode (*C. elegans*)	100,000,000	14,000
Fruit fly (*D. melanogaster*)	140,000,000	14,000
Mouse (*M. musca*)	2,500,000,000	30,000
Humans (*H. sapiens*)	3,000,000,0o00	25,000

Comparisons between these genomes show that differences in organisms can't be explained based on genome size alone. Mustard weed and corn are both plants, but the corn genome is much larger than that of the mustard weed. Fruit flies and nematodes have the same number of genes, but the fruit fly genome is much larger than that of nematodes. Likewise, humans and mustard weeds have the same number of genes, but the human genome is much larger.

Two main observations can explain the observed differences in genome size between organisms:

- Some organisms, such as certain plants, have acquired multiple sets of chromosomes during their evolution. Organisms that have many sets of chromosomes are *polyploid.* Polyploid organisms can have very large genomes.
- Some organisms, such as humans, have lots of repetitive sequences in their genomes. Repetitive sequences range in size from 150 to 300 base pairs and can be repeated thousands of time. For example, a 300 base pair sequence called *Alu* occurs more than 1.1 million times in humans.

What's Your Function?: Functional Genomics

Functional genomics tries to discover the functions of DNA sequences, including the relationships between genes and the interactions between gene products, such as RNAs and proteins. Functional genomics considers the entire group of gene products, or *transcriptome,* that is together in the cell at any one time and compares it to the transcriptome that is present at other times. Scientists can answer the following kinds of questions through functional genomics:

- **How much variation is there within a genome?** Scientists can compare the genomes of different individuals within species for nucleotide changes called *single nucleotide polymorphisms* (SNPs) to determine how much variation exists within the genome of a species.
- **When and where does gene expression occur?** Scientists can compare cells of different species or at different developmental stages to each other in order to determine patterns of gene expression between species and during development.
- **How do genes and the products of different genes interact with each other?** Relationships between genes are suggested when the same genes occur in a number of related species or when genes occur together in the same part of the genome. Relationships between gene products are proposed when they're produced at the same time in the cell.

Looking for open reading frames

Scientists hope to discover new proteins by searching or them within the genome. Genes are encoded within *open reading frames,* sections of DNA that begin with the code for a start codon and end with the code for a stop codon. (See Chapter 18 for more on codons.)

Computers search DNA sequences for patterns that correspond to open reading frames. Scientists can locate some types of genes more easily than others:

- **Prokaryotic genes** are relatively easy to locate. The sequences of prokaryotic promoters are well known, which helps to locate prokaryotic genes. (Also, prokaryotic genes do not contain introns, making the coding sequence easier to locate. See Chapter 18 for more on promoters and introns.)
- **Eukaryotic genes** are harder to identify. Less than 2 percent of the human genome is actually made of genes, which makes gene hunting a little like looking for a needle in a haystack. Also, eukaryotic genes contain introns that break up genes and make them harder to recognize.

Comparing gene expression with DNA microarrays

Cells turn genes on and off as they respond to signals, go through different developmental stages, or even become cancerous. One goal of functional genomics is to compare these changing patterns of gene expression (for more on gene expression, see Chapter 19). To determine which genes are expressed at any one time in the cell, geneticists track the presence of RNA sequences in cells using *DNA microarrays.*

A DNA microarray is a glass slide with many single-stranded copies of DNA stuck to it. The DNA sequences stuck to the slide are single-stranded copies of all the genes for a particular cell.

1. **Molecules of mRNA are isolated from the two cells to be compared.**

   The two cells are different in some way that you are interested in studying, such as two cells from different stages of development, cells from two different species, cells grown under different environmental conditions, or a normal cell and a cancer cell.

2. **Reverse transcription produces labeled cDNA copies of each population of mRNA.**

   One of the nucleotides included as building blocks for the formation of the cDNA in each population is labeled with a colored fluorescent tag so that all the cDNA molecules from one cell are tagged with a certain color. (See Chapter 20 for a description of how to make cDNA.)

3. **The cDNA molecules from each cell are washed over the surface of the DNA microarray.**

   Wherever the cDNA molecules find complementary sequences on the slide, they will stick. The cDNA molecules are acting as probes to scan the genome on the microarray and identify which sequences are being expressed in the cell from which the cDNAs were made.

4. **Use a laser to activate the fluorescent tags on the cDNA molecules and observe the color patterns to determine which DNA sequences are expressed in each cell type.**

   a. If a particular spot on the microarray is dark, then the sequence in that location on the microarray isn't expressed in either cell type.

   b. If a particular spot on the microarray is the color of one cell type or the other, then the sequence in that location on the microarray is expressed only in the corresponding cell type.

   c. If a particular spot on the microarray is the color of the combined fluorescent tags, then the sequence in that location on the microarray is expressed in both cell types.

Reaping the Rewards: Pharmacogenomics

Pharmacogenomics combines traditional pharmacology and genomics with the goal of tailoring drug therapies more precisely to suit particular diseases or an individual's genome. The kinds of questions that pharmacogenomics hopes to answer include

- **Which genes and proteins are associated with this disease?** When the genes and proteins that cause a particular disease are known, then scientists can develop very specific drugs that target just those genes and proteins. Drugs that are very specific will cause far fewer side effects than some of the medications in use today that affect normal proteins as well as disease-causing proteins.

- **Which drug is the best treatment for a person with a particular genotype?** Many people have adverse reactions to certain drugs. If certain adverse reactions are more common in people with a particular genotype, then it will be possible to use a person's genotype to predict how they'll react to a drug. Doctors can choose the best drug for a person right from the beginning of treatment based on their genotype, preventing additional illness from an adverse drug reaction.
- **For which diseases is a person with a particular genotype most at risk?** Some diseases have a genetic basis, while other diseases are more likely to affect some people than others. Genotype screening allows identification of any genetic diseases that a person inherited, as well as prediction of their risk factors for other diseases. Based on this information, people can make lifestyle choices to minimize their risk for disease.

I've Got a System: Systems Biology

The availability of sequenced genomes has had a huge impact on many branches of biology. Much of the genome information is stored in publicly accessible data bases — all you need to access this data is a computer with an Internet connection! Scientists all over the world are mining this information to develop new understandings in evolution, medicine, and development. Scientists are using genomic information and the techniques developed by genomics in many ways, as you can from the following list:

- **Evolutionary biologists** can study entire genomes, looking for evolutionary patterns to determine the relationships between organisms, the sources of new genes, and the effects of natural selection on genes.
- **Geneticists** share gene maps and probe sequences, facilitating the identification and location of disease-causing genes. Geneticists also look for open reading frames in genomes to discover new genes. (For more on this topic, see the section "Looking for open reading frames," earlier in this chapter.)
- **Developmental biologists** compare genomes of cells at different developmental stages using DNA microarrays in order to learn the expression patterns of genes during development. Developmental biologists also can use comparative genomics to screen genomes of different organisms for sequences known to be important in development, identifying sequences that have a common function across a group of organisms.
- **Biochemists** can predict the structure and function of proteins based on DNA sequences found in open reading frames. Biochemists can also look for drugs that target specific proteins discovered to be associated with particular diseases.

In addition to the impact of genomic information and techniques, the systems approach of genomics is spreading to other areas of biology. *Proteomics* is a new branch of biology that studies all the proteins present in a cell at one time, and *metabolomics* is a new branch of biochemistry that looks at the full set of reactions that happen in a cell at one time. These systems approaches are different from traditional approaches where a portion of a system is isolated and studied at one time. By considering an entire system of molecules or reactions at once, biochemists can better understand the interactions between components of the system and can consider the properties of the entire system as a whole.

The traditional central dogma of molecular biology is DNA → RNA → protein. In the era of genomics, the central dogma can be rewritten as genome → transcriptome → proteome.

Part VII
The Part of Tens

"The doctors think it's a result of all the shopping I did while I was pregnant."

In this part . . .

The amazing things that molecular and cellular biologists do today rests upon the genius and inspiration of past scientists. By studying and comparing cells through the years, scientists have discovered some fundamental laws that govern the way cells function.

This part presents some of the most important scientific principles that form the foundation of molecular and cellular biology. In addition, I share my top ten tips for making the grade in a college class on this subject.

Chapter 22

Ten Important Rules for Cells to Live By

In This Chapter

- Creating the Cell Theory
- Conserving matter and energy
- Evolving by natural selection
- Believing the Central Dogma

Laws that govern the energetics of the entire universe have implications for cells right here on Earth. And, by studying the behavior of all different types of cells, common themes emerge that suggest rules that all cells follow. In this chapter, I present ten of the rules that I find to be particularly important to cells.

The Cell Theory

The roots of the Cell Theory trace back to the first observations of cells by Robert Hooke and Antony vanLeeuwenhoek. Back around 1665, Robert Hooke looked at thin slices of cork, which is a tissue from the bark of certain trees, with a microscope. He saw what looked like tiny little rectangles all stuck together in a sheet. The little rectangles reminded him of the small rooms that monks lived in, which were called cells. The cells that Robert Hooke saw were the cell walls of the plant cells that had once made up the cork tissue. The cells were dead and gone, but their cell walls remained.

Antony vanLeeuwenhoek was a Dutch cloth merchant who read about Hooke's discovery and got curious about the world he saw through a microscope. He made his own tiny single lens microscopes and started exploring the world around him. (See Chapter 1 for more details on his discoveries.) In a tiny drop of pond water, he was amazed to find small swimming creatures that turned out to be the first living cells ever seen by a person! In 1673, vanLeewenhoek sent a letter about his discoveries to the Royal Academy of London, causing a shockwave of excitement throughout Europe.

People continued to look at the world through microscopes for another 200 years, making various observations. In 1833, the Scottish botanist Robert Brown discovered that plant cells had little spots in them that he named nuclei because they looked like little seeds. A German botanist, Matthias Jacob Schleiden, heard about Brown's discovery and started looking more closely at plant cells. By 1838, Schleiden decided that plants grew by producing new cells. He went to a dinner party where he met Theodor Schwann, a German zoologist who was studying animal tissues. Schleiden told Schwann about his observations and conclusions about plant cells, and the two of them got very excited about working together. Schwann invited Schleiden to visit his lab, and together the two scientists looked closely at animal cells. They discovered that animal cells had nuclei just like plant cells. Because plant and animal cells had this remarkable similarity, the scientists concluded that all living things are made of cells. This conclusion became the cornerstone of the modern Cell Theory:

- The cell is the fundamental unit of structure, function, and organization in all living organisms.
- All cells come from pre-existing cells.
- The cell is the smallest form of life.

The First Law of Thermodynamics

The First Law of Thermodynamics says that energy can't be created or destroyed. What the First Law of Thermodynamics means for cells is that because they can't make the energy they need, they have to get it from somewhere in their environment. Cells that do photosynthesis get their energy from the Sun and store it in the form of sugars that they make. Cells that can't make their own sugars must use sugars made by other cells for energy — in other words, they've got to eat somebody.

Single-cells organisms eat smaller single-celled organisms. Multicellular animals have digestive systems that are specialized for eating and breaking down other living things. Fungi and bacteria release enzymes onto dead things that turn them into molecule soup that the fungal and bacterial cells absorb for their matter and energy. All living things on Earth, including the cells that make their own food, have the ability to transfer energy from food to a form that is usable for cells.

Another consequence of the First Law of Thermodynamics for cells is that the energy that they use has to go somewhere; it doesn't just disappear. Cells transfer energy from food into movement, store it in molecules they build, or transfer it back out to their environment as heat. You can account for all the energy that enters the cell if you add up the energy within the cell and the energy that is transferred back to the environment. In keeping with the first law, the total energy in the universe remains the same.

The Second Law of Thermodynamics

The Second Law of Thermodynamics, phrased in a way that makes sense for cells, states that processes can happen spontaneously only if they increase the entropy, or disorder, in the universe. What the Second Law of Thermodynamics means to a cell is that the reactions that break molecules down or increase the amount of disorganization in the cell are easy to do, but anything that keeps the cell organized and maintained is going to cost the cell some energy. Because cells must stay organized and maintained to stay alive, cells must constantly transfer energy from the environment into the cell to pay for this expensive upkeep. If a cell runs out of energy, such as when no food is available, the cell will no longer be able to maintain itself and will die.

The Second Law also says that cells will never be able to access all the available energy that is stored in the food they break down. With every energy transfer, the potential energy of the system decreases (if you don't count any energy entering or leaving the system). For cells, what the second law means is that every time cells transfer energy from food molecules into cellular processes, some of the potential energy that was stored in the food is transferred to heat. The heat produced from these energy transfers leaves the cell and goes back out into the environment.

Any time you exercise, you can see the effect of the second law in action. Exercise requires energy, so your cells increase the speed of the reactions that transfer energy from food. With every transfer of energy from food to your cells, some heat energy is produced. You can feel this heat as your body begins to warm up and then starts to sweat to cool itself. The heat produced by the energy transfers in your cells is released from your body to the air and objects around you. Heat production by your cells is most dramatic during exercise, but it's actually occurring all the time.

The Theory of Evolution by Natural Selection

Things change, including cells. When groups of living things change over time, scientists call it *evolution.* You can find proof that evolution occurs in your daily newspaper in stories about the increase in numbers of antibiotic-resistant bacteria. Types of bacteria that were once killed by certain antibiotics now survive these same drugs because the bacteria have changed from what they used to be. Evidence for evolution is also found in the fossil record, which includes imprints of living things, such as dinosaurs that no longer exist. Living cells, from *E. coli* to your own cells, show that they're related to each other by using the same fundamental chemistry based on DNA and proteins. If cells on Earth are all related, then presumably they all have the same ancestors. If all cells have the same ancestor, then life today, in all of its varied forms, developed by evolution from a common source.

The Theory of Evolution by Natural Selection was proposed by Charles Darwin as a way to explain how cells evolve, or change, over time. He based his idea on observations of how plants and animals can change, including organisms bred for certain traits by farmers and the unique plants and animals that developed on the isolated islands in the Galapagos. Darwin's proposal had the following main points:

- **Living things inherit many of their traits from their parents.** Traits are inherited through the DNA that is passed from parents to offspring.
- **Living things make more offspring than can survive based on available resources in the environment.** From bunnies to bacteria, organisms typically make as many babies as they can. The number of offspring that will survive is limited by the availability of resources, such as food, water, and shelter.
- **Living things have differences from each other.** Changes in traits result from changes in the DNA called *mutations.* A small but significant number of mutations occur in the DNA of a cell every time the cell copies its DNA before it divides. Environmental agents can increase the rate of mutation in cells, resulting in faster changes. Basically, mutation happens. And when it does, individual cells and organisms can be slightly changed as a result.
- **Because of competition for survival, the living things that have the traits that are best suited for a particular environment are more likely to survive and reproduce**. In other words, it's *survival of the fittest.*
- **The survivors reproduce, passing their traits to their offspring. Thus, more of the individuals in the new generation have the traits that enable survival.** The population of living things has changed, or evolved, from what it used to be.

The change from susceptible bacteria that antibiotics can kill to resistant bacteria that antibiotics can't kill is an excellent example of evolution by natural selection. The rise of antibiotic-resistant bacteria follows all of Darwin's points:

- Bacteria copy their DNA before they divide so that each new cell is very similar to the original cell.
- Bacteria reproduce very quickly and can make thousands of new cells in just a few hours.
- Bacterial DNA is mutated slightly whenever the cells divide, creating slight differences in new cells.
- When humans use antibiotics to bacterial infections, the antibiotics will kill any individual bacteria that are susceptible. However, if some bacteria have traits that enable them to withstand the antibiotic, they will survive.
- Even if an antibiotic kills 99.9 percent of a bacterial population, bacteria reproduce so quickly that the 0.1 percent of survivors will be able to generate a whole new population in a very short period of time. This new population will receive its DNA from the survivors, inheriting the trait that enabled survival. This new population of bacteria is now resistant to the antibiotic that was used.

Evolution by natural selection acts on populations, or groups, not on individuals. In other words, individuals can't change their traits in order to survive and thus evolve into some sort of new being. When something limits survival, such as lack of food or the presence of an antibiotic (if you're a bacterium), individuals die or survive based on the traits they already have — traits they inherited from their parents. What changes over time is the *allele frequency* (the frequency of a trait) in a group of organisms the individuals belong to. When part of the population is killed, the survivors repopulate the group with their offspring, so the group changes from one generation to the next.

The Law of Conservation of Matter

The Law of Conservation of Matter (or Mass) says that matter can't be created or destroyed. The chemist Antoine Laurent Lavoisier originally formulated the law to explain what happens during chemical reactions. Lavoisier showed that even though materials may seem to disappear during chemical reactions, if you carefully track the mass of everything involved in the reaction, the mass of materials entering the reaction is the same as the mass of materials exiting the reaction. In other words, during chemical reactions atoms change partners, but no atoms are lost or gained.

According to the Law of Conservation of Matter, then, every atom in every food molecule taken in by a cell must be trackable. For example, in order to transfer energy out of food, cells break down food by cellular respiration. The summary reaction for the breakdown of glucose by cellular respiration is represented by this formula:

$$C_6H_{12}O_6 + 6\ O_2 \rightarrow 6\ CO_2 + 6\ H_2O$$

When cells break down glucose ($C_6H_{12}O_6$), it may seem that the glucose disappears. But in reality, the carbon and oxygen atoms in the glucose are bonded to each other to form the molecule carbon dioxide (CO_2). Carbon dioxide is a gas and is thus invisible. However, if you count all the atoms in the reactants and products of this reaction, you'll see that they're the same. If you were to trap all the carbon dioxide and water produced by cellular respiration, the mass would be the same as that of the sugar and oxygen that enters the reaction.

Another effect of the Law of Conservation of Matter on cells is that cells can't grow unless they take in matter from the environment. Cells can't just create molecules for growth out of nothing; they must use food molecules to provide the atoms needed for the construction of growth molecules. Cells break food molecules down into their building blocks and then use those building blocks to build the molecules of the cell. So, cells can either break food molecules all the way down to carbon dioxide and water, releasing the atoms from food back into the environment, or they can partially break down their food, keeping the atoms in the cell for cellular growth.

Exceptions to the Law of Conservation of Matter do exist. As objects travel at speeds approaching the speed of light, the distinction between matter and energy begins to blur. Also, matter may convert into energy during nuclear reactions. However, in the world of the cell, the Law of Conservation of Matter applies and is a useful way of thinking about how cells use food.

Nucleic Acids Pair in Antiparallel Strands

In DNA, two strands of nucleotides are held together by hydrogen bonds to form the double helix. The two strands of nucleotides run in opposite, or antiparallel, directions to each other. When nucleotide strands are oriented antiparallel to each other, the hydrogen bonding sites on the nitrogenous base adenine (A) line up with those on the nitrogenous base thymine (T), and the hydrogen bonding sites of guanine (G) line up with those of cytosine (C).

The nitrogenous bases are like puzzle pieces that fit together only if they're in just the right position next to each other — in this case, that means one base right side up, and the other base upside down.

The lesson of antiparallel base pairing demonstrated by DNA applies to all other interactions between nucleic acids, too. Any time two nucleic acid strands pair with each other, they must be oriented opposite to each other. Nucleic acids pair during several cellular processes and molecular techniques:

- **RNA pairs with DNA during transcription.** As the message in DNA is copied into RNA during transcription, the two nucleic acids are held together by hydrogen bonds between their bases. The growing RNA strand is built in the 5' to 3' direction, starting with its 5' end. This growing strand is hydrogen bonded to the DNA template strand that runs in the 3' to 5' direction.
- **mRNA pairs with tRNA during translation.** When tRNA molecules enter the ribosome, the tRNA anticodons bind to the mRNA codons. The mRNA molecule is oriented in the 5' to 3' direction, so the tRNA anticodons are oriented in the 3' to 5' direction.
- **snRNA pairs with pre-mRNA during splicing.** In eukaryotes, the primary RNA transcript must have its introns removed before translation. During splicing, mRNA and snRNA are held together by hydrogen bonds. The mRNA and snRNA are antiparallel to each other.
- **DNA probes bind to DNA in order to locate specific sequences.** DNA probes tagged with fluorescent or radioactive labels are used to search for specific sequences in a sample of DNA. The probe must be constructed so that its sequence is complementary and antiparallel to the sequence that is being looked for.
- **DNA primers bind to DNA in order to begin amplification of a DNA sequence during PCR.** DNA primers bind to single-stranded DNA on each side of the sequence to be amplified. The DNA primers must have sequences that are complementary and antiparallel to the sequences on either side of the sequence to be amplified.

Central Dogma

The Central Dogma of molecular biology describes the flow of information in the cell. Frances Crick first described the Central Dogma in 1958 as a way of representing the understanding at the time of how information flows from the language of nucleic acids to the language of proteins. Since then, evidence has supported and added to the understanding of the Central Dogma.

In its simplest form, the Central Dogma says that information flows from DNA to RNA to protein. In other words, the code in DNA is used to make RNA, and then the code in RNA is used to make protein. A more complete expression of the Central Dogma in cells has several components:

- **Information flows from DNA to DNA.** Each strand of DNA is used as the pattern for building complementary strands during DNA replication.
- **Information flows from DNA to RNA.** The DNA code in genes is used as the pattern for building complementary molecules of RNA. Several kinds of RNA exist in cells, including mRNA, rRNA, tRNA, and other small RNAs.
- **Information flows from mRNA to protein.** The code in mRNA is decoded during translation in order to determine the sequence of amino acids in proteins.

If information flow in viruses is included, then two more components must be added to the Central Dogma:

- **Information flows from RNA to RNA.** Some viruses have RNA for their genetic material. These viruses use RNA as a pattern to make more RNA strands during replication of the virus.
- **Information flows from RNA to DNA.** Some viruses that have RNA for their genetic material use the RNA as a pattern for the construction of a DNA molecules. These viruses are called *retroviruses.*

Protein Shape Is Essential to Their Function

Proteins are incredibly important to the normal functioning of cells (see Chapter 6). Proteins catalyze reactions, provide structure to cells, turn genes on and off, send and receive signals, and transport materials. In every job that proteins do, their shape is key to their function. Here's a list of just some of the ways protein shape is important to function:

- Enzymes have pockets called *active sites* that are just the right shape to bind the enzyme's substrate.
- Enzymes have regulatory sites called *allosteric sites* that are the right shape for regulatory molecules to bind to.
- DNA-binding proteins have pockets called DNA-binding sites that are the right shape to bind to particular sequences of DNA.

- Receptor proteins have to be a particular shape in order to bind to their ligand.
- Membrane transport proteins called carrier proteins have pockets that bind specifically to the substance they transport across membranes.
- Antibody proteins have antigen-binding sites that are just the right shape to bind to antigens from bacteria and viruses.

Cells use protein shape to regulate cell processes. Because protein shape is essential to function, if a protein's shape is changed, it will no longer be able to do a particular job — and a protein's shape changes any time something binds to it. So, regulatory molecules bind to proteins to change their shape and turn them on or off. Cells regulate proteins in order to control two of the most important activities in cells, metabolism and gene regulation:

- Allosteric inhibitors and activators bind to allosteric sites on enzymes to control enzyme activity. Allosteric inhibitors change enzyme shape to stop enzyme function and shut down metabolic pathways. Allosteric activators change enzyme shape to improve enzyme function and increase the rate of metabolic pathways.
- Inducers and co-repressors bind to DNA-binding proteins called *repressors* to regulate transcription. Inducers change repressor shape, making them inactive and unable to bind to DNA. When repressors are inactive, transcription occurs. Co-repressors change repressor shape, making them active and able to bind to DNA. When repressors are active, transcription is blocked.
- General transcription factors bind to each other and to RNA polymerase, changing the shape of RNA polymerase so that it can bind to promoters and initiate transcription.

Law of Segregation

Organisms that reproduce by sexual reproduction follow the Law of Segregation. Diploid cells, like body cells in humans, have two copies of every gene. During reproduction, some diploid cells undergo meiosis. Meiosis segregates each gene pair by separating the chromosomes of the cell. Thus, cells that have two copies of every gene give only one copy to their gametes (eggs and sperm). Gametes are haploid because they have only one copy of every gene. In order to make a new individual, two gametes need to get together, which creates a new diploid cell that again has two copies of every gene, one from each parent.

Law of Independent Assortment

Diploid cells have two of each kind of chromosome. Pairs of matching chromosomes are called homologous chromosomes. When cells undergo meiosis to create gametes (eggs and sperm), they separate their homologous chromosomes so that each gamete gets only one of each pair. The segregation of homologous chromosomes separates pairs of alleles for each gene. So, when meiosis happens over and over again in an individual to produce many gametes, each of those gametes will have a slightly different combination of alleles. The law of independent assortment increases the variation in the offspring of sexually reproducing organisms.

Chapter 23

Ten Ways to Improve Your Grade

In This Chapter

- Becoming an active participant in your learning
- Discovering the best strategies that work for you
- Figuring out how your instructor likes to test

Molecular and cellular biology is a tough subject, and you need to tackle it hard in order to succeed. The good news is that the skills you build to be successful in this class will be very useful skills in other classes and in careers in science and medicine. The bad news may be that you have to reserve a good chunk of time in your life to dedicate to this subject. In this chapter, I present ten tips for getting the most for your investment of time, whether in lecture or when studying on your own.

Keep Your Mind Alive During Lecture

Some instructors are good lecturers, others not so much. But whatever your circumstance, you can do a lot on your end to get the maximum benefit out of attending lecture:

- **Write notes in your own words**. Listen to what your instructor is saying and write your own notes. Writing your own notes is very different from just sitting there and writing down whatever the instructor says or writes on a board. If you're listening and writing things in your own words, you're processing the information as you go.
- **Take notes on interesting stories and anecdotes**. Instructors often tell stories and give examples to show the relevance of the information they're presenting. They don't usually write down these stories, however, so many students don't write them down either. If the instructor tells a good one that helps you grasp the concept they're talking about, jot down a few notes about it in the margin of your notes. When you're studying later, these side notes may help you recall the topic.

- **Sit in the best place for you in lecture**. Usually, the front is best. It's too easy to get distracted and tune out in the back. However, if you're someone who gets sleepy and might need to move around a little to wake up, then try an aisle seat. If you get sleepy, you can get up and take a short walk to the rest room. It's better to get up and move than to miss half of lecture because you took a nap.
- **Ask questions when you don't understand**. If you're prepared for class and following the lecture but something doesn't make sense to you, then ask about it. Chances are if you don't get it, someone else doesn't either.
- **Be in good physical shape for class**. Get enough sleep, exercise, and healthy food so that you're ready to participate.

Schedule Your Study Time

You'll get the best bang for your study time if you're smart about how you schedule it.

- **Plan for studying like you do for other important events**. Make studying a priority and figure out how to fit it into your life.
- **Schedule enough time each week**. Science classes require more study time than other types of classes. Each branch of science has its own language that you need to master, with tons of new terms. Also, science instructors typically expect you to memorize lots of detailed information, including how complex processes work. The general rule is two hours outside of class for every hour in class.
- **Schedule short blocks of time each day as well as longer blocks a few times a week**. Studies show that you get the most out of studying your lecture notes the sooner you review them after lecture. So, plan some study time every day so that you can do a review before you forget what you heard.

Be Active, Not Passive

You'll have the most success if you use a combination of approaches to studying molecular and cellular biology. Some activities, such as attending lecture, reading your text, and even reviewing your notes, are fairly passive because you're receiving information from another source.

If you understand what you're hearing and reading, great! Understanding is one of the first steps to learning the material. However, understanding what you hear and read isn't enough. And for most people, reading and re-reading the text isn't effective. You need to take active approaches to get yourself to the next levels.

- **Put things in your own words.** Memorizing a definition from the book or an explanation from an instructor will only get you so far. You can say the same thing many different ways, and your instructor is likely to use different ways of phrasing things in lecture and on exams. Practice the language you're learning by putting things in your own words rather than relying totally upon the words of others.
- **Recreate the information by yourself**. Practice drawing diagrams on blank pieces of paper, looking at your notes only when you have to. Tell your study partners or your dog about metabolic pathways, naming the essential steps out loud. The key here is that you're working with what is in your own brain as much as possible, over and over until you've got it down.
- **Practice applying the information to new situations**. Once you think you've got it, try solving every problem you can get your hands on. Solve the harder problems at the end of the chapter in the text, review any problems given in class, and do any homework that was assigned. If you really get it, you should be able to take it out of the box and use it in a new situation.

Give Your Brain a Well-Rounded Workout During Study Sessions

Some people learn best from pictures, some from sounds, and some from moving around. You may have one way that works best for you, but chances are you can learn from all these approaches. So, when you're studying, mix it up a bit. Draw pictures, talk out loud, act out processes using the objects on your desk, or get up and move around. Also, relate the information you're learning to your own life or things you've heard in the news. The more neurons you can engage while you're studying, the better your understanding is likely to be.

Get Creative with Memory Tricks

Molecular and cellular biology is full of new terms and processes you'll need to learn. There's really just no way around it. One way you can help yourself is to make up words or sayings to help trigger your memory of terms and events. For example, way, way, back in the fourth grade, I learned HOMES to represent the Great Lakes of the United States. To this day, I still know HOMES represents Huron, Ontario, Michigan, Erie, and Superior. You can make up your own memory tricks, borrow them from friends, or look online.

Recognize the Difference Between Levels of Understanding

If you're taking a course in molecular and cellular biology, then you're probably not an expert in the subject. To get from beginner to expert, you'll need to do a lot of learning on different levels. And, you need to be prepared to be tested at various levels of understanding. If you want to get a full overview of what the different levels are, look up Bloom's Taxonomy online at `www.odu.edu/educ/roverbau/Bloom/blooms_taxonomy.htm`. Here's my shorter version with some tips on how to recognize different types of questions:

- **Recall factual information**. You'll need to learn lots of terms for different structures and processes in molecular and cellular biology. Before you can move on to higher levels or even understand your instructor, you need to learn the language of the subject. You'll be tested at this level by questions that ask *what* things are. Key phrases to look for are "name," "what are," "state," and so on.
- **Describe processes**. Recalling processes is a little bit harder than just learning terms. You'll have to learn all the events, plus the order in which they occur. You'll be tested at this level by questions that ask *how* things work. Key phrases to look for are "describe," "explain how," or "identify the *mechanism*" (how something works).
- **Make connections to the big picture**. If you understand why something is important, you can relate it to the real world and other topics in biology. You'll be tested at this level by questions that ask you *why* something is the way it is. Key phrases to look for are "what is the significance," "why," or "what is the importance."
- **Apply information to solve new problems.** Application questions are often word problems, where you're given a scenario and asked to use what you've learned to solve a problem. Sometimes these questions will surprise you because they aren't phrased exactly like something you've heard in lecture. When tackling these questions, underline key information in the problem and ignore any extra details. (Key phrases to look for are "apply" and "solve.")

Remember the Supporting Material

Textbooks come with lots of supporting materials these days — CDs, Web sites, practice tests, student study guides, and more. (Not like the days when I was an undergrad and had to walk 10 miles barefoot in the snow to get to class and copy lecture notes down on a slate!) Some of these materials are very good and useful.

The supporting materials that came with your text were probably written by someone other than the textbook author and may not be at the same level as your course. For example, many practice tests seem heavy on recall-type questions and very skimpy on higher order questions. If a practice test is heavy on recall, but your instructor tests at a higher level, the practice test may give you a false sense of confidence before an exam. You should definitely take advantage of whatever materials come with your text, but you need to evaluate them carefully first!

Test Yourself Often

The big test day will come, and you'll get a permanent grade. Before you get to that crucial moment, make sure that you've put yourself through your paces in a less stressful situation. Test yourself during study sessions and get together with other students in the class if possible to test each other. Make sure that when you test yourself, you're working at all levels of understanding. (See the earlier section "Recognize the Difference Between Levels of Understanding.)

If your instructor releases old exams, use them as practice tests after you've studied the material and think you're ready.

In addition, keep these pointers in mind:

- **Flash cards are good for memorizing factual information**. Write a term or process on one side of the card and the meaning on the other side. Write things in your *own* words and practice saying things different ways. You can't count on your instructor to always use the exact same phrase!
- **Making diagrams and drawings are good ways to practice processes**. Make sure that you label your pictures to explain what is going on. Sketching your own diagrams on a blank piece of paper is much better practice than reviewing the same figure from the book over and over. Look online for animations, too!
- **Textbooks and instructors are usually pretty good at highlighting the big ideas**. The bold topics or questions at the beginning of sections usually point to the main concept being covered. Check your notes for things your instructor emphasized as "big ideas" or "really important." Compare any lab activities you did to what was said in lecture and make connections between lecture and lab. Also review any real-world connections you made while studying.

Use Your First Test as a Diagnostic Tool

You won't really know how your instructor tests until you take your first test, unless you've had him in the past (or know someone that has). Some instructors include recall questions in their tests, while others just write problems for you to solve. (Some instructors release old exams, which give you a great preview of what's coming!)

Your first test is a very valuable piece of information. When you get your graded test, go over it and look for the types of questions your instructor asked. At what level were you tested? Pay particular attention to any questions you missed and ask yourself why you missed it. What type of question was it — recall, concept, application? Where was the information? Does your instructor test only off notes? Or does he test on concepts in the book that he didn't cover in class? Is information learned in lab covered on exams?

Also look at how you did on the different types of questions. Did you do okay on recall questions, but miss questions that asked you how things worked? Or did you get any questions based on lecture, but miss those based on readings or labs?

The more you know about what to expect, the more you can tailor your study sessions to fit your instructor. Study smarter, not longer!

Get Help Sooner Rather Than Later

If you're doing your best to actively follow along in lecture and you're not getting it, go for help right away even if it's the first day of class! The longer you wait to get help, the greater the chance that you'll accumulate low grades on assignments. Even if you get help eventually, those low grades will continue to pull you down. Learning is usually cumulative, so if you don't understand early concepts, you won't understand later concepts that build on them. So, get help at the first sign of trouble, rather than at the end of the class! Your instructor and teaching assistants have office hours; don't be afraid to use them! Many schools have tutoring services as well.

Index

• Numerics •

• A •

• B •

• C •

• D •

• E •

• F •

• G •

• H •

• I •

• L •

• M •

• N •

• P •

• Q •

• R •

• S •

• T •

• U •

• V •

• W •

• Y •

• Z •